Manual of
LIPIDOLOGY

Manual of
LIPIDOLOGY

Editors

Sujoy Ghosh DM FRCP (London, Glasgow, Edinburgh) FACE FICP
Professor
Department of Endocrinology
Institute of Post Graduate Medical Education and Research
Kolkata, West Bengal, India

Banshi Saboo MD PhD MNAMS (Diabetes) FRCP (UK) FACE
Director and Chief Diabetologist
Department of Diabetes
Diabetes Care and Hormone Clinic
Ahmedabad, Gujarat, India

JAYPEE BROTHERS MEDICAL PUBLISHERS
The Health Sciences Publisher
New Delhi | London

 **Jaypee Brothers Medical Publishers (P) Ltd**

Headquarters
EMCA House
23/23-B, Ansari Road, Daryaganj
New Delhi 110 002, India
Landline: +91-11-23272143,
+91-11-23272703
+91-11-23282021, +91-11-23245672
E-mail: jaypee@jaypeebrothers.com

Overseas Office
JP Medical Ltd.
83, Victoria Street, London
SW1H 0HW (UK)
Phone: +44-20 3170 8910
E-mail: info@jpmedpub.com

Corporate Office
Jaypee Brothers Medical Publishers (P) Ltd.
4838/24, Ansari Road, Daryaganj
New Delhi 110 002, India
Phone: +91-11-43574357
Fax: +91-11-43574314
E-mail: jaypee@jaypeebrothers.com

EU GPSR Authorised Representative
Logos Europe, 9 rue Nicolas Poussin
17000, La Rochelle, France
Phone: +33 (0) 6 67 93 73 78
E-mail: Contact@logoseurope.eu

Website: www.jaypeebrothers.com
Website: www.jaypeedigital.com

© 2021, Jaypee Brothers Medical Publishers

The views and opinions expressed in this book are solely those of the original contributor(s)/author(s) and do not necessarily represent those of editor(s) of the book.

All rights reserved. No part of this publication may be reproduced, stored or transmitted in any form or by any means, electronic, mechanical, photocopying, recording or otherwise, without the prior permission in writing of the publishers.

All brand names and product names used in this book are trade names, service marks, trademarks or registered trademarks of their respective owners. The publisher is not associated with any product or vendor mentioned in this book.

Medical knowledge and practice change constantly. This book is designed to provide accurate, authoritative information about the subject matter in question. However, readers are advised to check the most current information available on procedures included and check information from the manufacturer of each product to be administered, to verify the recommended dose, formula, method and duration of administration, adverse effects and contraindications. It is the responsibility of the practitioner to take all appropriate safety precautions. Neither the publisher nor the author(s)/editor(s) assume any liability for any injury and/or damage to persons or property arising from or related to use of material in this book.

This book is sold on the understanding that the publisher is not engaged in providing professional medical services. If such advice or services are required, the services of a competent medical professional should be sought.

Every effort has been made where necessary to contact holders of copyright to obtain permission to reproduce copyright material. If any have been inadvertently overlooked, the publisher will be pleased to make the necessary arrangements at the irst opportunity. The **CD/DVD-ROM** (if any) provided in the sealed envelope with this book is complimentary and free of cost. **Not meant for sale.**
Inquiries for bulk sales may be solicited at: jaypee@jaypeebrothers.com

Manual of Lipidology / *Sujoy Ghosh, Banshi Saboo*

First Edition: 2021, Reprint: **2025**
ISBN: 978-93-5270-029-5

Contributors

Editors

Sujoy Ghosh DM FRCP (London, Glasgow, Edinburgh) FACE FICP
Professor
Department of Endocrinology
Institute of Post Graduate Medical Education and Research
Kolkata, West Bengal, India

Banshi Saboo MD PhD MNAMS (Diabetes) FRCP (UK) FACE
Director and Chief Diabetologist
Department of Diabetes
Diabetes Care and Hormone Clinic
Ahmedabad, Gujarat, India

Associate Editors

Sayantan Ray MD DM
Consultant Endocrinologist
Medica Superspeciality Hospital
Kolkata, West Bengal, India

Amritava Ghosh MBBS MD DM
Assistant Professor and Faculty In-Charge
Department of Endocrinology and Metabolism
All India Institute of Medical Sciences
Raipur, Chhattisgarh, India

Contributing Authors

Amarta Shankar Chowdhury MD DM
Consultant
Department of Endocrinology
The Mission Hospital
Durgapur, West Bengal, India

Ajitesh Roy MD DM
Assistant Professor
Department of Endocrinology
Vivekananda Institute of
Medical Science
Kolkata, West Bengal, India

Amitabh Sur MBBS MD DM
Consultant Endocrinologist
Peerless Hospital
Kolkata, West Bengal, India

Anshita Aggarwal MBBS MD DM
Assistant Professor
Department of Endocrinology
Atal Bihari Vajpayee Institute of Medical
Sciences (ABVIMS) and Dr Ram Manohar Lohia
(RML) Hospital
New Delhi, India

Avivar Awasthi MD
Senior Resident (Academic)
Department of Endocrinology and Metabolism
Medical College
Kolkata, West Bengal, India

Beatrice Anne MBBS MD DM
Assistant Professor
Department of Endocrinology
Nizam's Institute of Medical Sciences
Hyderabad, Telangana, India

Chitra Selvan MBBS MD DM MRCP
Associate Professor
Department of Endocrinology
MS Ramaiah Medical College
Bengaluru, Karnataka, India

Deep Dutta MBBS MD DM DNB MNAMS FRCP FACE
Consultant
Department of Endocrinology
CEDAR Superspeciality Clinics
Dwarka, New Delhi, India

Dibbendhu Khanra MD DM
Assistant Professor
Department of Cardiology
All India Institute of Medical Sciences (AIIMS)
Rishikesh, Uttarakhand, India

Hridish Narayan Chakravarti MBBS MD DM
Consultant Endocrinologist
Narayana Superspeciality Hospital
Howrah, West Bengal, India

Indira Maisnam MBBS MD DM FRCP FACE
Consultant Endocrinologist
RG Kar Medical College
Kolkata, West Bengal, India

Ipsita Ghosh MD DM
Senior Resident
Consultant Endocrinologist
IPGME&R and SSKM Hospital
Kolkata, West Bengal, India

Kaushik Biswas MBBS MD DM
Consultant Endocrinologist
Medica Superspeciality Hospital
Kolkata, West Bengal, India

Kaushik Sen MBBS MD DM
Assistant Professor
Department of General Medicine
Nil Ratan Sircar Medical College
Kolkata, West Bengal, India

Lohit Kumbar MBBS MD
Senior Resident
Department of Endocrinology
MS Ramaiah Medical College
Bengaluru, Karnataka, India

Mounam Chattopadhyay MD
Senior Resident (Academic)
Department of Endocrinology
Nil Ratan Sircar Medical College
Kolkata, West Bengal, India

Nodee Chowdhury MBBS MD
Senior Resident DNB-SS
Department of Neurology
Medanta – The Medicity
Gurugram, Haryana, India

Partha Pratim Chakraborty MD DM DNB FACE FICP
RMO cum Clinical Tutor
Department of Endocrinology and Metabolism, Medical College
Kolkata, West Bengal, India

Partha Sarathi Choudhury MBBS MD DM
Senior Resident
Department of Endocrinology and Metabolism, Medical College
Kolkata, West Bengal, India

Rajan Palui MD DM
Consultant
Department of Endocrinology
The Mission Hospital
Durgapur, West Bengal, India

Contributors

Rana Bhattacharjee MD DM MRCP FICP FACE
Assistant Professor and In-Charge
Hormone Assay Lab
Department of Endocrinology and Metabolism
IPGME&R and SSKM Hospital
Kolkata, West Bengal, India

Ranajit Bari MBBS MD DM
Assistant Professor
Department of Medicine
Murshidabad Medical College
Berhampore, West Bengal, India

Ruby Raphael MBBS MD
Senior Resident
Department of Endocrinology
Nizam's Institute of Medical Sciences
Hyderabad, Telangana, India

Satyam Chakraborty MBBS MD DM
Consultant Endocrinologist
Institute of Neurosciences
Kolkata, Fortis Hospitals
West Bengal, India

Shishir Soni MD
Senior Resident
Department of Cardiology
All India Institute of Medical Sciences (AIIMS)
Rishikesh, Uttarakhand, India

Soumik Goswami MD DM
RMO cum Clinical Tutor
Department of Endocrinology
Nil Ratan Sircar Medical College
Kolkata, West Bengal, India

Soumitra Kumar MD DM FCSI FACC FESC FSCAI FIAE FICC FICP
Professor and Head
Department of Cardiology
Vivekananda Institute of Medical Sciences
Kolkata, West Bengal, India

Subrata Chakrabarti MD DM
Consultant Endocrinologist
DESUN Hospital
Kolkata, West Bengal, India

Sunetra Mondal MD DM
Senior Resident
Department of Endocrinology
IPGME&R, Kolkata
Consultant Endocrinologist
HealthWorld Hospitals
Durgapur, West Bengal, India

Subhodip Pramanik MD DM DNB MRCP
Consultant Endocrinologist
Neotia Getwel Healthcare Hospitals
Siliguri, West Bengal, India

Tapas Chandra Das MBBS MD DM
RMO cum Clinical Tutor
Department of Endocrinology and Metabolism
IPGME&R and SSKM Hospital
Kolkata, West Bengal, India

Preface

Over the last two centuries, the role of lipids in the development of cardiovascular disease (CVD) has gained significant attention, and specifically, the role of cholesterol on the development of atherosclerosis. The first studies involving cholesterol and CVD way back to the early 1900s when Anichkov, a pathologist, provided rabbits with high cholesterol diets and observed stiffening of their artery walls. The first lipid lowering medication, niacin, was discovered coincidentally after it was tested on a rabbit schizophrenia model. One of the side effects observed was lowering of cholesterol. In the late 1970s, a Japanese microbiologist Akira Endo first discovered natural products with a potent inhibitory effect on cholesterol synthesis in a fermentation broth of *Penicillium citrinum*, during his search for antimicrobial agents. Concurrently, Brown and Goldstein (later earning the Nobel Prize for their work in 1985) demonstrated that HMG-CoA reductase inhibition represented the rate limiting step in cholesterol biosynthesis. These early basic and translational studies led to the production of statins. The first of many statin trials to come was the 4S study conducted in Scandinavian countries in the late 1980s and showed a significant reduction in CV events. Of note, many subsequent trials confirmed the benefits of statins in reducing risk of heart disease. On another front, the publication on the Mediterranean diet in the mid-90s transformed our understandings of the dietary components that protect against heart disease. In that study, participants randomized to olive oil, nuts survived longer following coronary bypass surgery than individuals kept on their regular diet. Several dietary studies have been conducted since confirming a distinct role for fats in the pathogenesis of heart disease.

Although the focus in treatment of lipid disorders is on reducing low-density lipoprotein (LDL) cholesterol concentrations, additional lipid-related independent risk factors, such as triglyceride (TG), high-density lipoprotein (HDL) cholesterol, and lipoprotein(a) levels should be used clinically to assess CV risk. Lipid treatment guidelines have continued to evolve as new evidence comes out from extensive research and clinical trials on lipid lowering medications. Clinicians need to be updated with all these emerging evidences when considering how to approach their patients with lipid disorders.

This book is published with the objective of providing both basic concepts and clinical approaches to understanding and managing lipid related disorders that is associated with an increased CV risk. To reach this aim, the focus is on up-to-date information on different aspects of *lipidology*. This has been possible through the contributions of endocrinologists and cardiologists who are expert in their respective fields. The book covers a wide range of topics from basic to clinic. The first section covers basic aspects of lipidology including genetics. Subsequent sections include updates from the recent clinical trials on lipid lowering medications, discussion on controversial areas in lipidology and current approaches in management. The last sections focus on management lipid disorders in special populations such as dyslipidemia in pregnancy, pediatric dyslipidemias, HIV induced dyslipidemia.

This book is intended for the postgraduate students, fellows, physicians, internists, cardiologists, endocrinologists, pediatricians and gynecologists with interest in lipids. The editors hope that by coupling of basic concepts and management approaches to lipid disorders, this book will assist the clinicians in making the best decisions in diagnosing and treating their patients. We have made every possible effort to maintain the quality and standard of "Manual of Lipidology". Despite our best efforts, there may be some unintentional errors. Kindly excuse us for the same.

We also express our thanks to each and every manuscript contributors, our associate editors and industry. Finally, we express our thanks to the efficient staff of Jaypee Brothers Medical Publishers Ltd, who actively participated in this project.

We hope that this book lives up to the expectations of everyone.

Sujoy Ghosh
Banshi Saboo

Contents

Section 1: Basic Lipidology

1. **Human Lipoprotein Metabolism: An Overview** 3
   Sayantan Ray

2. **Lipoproteins: Mechanisms for Atherosclerosis** 11
   Subrata Chakrabarti

3. **Apolipoprotein B and Cardiovascular Disease** 23
   Hridish Narayan Chakravarti

4. **Lipoprotein(a): An Update** 29
   Tapas Chandra Das

5. **Monogenic Lipoprotein Disorders: From Phenotype to Genotype** 37
   Sunetra Mondal, Subhodip Pramanik

Section 2: Clinical Lipidology

6. **Dietary Fats and Cardiovascular Disease** 53
   Ajitesh Roy

7. **Dyslipidemia: Relationship with Insulin Resistance, Fatty Liver, and Subclinical Atherosclerosis** 64
   Kaushik Sen

8. **Risk Assessment of Different Lipid Markers: Clinical Appraisal and Goal of Therapy** 72
   Amitabh Sur

9. **Rethinking "Good" Cholesterol: Understanding the Role of High-density Lipoprotein** 81
   Kaushik Biswas, Satyam Chakraborty

10. **Triglycerides, Remnant Cholesterol, and Atherosclerotic Cardiovascular Disease Risk** — 94
    Amritava Ghosh

11. **Non-HDL is a Better Marker than LDL for Prediction of ASCVD Risk** — 108
    Shishir Soni, Dibbendhu Khanra

12. **Approach to the Patient with Lipid Disorders** — 115
    Partha Pratim Chakraborty, Avivar Awasthi

13. **Laboratory Assessment of Lipoproteins: Principles and Pitfalls** — 127
    Rana Bhattacharjee

14. **Statins Diabetogenicity: Are All Same?** — 136
    Lohit Kumbar, Chitra Selvan

Section 3: Therapeutic Lipidology

15. **The Effects of Diet and Exercise in Hyperlipidemia** — 145
    Partha Sarathi Choudhury

16. **Nonalcoholic Fatty Liver Disease: The Lipid Disease of the Liver and the Effect of Lipid-lowering Drugs** — 154
    Anshita Aggarwal, Deep Dutta

17. **Beyond Statins: Who and When to Prescribe?** — 161
    Subhodip Pramanik

18. **Role of Triglyceride Lowering Drugs on Cardiovascular Outcomes: Focus On Omega-3 Fatty Acid** — 170
    Rajan Palui, Amarta Shankar Chowdhury

19. **HDL-targeting Therapies: Progress and Future** — 180
    Sayantan Ray

20. **Statin Intolerance: Impact on Statin Therapy—Assessment and Management** — 189
    Indira Maisnam

21. **Diabetic Dyslipidemia: Current Concepts and Guidelines** — 198
    Amritava Ghosh

22.	**Comparative Analysis of Major Cholesterol Guidelines** *Soumik Goswami, Mounam Chattopadhyay*	211
23.	**International Lipid Guidelines: Implications for Our Indian Patients—A Cardiologist's Viewpoint** *Soumitra Kumar, Nodee Chowdhury*	228
24.	**Lipoprotein Apheresis** *Satyam Chakraborty, Kaushik Biswas*	237
25.	**Evolving Targets of Therapies: PCSK9 and Beyond** *Subhodip Pramanik*	245

Section 4: Dyslipidemia in Special Population

26.	**Management of Dyslipidemia in Chronic Kidney Disease** *Ipsita Ghosh*	259
27.	**Dyslipidemia in Children and Adolescents: Screening, Diagnosis, and Management** *Sayantan Ray*	266
28.	**Dyslipidemia in Pregnancy** *Sayantan Ray*	278
29.	**Drug Treatment of Dyslipidemia in Older Adults** *Ruby Raphael, Beatrice Anne*	287
30.	**Dyslipidemia in Human Immunodeficiency Virus** *Ranajit Bari*	297

Index   305

Section 1

Basic Lipidology

CHAPTER 1

Human Lipoprotein Metabolism: An Overview

Sayantan Ray

ABSTRACT

Lipoproteins are macromolecular complexes, which carry triglycerides (TGs) and cholesterol in the circulation. The structure of lipoproteins is a hydrophobic neutral lipid core consisting of TGs and cholesteryl esters (CE), surrounded by a hydrophilic coat of phospholipids and specialized proteins called apolipoprotein. Larger lipoproteins hold more core lipid and are less dense than the smaller lipoproteins. The metabolism of the lipoproteins is the process by which hydrophobic lipids, namely cholesterol, and TGs are transported within the interstitial fluid and plasma. In the plasma, lipoprotein transport is facilitated by apolipoproteins. Apolipoproteins also play active roles in lipoprotein metabolism and act as ligands for lipoprotein receptors and cofactors for lipolytic enzymes and lipid transferases. Elevated levels of the apolipoprotein B (ApoB)-containing lipoproteins and low levels of the ApoA-I–containing lipoproteins are associated with cardiovascular disease (CVD).

INTRODUCTION

Plasma lipoproteins contain a hydrophobic nonpolar lipid core of cholesteryl esters (CE) and triglycerides (TGs) and are surrounded on the surface by a more polar, hydrophilic coat of apolipoproteins, phospholipids, and unesterified cholesterol. The compositions of the lipoproteins ascertain their size and structures. The size and density of lipoprotein are a direct function of neutral lipid content. The largest lipoprotein particles being least dense have the highest ratio of neutral to polar lipids (**Table 1**).[1] There are five major types of lipoproteins, classified on the basis of their density at which they are isolated, that is, as the high-, low-, intermediate-, and very low-density lipoproteins (HDLs, LDLs, IDLs, and VLDLs, respectively); The least dense and largest lipoproteins are the chylomicrons (CM), which are intestinally derived and composed mainly of dietary lipids and small amounts of protein. HDL appears in two subclasses, HDL2 and HDL3. Lipoprotein metabolism involves the transport of TG from liver and intestine to muscles and adipose tissue, plus the transport of cholesterol both from intestine and liver to peripheral

TABLE 1: Properties of plasma lipoproteins.								
Class	d (g/mL)	D (nm)	Composition (%)					
			Core		Surface			
			TG	CE	FC	PL	Pro	Major APOs
CM	<0.93	80–500	86	3	2	7	2	B-48, E, A-I, A-II, A-IV, C
VLDL	0.95–1.006	30–80	55	12	7	18	8	B-100, C-I, C-II, C-III, E
IDL	1.006–1.019	25–35	23	29	9	19	19	B-100, E
LDL	1.019–1.063	21.6	6	42	8	22	22	B-100
HDL2	1.063–1.125	10	5	17	5	33	40	A-I, A-II
HDL3	1.125–1.210	7.5	3	13	4	25	55	A-I, A-II
LP(a)	1.055–1.085	30	3	33	9	22	33	B-100, ApoA

[CM: chylomicrons; D: diameter; d: density; CE: cholesteryl ester; FC: free cholesterol; HDL: high-density lipoprotein; IDL: intermediate-density lipoprotein(s); LDL: low-density lipoprotein; Lp(a): lipoprotein(a); PL: phospholipid; Pro: Protein; TG: triglyceride(s); VLDL: very-low-density lipoprotein(s)]

tissues, and from periphery back to the liver as well. TGs are the key component of energy transport and metabolism, and cholesterol is a vital component of all cells and essential for steroidogenesis.[2] Correlations between coronary artery disease and the plasma concentrations of lipoproteins as well as their properties and compositions, have revealed mechanisms that eventually aided diagnosis and provided novel targets for pharmacologic treatment of dyslipidemia and atherosclerosis.[1]

APOLIPOPROTEINS

Apolipoproteins add structural stability and play a critical role in the recognition of lipoproteins. ApoB-containing lipoproteins include CMs, VLDL, VLDL remnants (also known as IDL), LDL, and lipoprotein(a) [Lp(a)]. ApoB-containing lipoproteins are lipid-rich and play an important role in carrying TGs and cholesterol in the blood. The majority of HDL lipoproteins have both ApoA-I and A-II. ApoA-containing lipoproteins are essential components of reverse cholesterol transport (RCT) and are the initial acceptors of cholesterol from peripheral tissues. HDL lipoproteins are much more complex than VLDL or LDL, and very heterogeneous.[3] ApoC and ApoE containing lipoproteins are "conductor" lipoproteins that can orchestrate the lipoprotein metabolism efficiently. Continuous exchange of these two classes of apolipoproteins occurs between HDL and VLDL/LDL particles after meals. ApoC lipoproteins regulate lipoprotein lipase (LPL) activity. The cholesterol efflux in the periphery is determined by ApoE lipoproteins.[3,4] **Table 2** illustrates the types and functions of different apolipoproteins.

LIPOPROTEIN METABOLISM

Lipoprotein metabolism has two pathways to maintain the movement of lipids from diet to blood to cells: (1) The exogenous pathway, and (2) The endogenous pathway.

TABLE 2: Types and functions of apolipoproteins.	
ApoA-I	HDL structural protein, it activates LCAT and participates in reverse cholesterol transport
ApoA-II	Forms HDL and activates hepatic lipase
ApoB-48	Structural component of chylomicrons. Binds to LDL receptor
ApoB-100	Structural component of all lipoproteins except HDL and chylomicrons. Binds to LDL receptor
ApoC-I	Inhibits lipoprotein binding to LDL receptor. Activates LCAT
ApoC-II	Activates lipoprotein lipase
ApoC-III	Inhibits lipoprotein lipase. Antagonizes ApoE, inhibiting liver VLDL uptake
Apo-E	LDL and LRP receptor ligand, and is essential component of reverse cholesterol transport and triacylglycerol clearance

[Apo: apolipoproteins; HDL: high-density lipoprotein; LCAT: lecithin-cholesterol acyl transferase; LDL: low-density lipoprotein; LRP: lipoprotein receptor-related protein; VLDL: very-low-density lipoprotein(s)]

Exogenous Lipid Transport Pathway

More than 90% of dietary lipids are TGs. The remaining consists of cholesterol, CEs, phospholipids, and fatty acids (FAs). TGs are not soluble in the blood and therefore are transported as CMs and VLDL particles.[5] Following consumption, dietary lipids are digested by lingual and gastric lipase in the stomach. Pancreatic lipase further converts TG to a mixture of 2-monoacylglycerol and free fatty acids (FFAs), while CEs are processed by cholesterol esterase to cholesterol and FFAs. These products are packed with bile salts and fat-soluble vitamins forming mixed micelles which then diffuse into intestinal mucosal cells. Inside the enterocytes, TG is reformed through the re-acylation of the 2-monoacylglycerols by monoacylglycerol acyltransferase and diacylglycerol acyltransferase,[6] while absorbed dietary cholesterol is esterified by cholesterol acyltransferase to form CE. The TGs and CEs are packaged within CMs, which are involved in the transport of exogenous (dietary) lipids from the intestine to the lymphatic system into the systemic circulation through the exogenous pathway (**Fig. 1**).[7] CMs are the largest in diameter with the highest TG content (TG to cholesterol ratio in CM is 8:1). These CMs carry TG and CE to the peripheral tissues including muscles and adipose tissues. Approximately 10–12 hours is required to clear the blood of CMs following a meal. Peak lipidemia is reached in about 3–5 hours. In the plasma, the ApoC-II on the CM surface activates LPL that is present at the capillary endothelial cells.[3] By the action of activated LPL, FFAs are released and undergo beta (β)-oxidation to be used for local metabolic needs or stored in the adipocytes. Through the action of cholesterol ester transfer protein (CETP), CM acquires CE from HDL in exchange for TG. Through these metabolic processes, the formation of smaller, cholesterol-enriched particles (chylomicron remnants) occurs. These particles are rapidly removed by the liver. In the lymphatic system, CMs exchange ApoA-I and A-II for ApoC and E from HDL. ApoC is necessary for the activation of the LPL and ApoE is required for the recognition of the CM remnants by the hepatic receptors.[3,8]

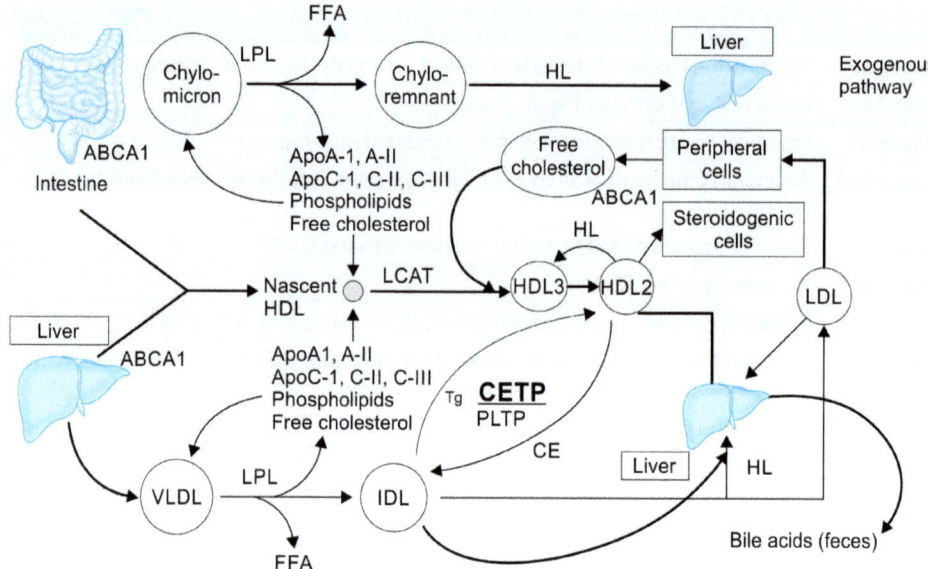

FIG. 1: Schematic representation of lipoprotein metabolism.
Source: Modified from Choi HY, Hafiane A, Schwertani A, Genest J. High-Density Lipoproteins: Biology, Epidemiology, and Clinical Management. Can J Cardiol. 2017;33:325-33.

Endogenous Lipid Transport Pathway

The first step in VLDL synthesis involves ApoB-100 synthesis on ribosomes attached to the endoplasmic reticulum. An enzyme called "microsomal triacylglycerol transfer protein (MTTP)" assembles triacylglycerols (TAG) and cholesterol with ApoB, E, and a phospholipid. These nascent particles contain more phospholipids and much less unesterified cholesterol in comparison to plasma VLDLs. VLDL is secreted from the liver into the plasma. TAGs form 50–60% of VLDL's weight and are the major fat that is transported from the liver into the blood. It contains a number of apolipoproteins, but ApoB-100 is necessary for its secretion from the liver. The TAG to cholesterol ratio in VLDL is 5:1 on average.[3] Alike the process with CM, ApoC, and ApoE are obtained from HDL where ApoC activates LPL that hydrolyzes TG in VLDL liberating FFAs that are taken up by the muscles for energy production or stored in adipose tissues. As with CM, through the action of CETP, VLDL exchange CE for TG with HDL resulting in the formation of relatively TG-depleted IDL which can be resorbed by the liver via the ApoE/remnant receptor or further hydrolyzed by hepatic lipase to form LDL (**Fig. 1**).[7] Phospholipid transfer protein (PLTP) mediates the transfer of phospholipids from VLDL to HDL. The main mechanism regulating VLDL secretion and uptake by the liver is under control of the Farsenoid X receptor (FXR) and SREBP-1c transcription factor.[9]

The main structural apolipoprotein of LDL is ApoB-100 that allows LDL to return to the liver for resorption through the LDL receptors with approximately one-third utilized by peripheral cells for membrane synthesis and steroid hormone production.[8] Unlike VLDLs, LDLs are mostly CEs with <10% TGs. Therefore, LDL carries more cholesterol per particle than other lipoproteins. The LDL receptor serves an essential mechanism for clearing both TGs and cholesterols from circulation. The liver is the primary organ responsible for removal of LDL and remnant lipoproteins.[3]

REVERSE CHOLESTEROL TRANSPORT PATHWAY

This pathway includes the mobilization of cholesterol from the cells along the arterial wall and the delivery of the cholesterol to the liver in the form of CE. The pathway begins with the secretion of a disk-shaped ApoA-I containing particle by liver and intestine, called nascent HDL. ApoA-I in this nascent HDL particle interacts with the adenosine triphosphate (ATP)–binding cassette (ABC) protein, ABCA1, on the peripheral cells such as macrophages. The ABCA1 protein mediates efflux of free cholesterol (FC) from the intracellular storage pools onto the HDL particle, thereby forming pre-β HDL. Lecithin-cholesterol acyl transferase (LCAT) and its cofactor, ApoA-I esterifies the FC to form CE on the surface of the HDL particle.[3] As it circulates, pre-β HDL particles are transformed into a more spherical α-HDL particle that contains CE in its core. Mature α-HDL converts into mature α-HDL subtypes, α-HDL2 and α-HDL3 and continue to receive FC from inside the cells, thus increasing the amount of cholesterol transported to the liver via the CE-rich α-HDL via both direct and indirect pathways. In the direct pathway, CE-rich α-HDL binds to scavenger receptor B1 (SRB1) that facilitates the selective uptake of α-HDL to the liver and excretion in bile. In the indirect pathway, CE-rich α-HDL exchanges CE for TG from the VLDLs and LDLs, a process that is facilitated by CETP. CEs are then taken by hepatocytes through LDL receptors, transformed to bile acids, and finally excreted via the biliary tree.[8] In physiological states, HDL particles are continuously shifting between larger and smaller particles that aid the transport of cholesterol, TG, and phospholipids between the different lipoproteins and are vital to the initial step of RCT.[3] An illustration of RCT is provided in **Figure 2**.

CHOLESTEROL METABOLISM

De novo synthesis and diet are the two sources of cholesterol for the human body. Cholesterol biosynthesis accounts for the majority of serum cholesterol even when individuals are on a high cholesterol diet. Intestinal absorption of cholesterol is the primary mechanism that regulates the contribution of dietary cholesterol to total cholesterol levels.[10] Our body produces around 700 mg of cholesterol daily. Although, any nucleated cell can synthesize cholesterol, the bulk of cholesterol in the blood comes from the liver. Consequently, regulating the capacity of the liver to synthesize or catabolize cholesterol is crucial to determining cholesterol levels in the circulation.

There are three key mechanisms to control the cellular cholesterol content: (1) De novo biosynthesis, (2) Cholesterol uptake and esterification, and (3) Cholesterol efflux. Brown and Goldstein showed an important mechanism by which the cell regulates its cholesterol content[11] via a cholesterol sensor system called the sterol-regulatory-element-binding protein (SREBP-2) cleavage-activating protein (SCAP). This system controls both cholesterol synthesis and its uptake by lipoproteins. The principal mechanism of cellular uptake of cholesterol is through the uptake of lipoproteins such as LDL and hydrolysis of their cholesterol by CE hydrolases. To maintain a critical level of FC inside the cell, it is stored as CE through the function of acyl-CoA cholesterol acyl transferase (ACAT), forming lipid droplets. The cholesterol released from the droplets is

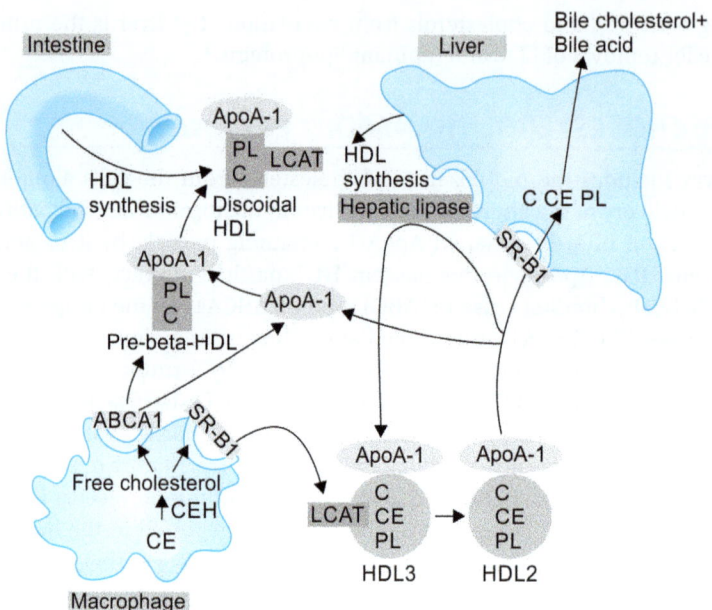

(CE: cholesteryl esters; HDL: high-density lipoproteins; LCAT: lecithin-cholesterol acyl transferase)

FIG. 2: Reverse cholesterol transport.

Source: Modified from Zhyvotovska A, Yusupov D, McFarlane SI. Introductory Chapter: Overview of Lipoprotein Metabolism. [online] Available from: https://www.intechopen.com/books/dyslipidemia/introductory-chapter-overview-of-lipoprotein-metabolism.

used for cell membranes and steroid hormone synthesis in steroidogenic cells.[3] At the peripheral level (macrophages), there are two major mechanisms for cellular cholesterol efflux and incorporation into lipoproteins for hepatic excretion: First, there is an active mechanism that involves the ABCA-1 transporter that generally gets activated after cholesterol loading of cells. The second one is a passive mechanism that depends on the cholesterol/phospholipid gradient between the cholesterol donor and acceptor. In the active pathway, the most avid cholesterol acceptor is ApoA-I. In the passive process, larger HDL particles with a large surface phospholipid to cholesterol ratio are the main cholesterol acceptors. Return of cholesterol back to the liver (CM in the fed state, VLDL and HDL in the fasting state), leads to activation of SREBP-1c to assist with cholesterol storage, efflux, or elimination in bile.[3]

BASICS TO CLINICS

Certain disease conditions interfere with lipid or lipoprotein metabolism pathways, leading to serious disorders. For example, in type 2 diabetes mellitus (T2DM), relative insulin resistance causes underutilization of VLDLs and CMs, ultimately leading to hypertriglyceridemia and increased small-dense LDLs which promotes atherosclerosis.[12] Insulin inhibits the release of FFAs from the adipocytes and suppresses hepatic VLDL production and secretion. These mechanisms come into play in the role of DM as a risk factor for cardiovascular disease (CVD). Additionally, the inability to suppress VLDL-TG kinetics has been implicated in the pathogenesis of the nonalcoholic fatty liver disease (NAFLD),[13] a serious complication that leads to CVD, liver fibrosis, and increases mortality.

CONCLUSION

Lipoprotein metabolism is an integrated process through which peripheral tissues exchange cholesterol, FAs, and phospholipids, and is mainly organized in the liver. In the fed state, the exogenous pathway predominates, whereas, the endogenous pathway predominates in times of fasting. When there is calorie surplus, lipids are packaged and stored in the adipose tissue. HDL metabolism plays a central role in RCT. These processes are well regulated in healthy states and are pretty abnormal in dyslipidemia. The increased exposure to cholesterol or oxidized lipids favors inflammation in the artery wall, ultimately leading to wall thickening, plaque formation, and rupture. A better understanding of the molecular mechanisms that facilitate lipid metabolism is important for designing appropriate therapies in order to address the CVD risk.

KEY POINTS

- Plasma lipoproteins are composed of neutral lipids, polar lipids, and a specialized protein called apolipoprotein.
- The apolipoproteins are important directors of lipoprotein metabolism.
- The liver is the central hub of lipoprotein metabolism.
- Lipoprotein metabolism serves the two major functions of providing TGs to adipose and muscle for storage or energy substrate and transport of cholesterol for cellular membrane and steroid hormone synthesis.
- A better understanding of the relationship between the structure and metabolism of lipoproteins and various pathologic states is important.

REFERENCES

1. Pownall HJ, Gotto AM. Human plasma lipoprotein metabolism. In: Ballantyne CM (Ed). Clinical Lipidology: A Companion to Braunwald's heart disease, 1st edition. Philadelphia: Elsevier; 2009. pp. 1-10.
2. Rader DJ. Lipoprotein Metabolism. In: Offermanns S, Rosenthal W (Eds). Encyclopedia of Molecular Pharmacology. Berlin, Heidelberg: Springer; 2008.
3. Yassine H, Tappin K, Sethi MJ. Basics in Lipoprotein Metabolism. In: Hussein Yassine (Ed). Lipid Management: From Basics to Clinic, 1st edition. Switzerland: Springer; 2015. pp. 1-16.
4. Mahley RW, Huang Y, Weisgraber KH. Putting cholesterol in its place: apoE and reverse cholesterol transport. J Clin Invest. 2006;116:1226-9.
5. Bakerman S. ABC's of Interpretive Laboratory Data, 4th edition. Scottsdale: Interpretive Laboratory Data Inc.; 2002.
6. Ferrier DR. Lippincott's illustrated reviews: biochemistry sixth edition. Baltimore: Lippincott Williams & Wilkins; 2014.
7. Choi HY, Hafiane A, Schwertani A, Genest J. High-Density Lipoproteins: Biology, Epidemiology, and Clinical Management. Can J Cardiol. 2017;33:325-33.
8. Zhyvotovska A, Yusupov D, McFarlane SI. Introductory Chapter: Overview of Lipoprotein Metabolism. [online] Available from: https://www.intechopen.com/books/dyslipidemia/introductory-chapter-overview-of-lipoprotein-metabolism. [Last accessed June, 2020]
9. Ferre P, Foufelle F. SREBP-1c transcription factor and lipid homeostasis: clinical perspective. Horm Res Paediatr. 2007;68:72-82.

10. Wilson JD, Lindsey C Jr. Studies on the influence of dietary cholesterol on cholesterol metabolism in the isotopic steady state in man. J Clin Invest. 1965;44:1805.
11. Brown MS, Goldstein JL. A proteolytic pathway that controls the cholesterol content of membranes, cells, and blood. Proc Natl Acad Sci U S A. 1999;96:11041-8.
12. McFarlane SI, Banerji M, Sowers JR. Insulin resistance and cardiovascular disease. J Clin Endocrinol Metab. 2001;86:713-8.
13. Poulsen MK, Nellemann B, Stødkilde-Jørgensen H, Pedersen SB, Grønbæk H, Nielsen S. Impaired Insulin Suppression of VLDL-Triglyceride Kinetics in Nonalcoholic Fatty Liver Disease. J Clin Endocrinol Metab. 2016;101:1637-46.

CHAPTER 2

Lipoproteins: Mechanisms for Atherosclerosis

Subrata Chakrabarti

ABSTRACT

Apolipoprotein B (ApoB) containing lipoproteins after passing through stages of entry and retention in the intima, oxidative modification, induction of vascular dysfunction, inappropriate inflammatory, apoptotic, and proliferative response, formation of foam cells ultimately lead to fatty streak formation which is the earliest lesion of atherosclerosis. This process is favored mostly by low-density lipoproteins (LDLs) but recent studies show triglyceride rich lipoproteins (TRLs) or plasma residual lipoproteins (PRPs) are equally responsible by sharing an almost equal set of actions. High-density lipoproteins (HDLs), the sole antiatherogenic particle acts against these processes but it suffers from a functional alteration in conditions of inflammation and dyslipidemia blunting its modulatory role in atherosclerosis.

INTRODUCTION

Atherosclerosis is considered as chronic inflammatory and metabolic disease characterized by the deposition of lipoproteins in the vessel walls, an influx of inflammatory cells, the formation of atheromatous plaques (consisting of cholesterol and macrophages) finally resulting in "stenosis," i.e., narrowing of the artery. Dyslipidemia is a prominent but modifiable risk factor and indicates an imbalance between elevated circulating levels of so-called good cholesterol in the form of apolipoprotein B (ApoB) containing lipoproteins including very low-density lipoprotein (VLDL), intermediate-density lipoproteins (IDLs), low-density lipoproteins (LDLs) and lipoprotein(a) [Lp(a)] versus subnormal levels of so-called bad cholesterol in the form of apolipoprotein A1 (ApoA-1) containing lipoprotein including high-density lipoproteins (HDLs).[1]

ROLE IN INITIATION OF ATHEROSCLEROSIS

Entry and Retention of Lipoproteins in the Intima

"Fatty streak" is the initial lesion of atherosclerosis which arises from local and focal increases in lipoproteins within specific regions of the intima—areas of abnormal shear stress and turbulence, nonlaminar or nonpulsatile flow (such as branching points of arteries). On a mechanistic basis, the accumulation of lipoproteins may not result simply from increased permeability or "leakiness," of the endothelium. On the contrary, penetration of the endothelial layer and the retaining of the lipoproteins in the intima occurs (**Flowchart 1**). Lipoprotein retention occurs by binding to constituents of the extracellular matrix (ECM) forming aggregates mostly by electrostatic interaction [interaction of the positively charged residues—arginine and lysine on ApoB with the negatively charged sulfate groups of sub-endothelial proteoglycans (SE-PG)]. While lipoprotein retention is initially related to direct binding of LDL to proteoglycans, after the infiltration of the intima by macrophages that secrete bridging molecules such as lipoprotein lipase (LPL) mediating indirect binding of lipoproteins to the intima, this retention is prolonged. These bridging molecules (SE-PG and LPL) work together in unison. This interaction slows the exit of the lipoproteins from the intima, increasing the residence time within the intima.[2,3]

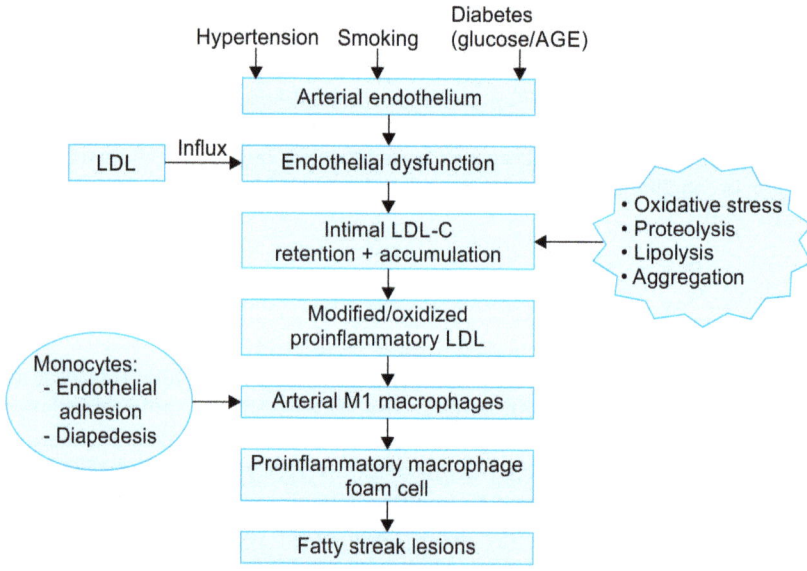

(AGE: advanced glycosylation end products; LDL-C: low-density lipoprotein cholesterol)

FLOWCHART 1: The initial stages of atherosclerotic lesion formation.[3]

Source: Adapted from Feng M, Rached F, Kontush A, Chapman MJ. Impact of Lipoproteins on Atherobiology: Emerging Insights. Cardiol Clin. 2018;36(2):193-201.

Modification of Lipoproteins in the Intima

After retention, LDL may be modified by aggregation, lipolysis, oxidation, or proteolysis. LDL by being sequestered from plasma antioxidants in the intima, becomes especially susceptible to oxidative modification, giving rise to hydroperoxides, lysophospholipids, oxysterols, and aldehydic breakdown products of fatty acids and phospholipids.[4]

Induction of Immune Response and Inflammation by Lipoproteins in the Intima

Modified LDL acts as a chronic stimulator of both innate and adaptive immune response causing activation of both endothelial cells and smooth muscle cells (SMCs); they then express surface adhesion molecules such as vascular cell adhesion molecule-1 (VCAM-1), intercellular adhesion molecule-1 (ICAM-1), chemo-attractants such as monocyte chemoattractant protein-1 (MCP-1), and growth factors such as macrophage colony-stimulating factor (M-CSF) and granulocyte-macrophage colony-stimulating factor (GM-CSF). They bind to receptors on circulating monocytes and stimulate their migration into the intima (homing of monocytes). These cells then differentiate into macrophages or dendritic cells.[5,6]

Formation of Foam Cells

Inside the intima, the monocytes transform into macrophages and become lipid-laden foam cells, process managed by the uptake of lipoprotein particles by receptor-mediated endocytosis. As the oxidation status of the lipoprotein increases, an affinity for the classical LDL receptor (LDLR) diminishes but its ability to enter cells increases because this action is mediated not by LDLR but by receptors of the LDL receptor-related protein (LRP) superfamily (so-called scavenger receptors) such as scavenger receptor-A (SR-A) and CD36. Modified LDL is next taken up by macrophages and SMCs through these scavenger receptors. A significant feature of these scavenger receptors and differing from the classic LDLR is that it escapes the usual homeostatic mechanisms shown by LDLR (i.e., feedback regulation by cellular cholesterol or reduced expression under conditions of cholesterol sufficiency). Thus, macrophages and SMCs internalize unregulated quantities of cholesterol and get engorged with droplets of lipoprotein-derived cholesteryl esters (CEs). They become known as foam cells (**Fig. 1**). Foam cells typically display a proinflammatory phenotype. This is important not only in the early fatty streak lesion, but also in determining the evolution of the plaque.[7]

Apart from the main role of oxidized LDL (oxLDL) in foam cell formation, nonoxidized, modified forms of LDL (small dense, electronegative, and especially desialylated forms) can participate in foam cell generation. Again, TG-rich ApoB-containing lipoproteins (remnant lipoproteins) do not require oxidative modification to be internalized by arterial macrophages. These remnant lipoproteins incite a more profound inflammatory response than do LDLs.

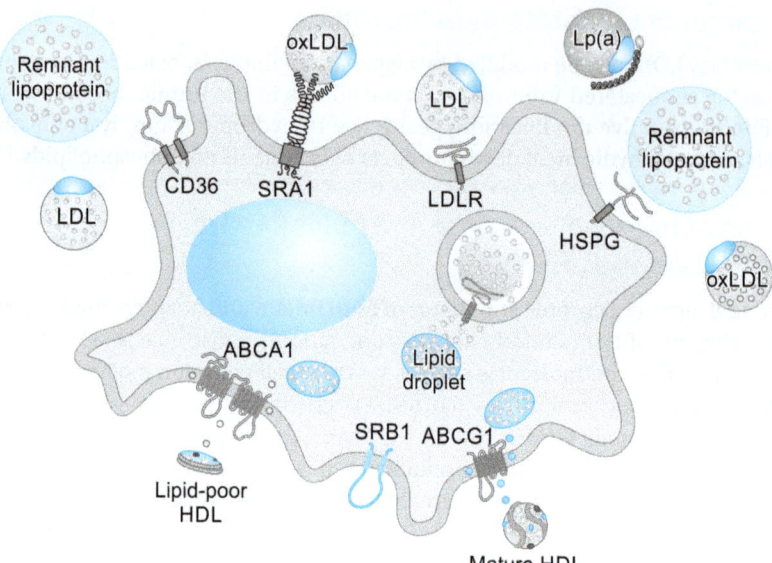

[ABCA1: ATP-binding cassette transporter A1; ABCG1: ATP-binding cassette transporter G1; HDL: high-density lipoprotein; HSPG: heparin sulfate proteoglycans; LDL: low-density lipoprotein; LDLR: low-density lipoprotein receptor; Lp(a): lipoprotein(a); oxLDL: oxidized low-density lipoprotein; SRA1: scavenger receptor A1; SRB1: scavenger receptor B1]

FIG. 1: Formation of foam cells.[2]

Source: Adapted from Shapiro MD, Fazio S. Apolipoprotein B-containing lipoproteins, and atherosclerotic cardiovascular disease. F1000Res. 2017;6:134.

ROLE IN PROGRESSION OF ATHEROSCLEROSIS

After the conversion of the monocyte-macrophage to foam cells, the intra-plaque inflammatory response starts. Foam cells augment inflammatory pathways by inducing the production of enzymes such as collagenases, elastases, cathepsins causing entry of more monocytes inside the intima and creating passages for the SMCs to reach from the media. This initial immune response can be treated as an appropriate and controlled attempt to get rid of the unwanted and hazardous debris from the intima (**Fig. 2**). But soon, the inflammatory response becomes chronic and maladaptive especially in advanced atherosclerosis. This is due to the functional failure of macrophages, causing failure to resolve the inflammation as foam cells lose mobility due to their lipid load and fail to exit the arterial wall.[2,4]

Macrophages and vascular smooth muscle cells (VSMCs) undergo programmed cell death or apoptosis forming the nidus of a lipid-rich or necrotic core of the atheroma. Foam cells secrete cytokines and growth factors. Macrophage necrosis leads to a more vigorous inflammatory response in a self-perpetuating cycle. This occurs by the following mechanism—macrophage engulfment of cholesterol crystals causes lysosomal destabilization and cathepsin B release, which generates a multimolecular signaling complex known as the nucleotide-binding leucine-rich repeat-containing pyrin receptor 3 (NLRP3) inflammasome. Activation of the NLRP3 inflammasome under influence of TH-1 cells, TH-2 cells, T-regulatory lymphocytes (Treg), and mast cells results in the secretion

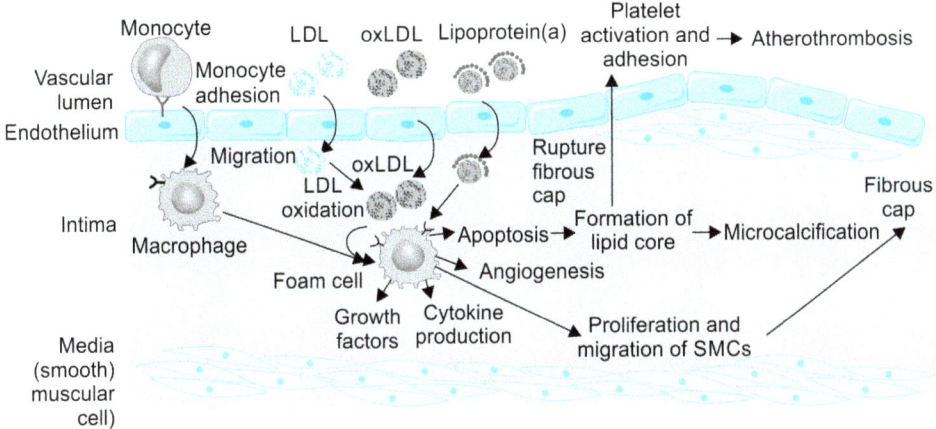

FIG. 2: The role of oxLDL in atherosclerosis.[9]
Source: Adapted from Arsenault BJ, Bourgeois R, Mathieu P. Do oxidized lipoproteins cause atherosclerotic cardiovascular diseases?. Can J Cardiol. 2017;33(12):1513-6.

of matrix-degrading enzymes such as matrix metalloproteinases (MMPs), reactive oxygen species (ROS), pro-inflammatory eicosanoids, cytokines [including interleukin-1 (IL-1) beta, interleukin-6 (IL-6)], all these factors favoring plaque progression and instability. Probably, it is the binding of modified LDL to pattern recognition receptors such as toll-like receptors (TLRs) on macrophages, which forms the trigger for the secretion of proinflammatory factors.[3]

During the initial stages of atherosclerosis, apoptotic cells are cleared by phagocytes in a process called efferocytosis and are mainly cleared. In advanced atherosclerosis, there is defective efferocytosis which causes a greater inflammatory response, necrotic core expansion, and plaque progression.

ROLE IN ATHEROTHROMBOSIS/ATHEROMA EVOLUTION AND COMPLICATIONS

Unstable plaques contain an infiltration of T-cells and macrophages, which can lead to plaque disruption and thrombosis by protease activation and matrix degradation. oxLDL by increasing monocyte-macrophage infiltration, which in turn generates plasmin and activates MMP activity, emboldens this process. LDL as well as residual lipoproteins (RLPs) may increase plasminogen activator inhibitor-1 (PAI-1) level increasing thrombosis chance.

Crystalline cholesterol within foam cells induce plaque rupture by physical disruption of the fibrous cap as the crystallization of cholesterol can result in sharp edges with the potential to penetrate biological membranes.

Oxidized LDL and RLPs also stimulate platelet adhesion and aggregation on the damaged vessel walls, promoting mural thrombus, i.e., thrombotic complications in advanced atherosclerosis (**Fig. 3**).[2,5]

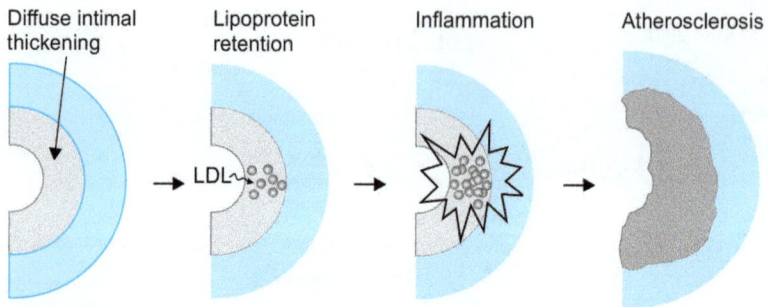

FIG. 3: The response-to-retention hypothesis of atherosclerosis.[11]

Source: Adapted from Fogelstrand P, Borén J. Retention of atherogenic lipoproteins in the artery wall and its role in atherogenesis. Nutr Metab Cardiovasc Dis. 2012;22(1):1-7.

ROLE OF INDIVIDUAL LIPOPROTEINS IN ATHEROSCLEROSIS

Role of Low-density Lipoprotein Cholesterol

Effects on Endothelial Nitric Oxide Synthase and Nitric Oxide Levels

Atherogenic concentrations of native LDL (nLDL) and even low concentrations of modified LDL lead to a decrease in nitric oxide (NO) bioavailability and action.

- Decrease endothelial NO synthase (eNOS) mRNA expression.
- Increase eNOS location in caveolae and its interaction with caveolin-1 (Cav-1) forming an inactive complex.
- Blocks acetylcholine-mediated eNOS activation.
- Induce nicotinamide adenine dinucleotide phosphate (NADPH) oxidase and xanthine oxidase causing an increase in superoxide radicals (O_2^-) that inactivates NO forming cytotoxic peroxynitrites.
- Increase plasma levels of asymmetric dimethylarginine (ADMA), an inhibitor of eNOS.

Effects on Vasoactive Compounds

- Cytochrome P450 (CYP)-monooxygenases produce epoxy fatty acids [endothelial-derived hyperpolarization factors (EDHF)] which cause vasodilation. oxLDL by negative modulation of these enzymes interferes with this action.
- Oxidized LDL and acetyl-LDL moieties but not native LDL induce endothelin-1 (ET-1) mRNA expression, a known vasoconstrictor.

Effect on Endothelial Adhesiveness

Oxidized LDL increases monocyte adhesion by an increase in VCAM-1, ICAM-1, E-selectin, and P-selectin, mainly mediated by LOX-1 (a receptor that mediates the uptake of oxLDL by endothelial cells).

Effect on Retention of Lipoproteins in the Intima

The size of LDL particles affects the binding of LDL to glycosaminoglycan (GAG) with small dense LDL showing the strongest interaction.

Effect on Extracellular Matrix Synthesis

Lipo-oxygenase (LO) controls collagen III transcription; reductions of LO activity decreases collagen levels. LO downregulation by oxLDL indirectly leads to disruption of ECM.

Regulation of Matrix Metalloproteinases and Tissue Inhibitor of Metalloproteinase-1

Oxidized LDL increases MMP-1 and MMP-3 and decreases tissue inhibitor of metalloproteinase-1 (TIMP-1) levels in endothelial cells.

Effect on Vascular Smooth Muscle Cells

Low-density lipoprotein induces proliferation of VSMC; lesser but similar effect of LDL on endothelial cells promotes apoptosis in both.[8]

Oxidation Theory of Atherosclerosis: Role of Lipid Oxidation Products

Lipoproteins (especially LDL) contain lipid oxidation product (LOP) in the form of hydroxides, hydroperoxides, epoxides, ketones, aldehydes, and isoprostanes. Enzymes involved in oxLDL generation include lipoxygenases, cyclooxygenase, phospholipases, whereas free radicals might generate oxLDLs in a nonenzymatic manner.

Low-density lipoprotein, like carrying cholesterol, transports LOP to peripheral tissues giving rise to atherosclerosis. HDL, conversely transports these LOP to the liver (place of final disposal), explaining in part the antiatherogenic action of HDL. Ox lipids are an independent risk factor for atherosclerosis, and also for insulin resistance, fatty liver, hypertension, and cancer. This is known as the oxidation theory of atherosclerosis.[9,10]

The Response-to-retention Hypothesis of Atherosclerosis

Williams and Tabas (in 1995) mentioned the response-to-retention hypothesis of atherogenesis, which states that retention of atherogenic lipoproteins within the intima is the inciting event of atherosclerosis. Retained lipoproteins lead to an inflammatory response that ultimately results in atherosclerotic plaques.[11,12]

Role of Triglyceride, Triglyceride Rich Lipoproteins and Plasma Remnant Lipoproteins

Postprandial hyperlipidemia [nonfasting triglyceride (TG)] is found to be an independent risk factor for atherosclerosis, apart from LDL. Despite satisfactory LDL lowering with statin therapy, substantial remaining cardiovascular risk often remains. This "residual risk" of atherosclerosis is attributed to plasma remnant lipoproteins (PRPs) or RLPs or triglyceride rich lipoproteins (TRLs) (**Fig. 4**). TRLs include VLDL and VLDL remnants, as well as chylomicron (CM) remnants. PRPs/RLPs result from partial hydrolysis by LPL of TRLs of hepatic and intestinal origin. TRLs promote atherogenesis independently of LDL. Intact CMs are not regarded as atherogenic as they are too big to pass through vascular spaces. On the other hand, CM remnants play atherogenic role by virtue of their relatively small size.[13]

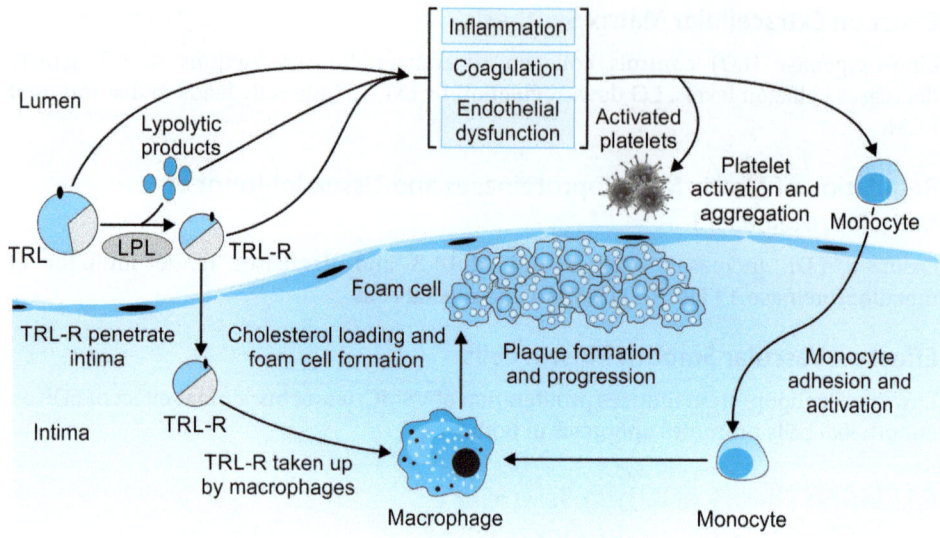

(LPL: lipoprotein lipase; TRL: triglyceride-rich lipoproteins; TRL-R: triglyceride-rich lipoprotein remnants)

FIG. 4: The role of TRLs in atherosclerosis.

Source: Adapted from Watts GF, Ooi EM, Chan DC. Demystifying the management of hypertriglyceridemia. Nat Rev Cardiol. 2013;10(11):648-61.

Moderately elevated plasma triglycerides indicate increased risk for cardiovascular diseases (CVD), and extremely elevated TGs (>500 mg/dL) signalize increased risk for pancreatitis. Now what is the link between hypertriglyceridemia and CVD? This is explained by the postprandial hypothesis of atherosclerosis or residual risk theory. Plasma TG levels also correlate with the levels of TRLs and their remnants, i.e., PRPs.[14]

The Postprandial Hypothesis of Atherosclerosis/Residual Risk Theory

Zilversmit was the first to forward the idea that delayed clearance of PRLs in the postprandial state can lead to premature atherosclerosis, the process commonly operational in obesity, metabolic syndrome. This is known as "post-prandial hypothesis of atherosclerosis."[15]

Plasma TRLs whose levels increase significantly after fat load, contribute to atherosclerosis. It is primarily the cholesterol content of TRLs that leads to atherosclerosis, rather than the TGs themselves. They directly contribute to intimal cholesterol deposition and cause activation of several proinflammatory, proapoptotic, and procoagulant pathways. In this regard, they share many proatherogenic features of oxLDL:

Both participate in macrophage foam cell formation. TG rich lipoprotein can directly enter vascular space and due to their larger size, have the potential to carry more cholesterol than LDL, leading to a hypothesis that its not LDL, but TRLs are the primary drivers in atherogenesis.

- Both are chemotactic for monocytes, T cells.
- Both enhance the expression of adhesion molecules on monocyte and endothelium.

- Both are mitogenic for VSMCs.
- Both accentuate proinflammatory gene expression in the intima.
- Both induce prothrombogenic molecules expression and platelet aggregation; assembly of the prothrombinase complex, PAI-1 and tissue factor.
- Both impair endothelial relaxation by decreasing NO; increase ROS.
- Both are cytotoxic and could induce apoptosis in endothelial and VSMCs.[16,17]

Role of ApoC-3

Atherogenicity of TRLs is related to their ApoC-3 content. VLDL and LDL bearing ApoC-3 stimulate most processes in atherosclerosis.[18]

Role of Lipoprotein Lipase

Lipoprotein lipase can attach to TRLs during lipolysis in endothelium and enhance their binding. LPL itself has proatherogenic effects independent of plasma lipoproteins and the resulting lipolytic products of TRLs such as oxidized free fatty acids (FFAs) activate several proinflammatory, procoagulant, and proapoptotic pathways.[19]

Role of High-density Lipoprotein

High-density lipoprotein is considered as antiatherogenic because of its role in cellular cholesterol extraction and reverse cholesterol transport. It shows functions beyond lipid metabolism but its role in reverse cholesterol transport is the most important about decreasing the chance of plaque development (**Fig. 5**). HDL can contribute to atheroprotection by:
- *Effluxing cellular cholesterol*: Cholesterol efflux by HDL is the key step in the reverse cholesterol transport from peripheral cells to the liver for excretion into the bile. It can remove cholesterol from macrophages and macrophage-derived foam cells.
- Attenuating vascular constriction by activation of NO by binding via SR-BI or ABCG1.
- Reducing vascular inflammation by decreasing adhesion molecule expression and attenuating monocyte activation.
- Cytoprotective action, protecting both macrophages and endothelial cells from apoptosis induced by oxidized LDL or tumor necrosis factor alpha (TNF-α), complement proteins.
- Protecting cells from oxidation by decreasing ROS.
- Attenuating platelet activation.
- Inhibits oxidative modification of LDL and blocks the effects of oxLDL by promoting the activity of antioxidant enzymes like platelet activation factor, acetylhydrolase and paraoxonase.[20,21]

The attenuated atheroprotective properties of HDL often seen in dyslipidemias and inflammatory states raises the possibility of indirect proatherogenic effects of such dysfunction. Defective biological activity of HDL is mainly due to the action of inflammatory mediators and acute phase reactants which rise in inflammation and commonly accompanied by attenuated cholesterol efflux capacity of HDL [impaired reverse cholesterol transport by cholesterol ester transfer protein (CETP)]. Dyslipidemic conditions are also characterized by defective antioxidative activity, the deficient anti-inflammatory activity of HDL, resulting in elevated oxidative stress.[22]

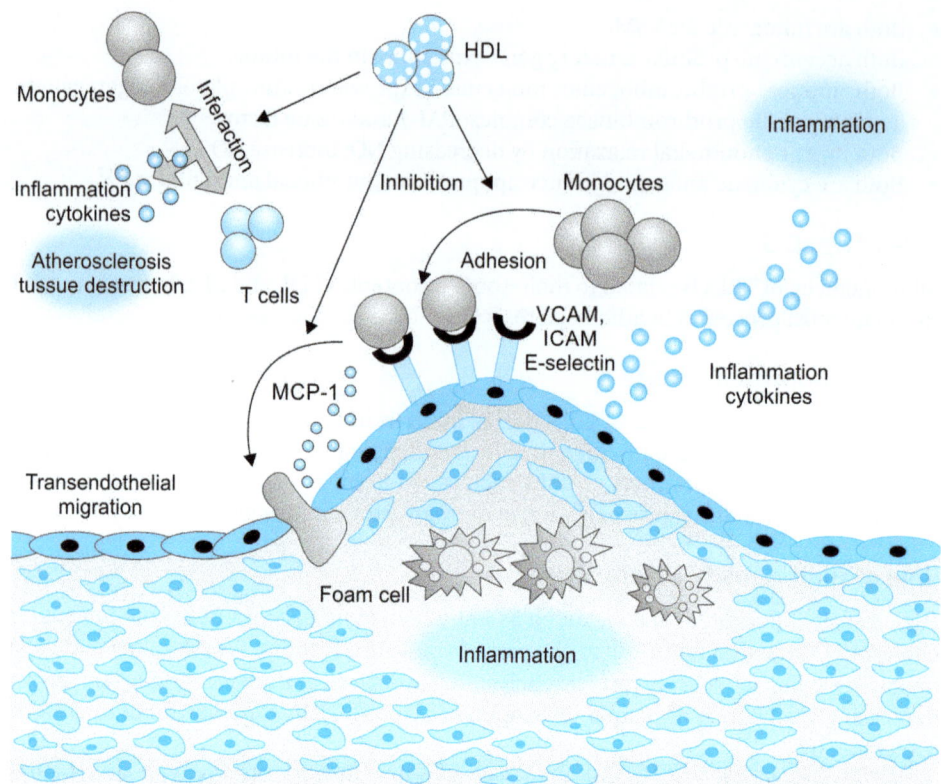

(ICAM: intercellular adhesion molecule; MCP-1: monocyte chemoattractant protein-1; VCAM: vascular cell adhesion molecule)

FIG. 5: Potential anti-inflammatory mechanisms of high-density lipoprotein (HDL) in atherosclerosis.[20]

Source: Adapted from Hu J, Xi D, Zhao J, Jinzhen Zhao, Tiantian Luo, Jichen Liu, et al. High-density lipoprotein and inflammation and its significance to atherosclerosis. Am J Med Sci. 2016;352(4):408-15.

Not all HDL particles have the same antiatherogenic potential. In the HDL spectrum, small dense protein rich particles display maximum antiatherogenic and anti-inflammatory properties. Incidentally, this population of HDL particles are most affected in metabolic syndrome.[23]

Role of Lipoprotein(a)

Lipoprotein(a) is both proatherogenic and prothrombotic. Its direct deposition occurs on the arterial wall such as LDL. Lp(a) is more likely to undergo oxidation than LDL itself and this facilitates uptake by macrophages via scavenger receptors. It induces endothelial dysfunction and expression of inflammatory cytokines such as TNF-α, TGF-β, IL-6, MCP-1. Due to the homology between ApoA and plasminogen, Lp(a) interferes with the fibrinolytic system by competing with plasminogen for binding sites on endothelial cells promoting thrombosis.[24]

SPECIAL SITUATIONS

Type 2 Diabetes Mellitus

Insulin resistance to characteristic dyslipidemia, in the form of high TG, low HDL levels with high VLDL and LDL levels. Apart from this, lipoproteins can undergo functional modification by forming advanced glycosylation end products (AGE) as AGE receptors are present on macrophages. Oxidative stress in type 2 diabetes mellitus (T2DM) is generated by ROS which promotes inflammation. Glycated lipoproteins along with ROS promote action and level of pro-inflammatory cytokines and impair endothelial function in the arterial endothelium. Modification in the form of glycated and oxidized lipoproteins along with oxidative stress in T2DM ultimately promotes atherosclerosis.

Obesity

High visceral fat leads to more FFAs manifesting in elevated levels of VLDL and decreased HDL. Simultaneously, there are elevated levels of inflammatory cytokines such as TNF-α and IL-6, leptin, and resistin produced from adipose tissue which can augment inflammation. High lipoprotein levels along with their altered form and inflammatory cytokines accelerate atherosclerosis.[25]

KEY POINTS

- Lipoproteins have a more biological role in the early stages of atherosclerosis; monocyte migration and attachment into the intima and maturation to form lipid-laden foam cells represent key steps in the formation of the fatty streak, the precursor to advanced atherosclerotic plaques.
- ApoB lipoproteins such as LDL, VLDL, and Lp(a) are highly atherogenic particles responsible for most cases of atherosclerosis.
- Low-density lipoprotein mediates proatherogenic function by inducing vascular dysfunction, proliferation, inflammation, apoptosis, oxidative stress, procoagulant effects; this role is exacerbated by oxidative modification of LDL (oxLDL).
- Postprandial remnant lipoproteins or TRLs have been found to explain the "residual risk" in atherosclerosis.
- High-density lipoprotein is classically the only antiatherogenic particle. But the complete or partial loss of the functional capacity of HDL in conditions of inflammation and oxidative stress and dyslipidemic states potentially amplify the role of modified LDL and may convert HDL itself into a proatherogenic one.

REFERENCES

1. Badimon L, Martínez-González J, Llorente-Cortés V, Rodríguez C, Padró T. Cell biology and lipoproteins in atherosclerosis. Curr Mol Med. 2006;6(5):439-56.
2. Shapiro MD, Fazio S. Apolipoprotein B-containing lipoproteins and atherosclerotic cardiovascular disease. F1000Res. 2017;6:134.
3. Feng M, Rached F, Kontush A, Chapman MJ. Impact of lipoproteins on atherobiology: Emerging insights. Cardiol Clin. 2018;36(2):193-201.
4. Ference BA. Causal effect of lipids and lipoproteins on atherosclerosis: Lessons from genomic studies. Cardiol Clin. 2018;36(2):203-11.

5. Morita SY. Metabolism and modification of apolipoprotein b-containing lipoproteins involved in dyslipidemia and atherosclerosis. Biol Pharm Bull. 2016;39(1):1-24.
6. Pirillo A, Bonacina F, Norata GD, Catapano AL. The interplay of lipids, lipoproteins, and immunity in atherosclerosis. Curr Atheroscler Rep. 2018;20(3):12.
7. Chistiakov DA, Melnichenko AA, Myasoedova VA, Grechko AV, Orekhov AN. Mechanisms of foam cell formation in atherosclerosis. J Mol Med (Berl). 2017;95(11):1153-65.
8. Kattoor AJ, Pothineni NVK, Palagiri D, Mehta JL. Oxidative Stress in Atherosclerosis. Curr Atheroscler Rep. 2017;19(11):42.
9. Arsenault BJ, Bourgeois R, Mathieu P. Do Oxidized Lipoproteins Cause Atherosclerotic Cardiovascular Diseases?. Can J Cardiol. 2017;33(12):1513-6.
10. Ahotupa M. Oxidized lipoprotein lipids and atherosclerosis. Free Radic Res. 2017;51(4):439-47.
11. Fogelstrand P, Borén J. Retention of atherogenic lipoproteins in the artery wall and its role in atherogenesis. Nutr Metab Cardiovasc Dis. 2012;22(1):1-7.
12. Borén J, Williams KJ. The central role of arterial retention of cholesterol-rich apolipoprotein-B-containing lipoproteins in the pathogenesis of atherosclerosis: a triumph of simplicity. Curr Opin Lipidol. 2016;27(5):473-83.
13. Peng J, Luo F, Ruan G, Peng R, Li X. Hypertriglyceridemia and atherosclerosis. Lipids Health Dis. 2017;16(1):233.
14. Talayero BG, Sacks FM. The role of triglycerides in atherosclerosis. Curr Cardiol Rep. 2011;13(6):544-52.
15. Zilversmit DB: Atherogenesis: a postprandial phenomenon. Circulation. 1979;60(3):473-85.
16. Nakajima K, Tanaka A. Postprandial remnant lipoproteins as targets for the prevention of atherosclerosis. Curr Opin Endocrinol Diabetes Obes. 2018;25(2):108-17.
17. Nakajima K, Tanaka A. Atherogenic postprandial remnant lipoproteins; VLDL remnants as a causal factor in atherosclerosis. Clin Chim Acta. 2018;478:200-15.
18. Botham KM, Wheeler-Jones CP. Postprandial lipoproteins and the molecular regulation of vascular homeostasis. Prog Lipid Res. 2013;52(4):446-64.
19. Toth PP. Triglyceride-rich lipoproteins as a causal factor for cardiovascular disease. Vasc Health Risk Manag. 2016;12:171-83.
20. Hu J, Xi D, Zhao J, Jinzhen Zhao, Tiantian Luo, Jichen Liu, et al. High-density lipoprotein and inflammation and its significance to atherosclerosis. Am J Med Sci. 2016;352(4):408-15.
21. Wang HH, Garruti G, Liu M, Portincasa P, Wang DQ. Cholesterol and lipoprotein metabolism and atherosclerosis: Recent advances in reverse cholesterol transport. Ann Hepatol. 2017;16(Suppl 1: s3-105.):s27-42.
22. Orekhov AN. The role of modified and dysfunctional lipoproteins in atherogenesis. Curr Med Chem. 2019;26(9):1509-11.
23. Orekhov AN, Sobenin IA. Modified and dysfunctional lipoproteins in atherosclerosis: Effectors or biomarkers? Curr Med Chem. 2019;26(9):1512-24.
24. Maranhão RC, Carvalho PO, Strunz CC, Pileggi F. Lipoprotein(a): structure, pathophysiology, and clinical implications. Arq Bras Cardiol. 2014;103(2):76-84.
25. Libby P, Buring JE, Badimon L, Hansson GK, Deanfield J, Bittencourt MS, et al. Atherosclerosis. Nat Rev Dis Primers. 2019;5(1):56.

CHAPTER
3

Apolipoprotein B and Cardiovascular Disease

Hridish Narayan Chakravarti

ABSTRACT

Traditional teaching on lipid metabolism since time immemorial has driven home the point that cholesterol is the principal culprit in atherogenesis. Low-density lipoprotein cholesterol (LDL-C) has been the focus of most lipid guidelines in primary or secondary prevention of cardiovascular disease. As more researches in lipidology pour in, we are becoming increasingly aware of the role of apolipoprotein B (ApoB) becoming the center of attention in atherosclerosis. Better understanding of the pathobiology is a prerequisite for passing on the benefits of modern research in cardiovascular disease prevention and treatment to the patients.

INTRODUCTION

Atherosclerosis is a process in the vascular tree that starts from fatty streaks and may end up with a clinical event, if goes unchecked.[1] The understanding of this continuum from streaks to strokes (brain, heart) is a central issue of preventive clinical cardiology.

The parameters that we examine in a standard lipid profile panel include total cholesterol (TC), triglycerides (TG), low-density lipoprotein cholesterol (LDL-C), and high-density lipoprotein cholesterol (HDL-C). The central dogma of atherosclerosis has been that LDL-C is bad and HDL-C is good.[2] However, this is an over-simplification. Though most guidelines force the physician to first look at LDL-C value in the lipid profile report, there is much more than that meets the eye. It is well accepted that LDL-C targeting is of prime importance, but other factors such as non HDL-C also have gained importance in recent times.[3] TG is considered to be a weak independent atherosclerotic risk factor. More recent studies have pointed figures at lipoprotein a [Lp(a)] and small dense LDL (sdLDL) as upcoming risk factors.[4,5]

WHAT ARE LIPOPROTEINS?

The food that we consume has fat in them. The fats are packaged, transported, metabolized, and assimilated through a complex cascade of biochemical reactions. As fat is sparsely soluble in water, they need a vehicle to get transported. Lipoproteins, as the name suggests, perfectly fill in this gap, and they help to transport lipids from one corner of the body to another. A lipoprotein is basically a biochemical assembly which helps to ferry hydrophobic lipid molecules in an aqueous environment.[6] An apolipoprotein (Apo) is the protein that binds to fats to create a lipoprotein. Most apolipoproteins are produced in the liver and intestine, but minute amounts of ApoB can be made in cardiac tissue and ApoE in macrophages.

Types of Lipoproteins

Based on their densities, lipoproteins can be classified in ascending order as:-

Chylomicrons (0.94 g/mL) < VLDL (0.94–1.006 g/mL) < IDL (1.006–1.019 g/mL) < LDL (1.019–1.063 g/mL) < HDL (1.063–1.2 g/mL)

As we move from left to right, their size decreases.

Types of Apolipoproteins

The major ones are A to E.

ApoB: It exists in two major types: ApoB-100 and ApoB-48.

Human ApoB-100 is a secretary glycoprotein with 4,536 amino acids. It has a major role in hepatic VLDL assembly. A truncated version, ApoB-48, is seen in the small intestine due to unique m-RNA editing and is useful in chylomicron formation. ApoB is water insoluble (hydrophobic) and it remains as a scaffold providing the structural integrity to the lipoprotein molecule.[7,8] There is unit to unit stoichiometry, i.e., 1 LDL particle will have 1 ApoB molecule. The ApoB stays with the particle right from formation to metabolism of the lipoprotein. ApoB helps to bind to LDL-R (LDL-Receptor) but not to other members of the LDL-R family.

The ApoB particles include chylomicrons, chylomicron remnants, VLDL, IDL, LDL, and Lp (a): The first two are associated with ApoB-48 while the rest have ApoB-100. As each particle has one molecule of ApoB, so total ApoB indicate the total number of atherogenic particles that can penetrate the arterial walls (**Fig. 1**).

Role in Atherosclerosis

Atherosclerosis is a multifactorial disease. There are hemodynamic factors such as areas in the arterial tree with low shear stress, nonpulsatile, nonlaminar flow, etc. Areas of arterial branching and vessel curvature may be particularly susceptible. These specific areas may interact in a way with systemic factors that are conducive to atherogenesis. As for example, low shear areas may demonstrate increased activation of endothelial cells, and expression of adhesion molecules ultimately facilitating inflammatory cells recruitment to the subendothelial space. Additionally, these susceptible areas may thwart or impair endothelio-protective functions. Increased affinity for LDL at these sites may be responsible for enhanced atherosclerosis susceptibility.[9]

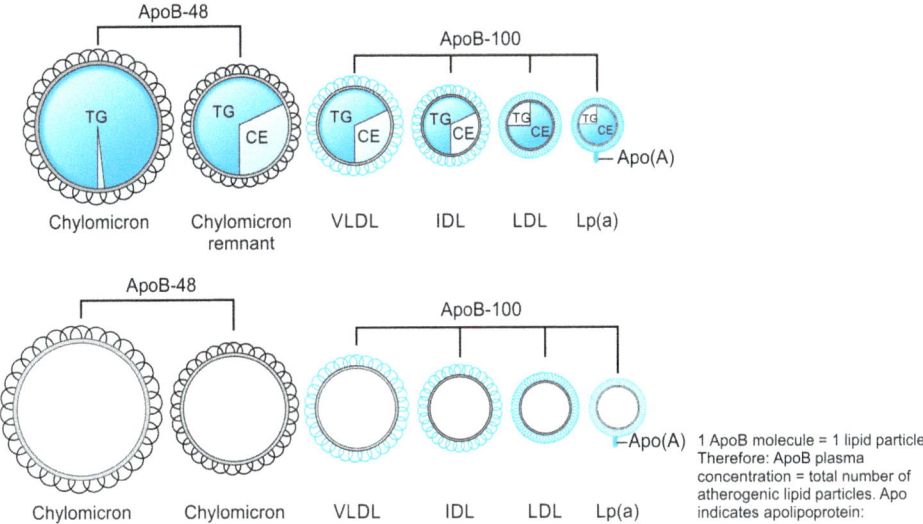

[CE: cholesterol ester; IDL: intermediate-density lipoprotein; LDL: low-density lipoprotein; Lp(a): lipoprotein(a); TG: triglyceride; VLDL: very low-density lipoprotein]

FIG. 1: Apolipoprotein B particles and cardiovascular disease.
Source: Sniderman AD, Thanassoulis G, Glavinovic T, Navar AM, Pencina M, Catapano A. Apolipoprotein B particles and cardiovascular disease: A Narrative Review. JAMA Cardiol. 2019;4(12):1287-95.

The process of atherosclerosis starts with percolation and subsequent subendothelial retention of ApoB particles [VLDL, LDL, Lp(a)].[10-13]

How much deposition occurs is directly correlated with arterial luminal exposure of cholesterol rich lipoproteins.[10] About 85% of lipoprotein delivery to subendothelial occurs by transcytosis while a small amount utilizes gap junction to enter.[14]

Any ApoB containing particles which is >70 nm in diameter cannot enter due to size restriction.[15]

The extracellular matrix has an important role in retaining the lipoproteins that have creeped into the subendothelium.[16] The proteoglycans (PG) of the extracellular matrix are heparan sulfate (HS), dermatan sulfate (DS), and chondroitin sulfate (CS).[17]

The CS component is particularly notorious in retaining the atherogenic lipoproteins. The bonhomie between the lipoproteins and the PG components is driven by an ionic charge-related interaction. Acting as a catalyst in this process is the lipoprotein lipase (LPL) that creates a bridge between the PGs and lipoproteins. LPL here is generated by macrophages and smooth muscle cells.[18,19]

In the subendothelium, lipoprotein particles get modified and aggregated. These bigger aggregated particles now tend to bind with more affinity to the PGs and the large size disallows them to come out of the subendothelial space.

Hence, it is a one way movement for these lipoprotein particles. The lipoproteins now undergo oxidation and it leads to a cascade where there is upregulation of cell adhesion molecules; the blood borne monocytes are chemoattracted, they convert to macrophages. These macrophages and smooth muscle cells then devour the excess cholesterol in lipoproteins and get transformed into foam cells storing large amounts of cholesteryl esters. The cholesterol in these ApoB particles (LDL) behaves as "passengers", but the intact particles drive atherogenesis.

From the above discussion, it becomes apparent that ApoB particles have a pioneering role in atherogenesis.

ApoB Particles Model of Atherogenesis

The question that baffled researchers is that "Is the ApoB related atherogenic risk more due to the mass of cholesterol going into the arterial wall or related to the number of ApoB particles trapped in the arterial wall?"

If mass gains more importance, the LDL-C and non HDL-C are better markers, while if particle number is important, then ApoB is a more accurate marker.

The ACC/AHA guidelines acknowledge that non HDL-C and ApoB are more accurate markers of CV risk than LDL-C.[20]

Several, but not all, epidemiological studies mention that ApoB is more predictive than non HDL-C.

Studies of ApoB

In Primary Prevention

A meta-analysis by Thompson et al. showed ApoB as a clear risk predictor of coronary heart disease (CHD).

The AMORIS study provided more compelling evidence; ApoB remained a significant predictor of MI. Here, LDL-C was an insignificant risk predictor in women and only modestly associated with MI in men.

Most of the published prospective studies of ApoB showed significant CAD association, even after adjustment for nonlipid risk factors.[21]

In Secondary Prevention

Baseline ApoB was a significant marker of repeat CV events in 4S, LIPID, and other studies.[21]

Low-density Lipoprotein Cholesterol Lowering and Residual Risk

Despite LDL-C targeting and lowering, it seems a residual risk persists that might suggest that a reduction in ApoB (or LDL particles), rather than LDL-C may be a better target.

Statins may be successful in targeting LDL-C, but may not be able to reduce LDL-particles concentration, to such a level where residual risk can be eliminated.

FUTURE DIRECTION

Mipomersen, an ApoB antisense oligonucleotide reduces LDL-C, ApoB, and Lp(a).[22] Lomitapide, a microsomal triglyceride transfer protein inhibitor, reduces LDL-C, ApoB, and TG.[23] The detailed discussion of these newer molecules are beyond the scope of this chapter.

From the preceding discussion, it emerges as a fact that ApoB may be a strong candidate to gain entry into standard lipid profile parameters. Its estimation does not mandatorily require fasting state and its estimation may be a better predictor of CVD than the standard parameters examined at present.

CONCLUSION

The concept of LDL cholesterol targeting in therapeutic lipidology is seeing a gradual shift. Though clinicians look at a standard lipid profile report through the prism of LDL cholesterol primarily, enough data is emerging to shift focus on ApoB as an important marker of atherosclerotic cardiovascular disease (ASCVD).

> **KEY POINTS**
> - Established concept that cholesterol as a whole is the sole culprit in atherogenesis is rapidly changing.
> - Increased plasma concentrations of cholesterol-rich ApoB-containing lipoproteins are causatively linked to ASCVD.
> - Incorporating ApoB as an investigative marker and targeting it appropriately for treatment may mitigate a lot of the burden of CVD in the society.

REFERENCES

1. Borén J, Williams KJ. The central role of arterial retention of cholesterol-rich apolipoprotein-B containing lipoproteins in the pathogenesis of atherosclerosis: a triumph of simplicity. Curr Opin Lipidol. 2016;27(5):473-83.
2. Elshourbagy NA, Meyers HV, Abdel-Meguid SS. Cholesterol: the good, the bad, and the ugly - therapeutic targets or the treatment of dyslipidemia." Med Princ Pract. 2014;23(2):99–111.
3. Soran H, Dent R, Durrington P. Evidence-based goals in LDL-C reduction. Clin Res Cardiol. 2017;106(4):237-48.
4. Nordestgaard BG, Chapman MJ, Ray K, Borén J, Andreotti F, Watts GF, et al. Lipoprotein(a) as a cardiovascular risk factor: current status. Eur Heart J. 2010;31(23):2844-53.
5. Hirayama S, Miida T. Small dense LDL: An emerging risk factor for cardiovascular disease. Clin Chim Acta. 2012;414:215-24.
6. Lent-Schochet D, Jialal I. Biochemistry, Lipoprotein Metabolism. Treasure Island (FL): StatPearls Publishing; 2020.
7. Chan L. Apolipoprotein B, the major protein component of triglyceride-rich and low density lipoproteins. J Biol Chem. 1992;267:25621-4.
8. Young SG. Recent progress in understanding apolipoprotein B. Circulation. 1990;82(5):1574-94.
9. Shapiro MD, Fazio S. Apolipoprotein B-containing lipoproteins and atherosclerotic cardiovascular disease. F1000Res. 2017;6:134.
10. Proctor SD, Vine DF, Mamo JC. Arterial retention of apolipoprotein B(48)- and B(100)-containing lipoproteins in atherogenesis. Curr Opin Lipidol. 2002;13(5):461-70.
11. Skålén K, Gustafsson M, Rydberg EK, Hultén LM, Wiklund O, Innerarity TL, et al. Subendothelial retention of atherogenic lipoproteins in early atherosclerosis. Nature. 2002;417(6890):750-4.
12. Tabas I, Williams KJ, Borén J. Subendothelial lipoprotein retention as the initiating process in atherosclerosis: update and therapeutic implications. Circulation. 2007;116(16):1832-44.
13. Kinnunen PK, Holopainen JM. Sphingomyelinase activity of LDL: a link between atherosclerosis, ceramide, and apoptosis? Trends Cardiovasc. Med. 2002;12:37-42.
14. Simionescu M, Simionescu N. Proatherosclerotic events: pathobiochemical changes occurring in the arterial wall before monocyte migration. FASEB J. 1993;7:1359-66.
15. Nordestgaard BG, Wootton R, Lewis B. Selective retention of VLDL, IDL, and LDL in the arterial intima of genetically hyperlipidemic rabbits in vivo. Molecular size as a determinant of fractional loss from the intima-inner media. Arterioscler Thromb Vasc Biol. 1995;15:534-42.
16. Öörni K, Pentikäinen MO, Ala-Korpela M, Kovanen PT. Aggregation, fusion, and vesicle formation of modified low density lipoprotein particles: molecular mechanisms and effects on matrix interactions. J Lipid Res. 2000;41:1703-14.

17. Pillarisetti S. Lipoprotein modulation of subendothelial heparin sulfate proteoglycans (perlecan) and atherogenicity. Trends Cardiovasc Med. 2000;10:60-5.
18. Ylä-Herttuala S, Lipton BA, Rosenfeld ME, Goldberg IJ, Steinberg D, Witztum JL. Macrophages and smooth muscle cells express lipoprotein lipase in human and rabbit atherosclerotic lesions. Proc Natl Acad Sci USA. 1991;88:10143-7.
19. O'Brien KD, Gordon D, Deeb S, Ferguson M, Chait A. Lipoprotein lipase is synthesized by macrophage-derived foam cells in human coronary atherosclerotic plaques. J Clin Invest. 1992;89:1544-50.
20. Grundy SM, Stone NJ, Bailey AL, Beam C, Birtcher KK, Blumenthal RS, et al. 2018 AHA/ACC/AACVPR/AAPA/ABC/ACPM/ ADA/AGS/APhA/ASPC/NLA/PCNA guideline on the management of blood cholesterol. Circulation. 2019;139:e1082–e1143.
21. Contois JH, McConnell JP, Sethi AA, Csako G, Devaraj S, Daniel M, et al. Apolipoprotein B and Cardiovascular Disease Risk: Position Statement from the AACC Lipoproteins and Vascular Diseases Division Working Group on Best Practices. Clin Chem. 2009;55(3):407-19.
22. Wong E, Goldberg T. Mipomersen (kynamro): a novel antisense oligonucleotide inhibitor for the management of homozygous familial hypercholesterolemia. P T. 2014;39(2):119-22.
23. Alonso R, Cuevas A, Mata P. Lomitapide: a review of its clinical use, efficacy, and tolerability. Core Evid. 2019;14:19-30.

CHAPTER 4

Lipoprotein(a): An Update

Tapas Chandra Das

ABSTRACT

Lipoprotein(a) is a unique lipoprotein which has causal, independent, intermediate, genetic risk factor for atherosclerotic cardiovascular disease (ASCVD). Plasma level of Lp(a) >30 mg/dL is atherogenic. Lp(a) level should be done at least once in a lifetime for cardiovascular risk assessment. There is no approved therapy for elevated Lp(a). As it is a genetic disease RNA directed therapy, antisense oligonucleotide (ASO), acting on the liver is the emerging therapy for elevated Lp(a).

INTRODUCTION

The traditional lipid profile has served as mainstay of atherosclerotic cardiovascular disease (ASCVD) risk assessment for many decades and targeted therapies such as statins, ezetamib and PCSK9 inhibitors have become available. In spite of these therapies, ongoing risk of many patients needs to search and investigate others additional risk factors. Lipoprotein(a) [Lp(a)] is a promising biomarker to help refine current strategies of atherosclerotic cardiovascular disease (ASCVD) risk assessment and aortic stenosis. Lp(a) level is estimated to be elevated in 20% of world population. Although not yet approved specific therapy for elevated Lp(a), research on Lp(a) suggested its role in preventive medicine and it is time for cardiologist and cardiovascular team member to perform routine testing of blood Lp(a) level in clinical practice.[1]

LIPOPROTEIN(A) STRUCTURE AND FUNCTION

Lipoprotein(a) is a lipoprotein by its name and often described as low-density lipoprotein (LDL) particle with an extra; the extra is apolipoprotein A that has been covalently linked to ApoB 100. ApoA has substantial protein homology to plasminogen, required for endogenous thrombolytic response, and contain multiple copies of plasminogen Kringle IV, a single copy of plasminogen Kringle V and inactive protease domains.[2] Total concentration of Lp(a) particles instead of apoA isoform size, mediates the relationship between Lp(a) excess and ASCVD risk.[3]

MECHANISM AFFECTING ATHEROSCLEROTIC CARDIOVASCULAR DISEASE RISK

Lp(a) excess may promote atherosclerosis as well as thrombosis by:
- Interfering with fibrinolysis. ApoA has close structure with plasminogen and Lp(a) inhibits fibrinolysis by contest with plasminogen to the molecules (**Fig. 1**), this blunts plasminogen action and the process of fibrinolysis.[4,5]
- Binding to the macrophage. Lp(a) also binds to macrophages by high affinity receptor that promotes foam cell formation and deposition of cholesterol in atherosclerotic plaque (**Fig. 2**). Lp(a) is ubiquitous in human coronary atheroma, colocalizes with plaque macrophages, and is detected in large amounts in tissue from culprit lesions in patients with unstable compared with stable coronary diseases.[6]
- Binding to the endothelium and components of extracellular matrix, this may interfere with normal endothelial function.[7]
- Increased expression of intercellular adhesion molecule-1 resulting in the recruitment monocytes to the arterial wall and binding to the macrophage, promoting foam cell formation and the localization of Lp(a) in the atherosclerotic plaque.[8]
Enhancing susceptibility of LDL to oxidative modification.[9]

GENETICS

Serum Lp(a) levels are genetically determined, hugely variable and largely unaffected by environmental factors. Lp(a) level differences exist between populations, suggesting ethnics difference in the control of Lp(a). African population, black Americans, and Asian

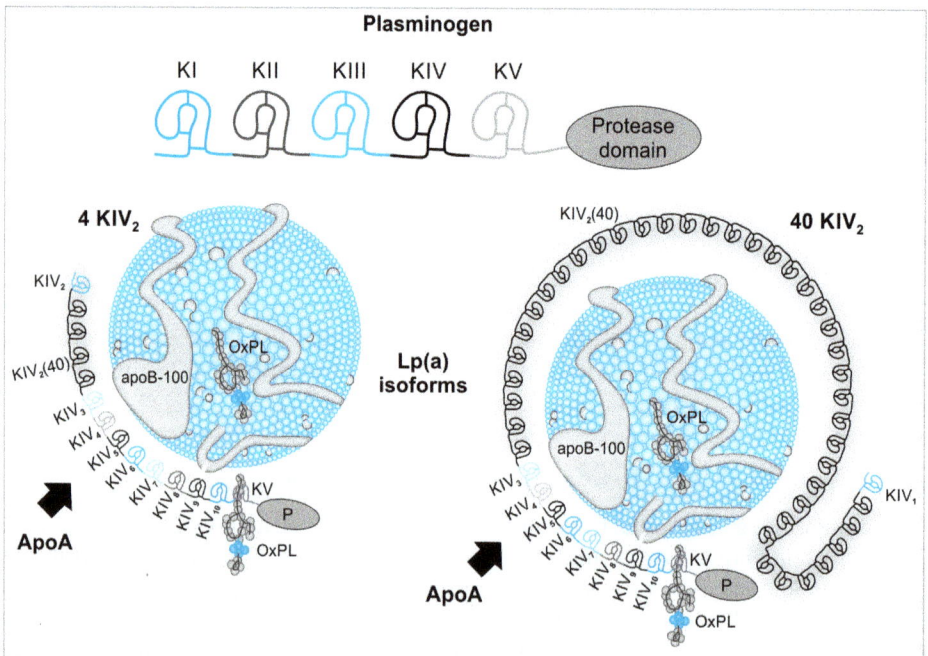

FIG. 1: Lipoprotein(a) and plasminogen structure.[4]

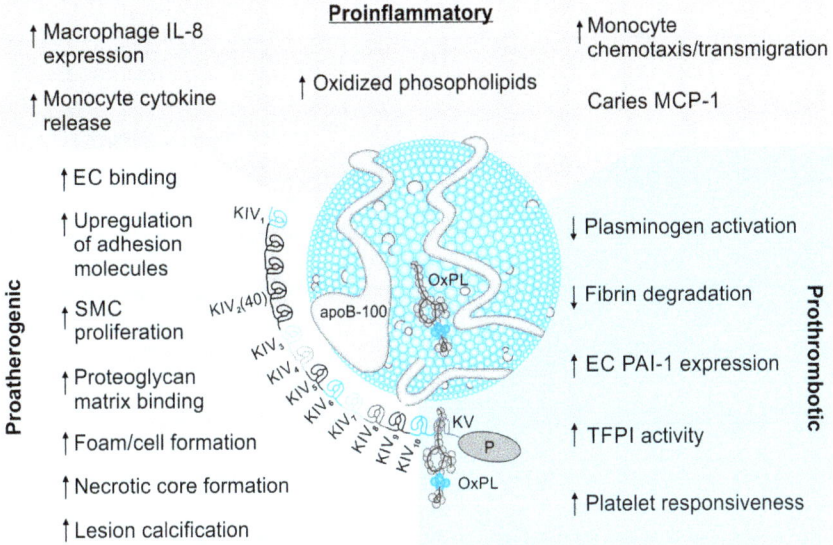

FIG. 2: Mechanism of atherosclerosis by lipoprotein(a) [Lp(a)].[4]

Indians have higher mean Lp(a) levels as compared with Caucasians and other races (**Fig. 3 and Table 1**). *Heterogeneous ApoA* gene is responsible for much of its variations.[10] Strong inverse relationship between the size of ApoA isoforms and Lp(a) concentration exists.[11-13] The molar concentration concentrations of Lp(a), not the ApoA size determine the risk of coronary artery disease.[3]

MEASUREMENT OF SERUM LP(A) CONCENTRATIONS

Measurement and reporting of Lp(a) levels are complicated and widely misunderstood. Several Lp(a) assays for clinical use, mostly measure Lp(a) mass but some estimate Lp(a) cholesterol. Some assays measure Lp(a) mass concentration (mg/dL) by *nephelometry* but these measure vary by the number of ApoA kringle repeats.[14] Newer assays measure Lp(a) mass concentration (nmol/L) by enzyme-linked immunosorbent assay (ELISA) which is not influenced by variability in ApoA isoforms. Lp(a) mass by ELISA better predicts atherosclerotic cardiovascular disease than measures using *nephelometry*.[15] Measurements cannot be converted accurately from mg/dL to nmol/L and vice versa. One conversion is to multiply the Lp(a) mass by 2.4 to provide a rough estimate of Lp(a) concentrations; it varies by the molecular mass of the typical Lp(a) in a population and is not recommended for clinical use.[16,17] Lp(a) levels are stable in healthy individual over times, though age-related increases in Lp(a) concentrations due to sex steroid deficiency and reduction of renal functions can be observed. High quality evidence supports a link between Lp(a) and variable cardiovascular outcome. One meta-analysis showed an increased risk of chronic heart disease (CHD) and myocardial infarction (MI) with concentration >30 mg/dL (62 nmol/L); INTERHEART trial showed Lp(a) >50 mg/dL is associated with increased risk of MI.

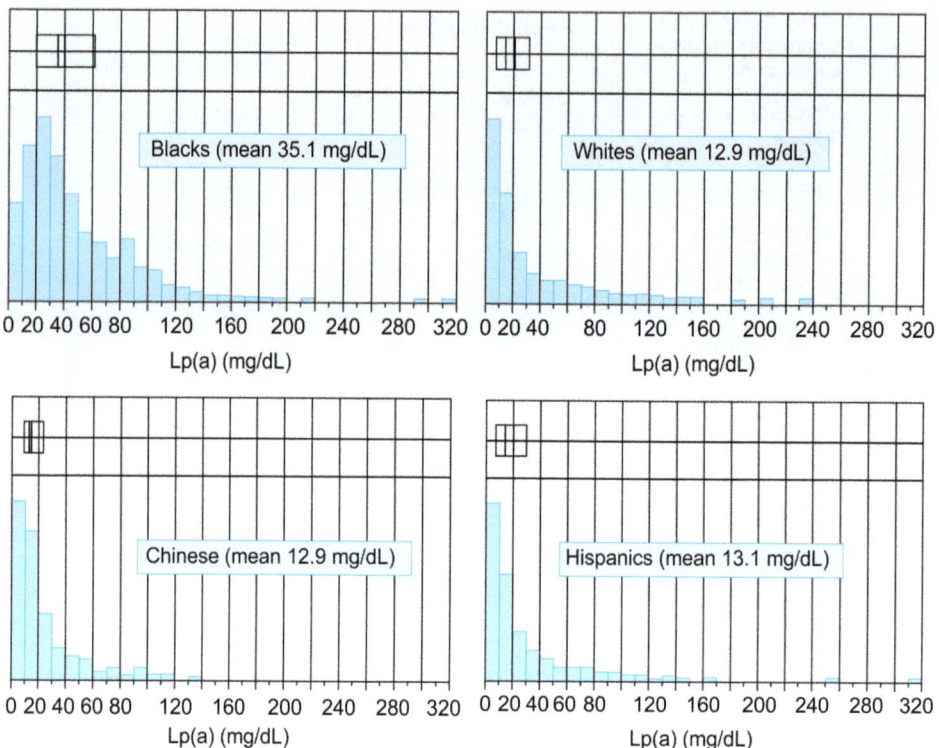

FIG. 3: Racial/ethnic differences of lipoprotein(a) [Lp(a)] level.[13]

TABLE 1: Prevalence of elevated Lp(a): US and globally.[13]				
Prevalence	Top 20%	Top 10%	Top 5%	Top 1%
Lp(a) level	60 mg/dL	90 mg/dL	116 mg/dL	180 mg/dL
Number (USA)	64 million	30 million	16 million	3.2 million
Number (EU)	150 million	75 million	37.5 million	7.5 million
Number globally	1.4 billion	700 million	350 million	7 million

ATHEROSCLEROTIC CARDIOVASCULAR DISEASE RISK DUE TO LP(A) EXCESS

Lipoprotein(a) excess is an independent risk factor for ASCVD and ASCVD events. Genetic and experimental data suggest it contributes to the pathogenesis of the disease.[18]

A 2018 meta-analysis evaluated the impact of elevated Lp(a) on the risk of ASCVD events in 29,029 individuals with high baseline ASCVD risk (including established ASCVD) using patient-level data from seven placebo-controlled, randomized statin trials.[18] Hazard ratios (HRs) for cardiovascular disease (CVD) events (e.g., fatal and nonfatal coronary heart disease, stroke, or revascularization procedures) were

calculated across Lp(a) concentration groups of normal, borderline elevated, and elevated. This analysis showed that:
- Lipoprotein(a) level on statin treatment with CVD risk were linear with extension of risk with Lp(a) value >30 mg/mL from base line value as compared with >50 mg/dL on statin values.
- After multivariable adjustment, the baseline HRs in individuals with Lp(a) of 15–<30 mg/dL, 30–<50 mg/dL, and ≥50 mg/dL were 1.04 (95% CI 0.91–1.20), 1.13 (95% CI 1.02–1.25), and 1.36 (95% CI 1.11–1.66), compared with Lp(a) <15 mg/dL, respectively. In statin-treated patients, respective HRs were 0.95 (95% CI 0.82–1.11), 1.08 (95% CI 0.95–1.23), and 1.42 (95% CI 1.16–1.74).

In an analysis of statin-treated CVD patients enrolled in the FOURIER study, which compared evolocumab with placebo, patients with baseline Lp(a) in the highest quartile (>165 nmol/L) had a higher risk of coronary heart disease death, myocardial infarction, or urgent revascularization compared with those in the lowest quartile (HR 1.22, 95% CI 1.01–1.48) (**Fig. 4**).[19] In post-acute-coronary-syndrome patients enrolled in ODYSSEY Outcomes, a 1 mg/dL reduction in Lp(a) mass was associated with an HR of 0.994 (95% CI 0.990–0.999) for ASCVD events.[19]

In a systematic review including six studies,[20] the relative risk for ischemic stroke related to Lp(a) excess was 2.14 (95% CI 1.85–2.97). Lp(a) levels[21] are associated with cerebrovascular disease; however, the risk estimates are widely variable, depending on the population studied and modeling strategy; risk may be stronger in men than in women.

In 2019, a large observational study reported that Lp(a) levels were associated with an increased risk of mortality. Having an Lp(a) >93 mg/dL (199 nmol/L) was associated with an HR of 1.5 (95% CI 1.28–1.76) for cardiovascular mortality and 1.20 (95% CI 1.10–1.30) for all-cause mortality.[15]

SCREENING

The National Lipid Association stated that Lp(a) testing is reasonable to refine ASCVD risk in adults with:[1]
- First-degree relatives with premature ASCVD (<55 years of age in men; <65 years of age in women).
- A personal history of premature ASCVD.

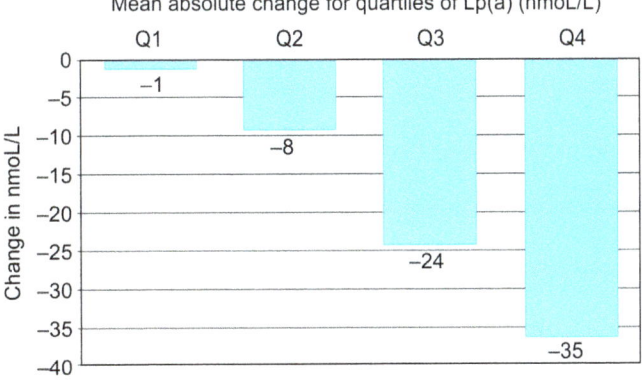

FIG. 4: Fourier effect of evolocumab on lipoprotein(a).[4]

♦ Primary severe hypercholesterolemia (LDL-C ≥190 mg/dL) or suspected familial hypercholesterolemia. The prevalence of Lp(a) excess is higher in people with heterozygous familial hypercholesterolemia than those without it and further increases ASCVD risk.[22]

Lipoprotein(a) testing may be reasonable in adults:
♦ Recurrent CVD despite statins.
♦ ≥3%, 10 years risk of fatal CVD.
♦ ≥10%, 10 years risk of fatal/nonfatal CHD.

MANAGEMENT

Lifestyle changes have minimal effects on reducing elevated Lp(a). Statins do not lower Lp(a) and may cause slight increase that do not appear to mitigate their beneficial effects on ASCVD events reduction. Ezetimibe does not lower Lp(a). Estrogen replacement reported to lower Lp(a) by 37%, does not recommend for CVD risk reduction.[25] Niacin 1–3 g/day lower Lp(a) by 30–40% but not accompanied by improved clinical outcome.[23] PCSK9 inhibitors[19] modestly lower Lp(a) by 25% but it is costly. Lipoprotein apheresis can significantly reduce Lp(a) level as much as 75%, may decrease events but is cumbersome.[24]

EMERGING THERAPIES FOR LOWERING OF LP(A)

Antisense therapy as elevated Lp(a) is genetically mediated. ASO is a promising therapy that reduces Lp(a) by inhibiting production of ApoA. ApoA-L$_{RX}$, is a newer antisense oligonucleotide that lowered Lp(a) in a dose dependent fashion by as much as 80% and also reduced oxidized phospholipid, leading to a more than 10 times lower dose and improved tolerability compared to native ASO (**Fig. 5**).[26,27]

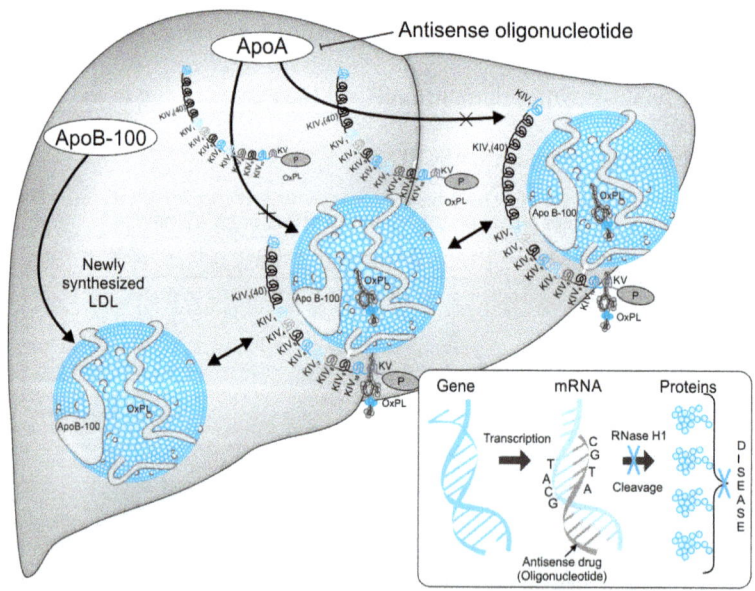

FIG. 5: Mechanism of action of antisense oligonucleotide.[27]

CONCLUSION

Lipoprotein(a) is a strongest, single inherited risk factor for ASCVD events. The risk of ASCVD increases linearly with Lp(a) concentration. ASCVD increases are clinically relevant as Lp(a) concentrations exceed 30–50 mg/dL and especially when exceed 180 mg/dL. Based on available evidence, Lp(a) level is usually estimated on targeted screening. No clinical trials have adequately proved that Lp(a) reduction reduces future ASCVD events. Patients with elevated Lp(a) who have reached their LDL-C target or who had all recommended therapies to lower LDL-C attempted, additional therapies to lower Lp(a) are approached, but still not recommended. Therapy such as antisense oligonucleotide is under clinical trial, and if successful they will become the important mode of interventions to lower Lp(a).

KEY POINTS

- Lipoprotein(a) is causal independent risk factors for ASCVD.
- There is no broad agreement on when to screen for Lp(a) excess; only targeted screening is done.
- None of the clinical trials have adequately tested the hypothesis that Lp(a) reduction reduces the incidence of future ASCVD events.
- As it is a genetic disease gene silencing approach, antisense oligonucleotide may be the promising therapy for elevated Lp(a).

REFERENCES

1. Wilson DP, Jacobson TA, Jones PH, Koschinsky ML, McNeal CJ, Nordestgaard BG, et al. Use of Lipoprotein(a) in clinical practice: a biomarker whose time has come. A Scientific statement from national lipid Association. J Clin Lipidol. 2019;13:374-92.
2. Steyrer E, Durovic S, Frank S, Giessauf W, Burger A, Dieplinger H, et al. The role of lecithin: cholesterol acyltransferase for lipoprotein(a) assembly. Structural integrity of low density lipoproteins is a prerequisite for Lp(a) in human plasma. J Clin Invest. 1994;94:2330-40.
3. Gudbjartsson DF, Thorgeirsson G, Sulem P, Helgadottir A, Gylfason A, Saemundsdottir J, et al. Lipoprotein(a) Concentration and Risks of Cardiovascular Disease and Diabetes. J Am Coll Cardiol. 2019;74:2982-94.
4. Tsimikas S. Lipoprotein(a): Diagnosis, Prognosis, Controversies, and Emerging Therpies. J Am Coll Cardiol. 2017;69:692-711.
5. Palabrica TM, Liu AC, Aronovitz MJ, Furie B, Lawn RM, Furie BC. Antifibrinolytic activity of apolipoprotein(a) in vivo: human apolipoprotein(a) transgenic mice are resistant to tissue plasminogen activator-mediated thrombolysis. Nat Med. 1995;1:256-9.
6. Dangas G, Mehran R, Harpel PC, Sharma SK, Marcovina SM, Dube G, et al. Lipoprotein(a) and inflammation in human coronary atheroma: association with the severity of clinical presentation. J Am Coll Cardiol. 1998;32: 2035-42.
7. Salonen EM, Jauhiainen M, Zardi L, Vaheri A, Ehnholm C. Lipoprotein(a) binds to fibronectin and has serine proteinase activity capable of cleaving it. EMBO J. 1989;8:4035-40.
8. Takami S, Yamashita S, Kihara S, Ishigami M, Takemura K, Kume N, et al. Lipoprotein(a) enhances the expression of intercellular adhesion molecule-1 in cultured human umbilical vein endothelial cells. Circulation. 1998;97: 721-8.
9. Riis Hansen P, Kharazmi A, Jauhiainen M, Ehnholm C. Induction of oxygen free radical generation in human monocytes by lipoprotein(a). Eur J Clin Invest. 1994;24:497-9.
10. Boerwinkle E, Leffert CC, Lin J, Lackner C, Chiesa G, Hobbs HH. Apolipoprotein(a) gene accounts for greater than 90% of the variation in plasma lipoprotein(a) concentrations. J Clin Invest. 1992;90:52-60.

11. Bowden JF, Pritchard PH, Hill JS, Frohlich JJ. Lp(a) concentration and apo(a) isoform size. Relation to the presence of coronary artery disease in familial hypercholesterolemia. Arterioscler Thromb. 1994;14:1561-8.
12. Guan W, Jing C, Steffen BT, Post WS, Stein JH, Tattersall MC, et al. Race is the key variable in assigning Lipoprotein(a) cut off value for CAD risk assessment. Arterioscler Thromb Vasc Biol. 2015;35(4):996-1001.
13. Steve Varbel, McConnell JP, Tsimikas S. Prevalence of Lp(a) mass level and patients threshold in 532359 US patients. Arterioscler Thromb Vasc Biol. 2016;36:2239-45.
14. Albers JJ, Kennedy H, Marcovina SM. Evidence that Lp[a] contains one molecule of apo[a] and one molecule of apoB: evaluation of amino acid analysis data. J Lipid Res. 1996;37:192-6.
15. Langsted A, Kamstrup PR, Nordestgaard BG. High lipoprotein(a) and high risk of mortality. Eur Heart J. 2019;40:2760-70.
16. Kamstrup PR, Tybjaerg-Hansen A, Steffensen R, Nordestgaard BG. Genetically elevated lipoprotein(a) and increased risk of myocardial infarction. JAMA. 2009;301:2331-9.
17. Marcovina SM, Albers JJ. Lipoprotein(a) measurements for clinical application. J Lipid Res. 2016;57:526-37.
18. Willeit P, Ridker PM, Nestel PJ, Simes J, Tonkin AM, Pedersen TR, et al. Baseline and on-statin treatment lipoprotein(a) levels for prediction of cardiovascular events: individual patient-data meta-analysis of statin outcome trials. Lancet. 2018;392:1311-20.
19. O'Donoghue ML, Fazio S, Giugliano RP, Stroes ESG, Kanevsky E, Gouni-Berthold I, et al. Lipoprotein(a), PCSK9 Inhibition, and Cardiovascular Risk. Circulation. 2019;139:1483-92.
20. Bittner VA, Szarek M, Aylward PE, Bhatt DL, Diaz R, Edelberg JM, et al. Effect of Alirocumab on Lipoprotein(a) and Cardiovascular Risk After Acute Coronary Syndrome. J Am Coll Cardiol. 2020;75:133-44.
21. Erqou S, Thompson A, Di Angelantonio E, Saleheen D, Kaptoge S, Marcovina S, et al. Apolipoprotein(a) isoforms and the risk of vascular disease: systematic review of 40 studies involving 58,000 participants. J Am Coll Cardiol. 2010;55:2160 -7.
22. Alonso R, Andres E, Mata N, Fuentes-Jiménez F, Badimón L, López-Miranda J, et al. Lipoprotein(a) levels in familial hypercholesterolemia: an important predictor of cardiovascular disease independent of the type of LDL receptor mutation. J Am Coll Cardiol. 2014;63:1982-9.
23. Guyton JR, Blazing MA, Hagar J, Kashyap ML, Knopp RH, McKenney JM, et al. Extended-release niacin vs gemfibrozil for the treatment of low levels of high-density lipoprotein cholesterol. Niaspan-Gemfibrozil Study Group. Arch Intern Med. 2000;160:1177-84.
24. Keller C. Apheresis in coronary heart disease with elevated Lp(a): a review of Lp(a) as a risk factor and its management. Ther Apher Dial. 2007;11:2-8.
25. Sacks FM, McPherson R, Walsh BW. Effect of postmenopausal estrogen replacement on plasma Lp(a) lipoprotein concentrations. Arch Intern Med. 1994;154:1106-10.
26. Tsimikas S, Karwatowska-Prokopczuk E, Gouni-Berthold I, Tardif J-C, Baum SJ, Steinhagen-Thiessen E, et al. Lipoprotein(a) Reduction in Persons with Cardiovascular Disease. N Engl J Med 2020; 382:244-55.
27. Tsimikas S, Viney NJ, Hughes SG, Singleton W, Graham MJ, Baker BF, et al. Antisense therapy Targeting Apoliprotein(a). Lancet. 2015;386:1472-83.

CHAPTER 5

Monogenic Lipoprotein Disorders: From Phenotype to Genotype

Sunetra Mondal, Subhodip Pramanik

ABSTRACT

Inherited causes of dyslipidemia can be polygenic or monogenic due to deficiencies of enzymes or apo-proteins involved in lipoprotein metabolism. Although less common than secondary causes, the inherited dyslipidemias exhibit characteristic clinical and biochemical picture and each disorder has unique therapeutic and prognostic implication. Genetic testing is available but often a systematic approach using only the biochemical and clinical picture can lead to the diagnosis. This review summarizes the existing knowledge on these disorders, provides algorithms for systematic approach to these disorders and focuses on their unique management issues.

INTRODUCTION

Disorders of lipoprotein metabolism, collectively termed as "dyslipidemias" mostly arise from a combination of genetic (often polygenic), environmental factors such as diet, lifestyle, and medical conditions such as diabetes mellitus, renal disease, or thyroid illnesses. However, specific molecular defects causing the deficiency or altered activity of multiple enzymes or cofactors involved in the different pathways of lipoprotein metabolism can lead to a spectrum of monogenic disorders.

CLASSIFICATION

The first classification of lipoprotein disorders was given by Fredrickson, Lees, and Levy (1967) based on the measurement of total plasma cholesterol and triglycerides (TGs) and they analyzed lipoprotein patterns after separation by electrophoresis (**Table 1**).[1] However, despite providing a useful conceptual framework, it has several drawbacks like not including high-density lipoprotein cholesterol (HDL-C) and not differentiating severe monogenic disorders from the more common polygenic disorders.

A more elaborate classification based on the chief biochemical abnormality and their corresponding molecular defect and clinical findings is given in **Table 2**.

TABLE 1: Fredrickson and Levy classification of lipoprotein disorders.

Type	Lipoprotein elevated	Lipid elevation
I	Chylomicrons	TG
IIa	LDL	Chol
IIb	LDL, VLDL	Chol > TG
III	IDL	Chol, TG
IV	VLDL	TG
V	Chylomicrons, VLDL	TG

(Chol: cholesterol; IDL: intermediate-density lipoprotein; LDL: low-density lipoprotein; TG: triglyceride; VLDL: very-low-density lipoprotein)

TABLE 2: Classification of monogenic lipoprotein disorders.

Genetic disorder	Protein (gene) defect	Lipoproteins elevated	Chief clinical findings
Hypertriglyceridemia (familial chylomicronemia syndrome)		Chylomicrons, VLDL	Eruptive xanthomas, pancreatitis, hepatosplenomegaly +/-
Lipoprotein lipase (LPL) deficiency	LPL (*LPL*)		
Familial ApoC II deficiency	Apo C-II (*APOC2*)		
ApoA-V deficiency	ApoA-V (*APOA5*)		
GPIHBP1 deficiency	GPIHBP1		
Hypercholesterolemia without hypertriglyceridemia		LDL	Tendon xanthomas, coronary artery disease
Familial hypercholesterolemia	LDL–receptor (*LDLR*)		
Autosomal dominant hypercholesterolemia type 2	ApoB-100 (*APOB*)		
Autosomal dominant hypercholesterolemia type 3	PCSK9 (*PCSK9*)		
Autosomal recessive hypercholesterolemia	LDL receptor adaptor protein (*LDLRAP*)		
Sitosterolemia	ABCG5 or ABCG8		
Combined hyperlipidemia (elevated cholesterol and triglycerides)		Remnant lipoproteins	
Familial dysbetalipoproteinemia	ApoE (*APOE*)	Chylomicron remnants, VLDL remnants	CHD, peripheral vascular disease, palmar and tuberoeruptive xanthomas
Familial hepatic lipase deficiency	Hepatic lipase (*LIPC*)	VLDL remnants, HDL	Pancreatitis, CHD

Continued

Continued

TABLE 2: Classification of monogenic lipoprotein disorders.

Genetic disorder	Protein (gene) defect	Lipoproteins elevated	Chief clinical findings
Low ApoB containing lipoproteins			
Abetalipoproteinemia	Microsomal TG transfer protein (*MTP*)		
Familial hypobetalipoproteinemia	Truncated ApoB protein (*APOB*)		
Familial combined hypolipidemia	Angiopoietin like 3 (*ANGPTL3*)		
PCSK9 deficiency	Proprotein convertase subtilisin/kexin type 9 (*PCSK9*)		
Low HDL-C			
APOA1 deficiency	*APOA1* gene mutations or gene deletions in APOA5-A1-C3-A4 locus		
Tangier disease	ABCA1 deficiency		
Familial LCAT deficiency	Complete or partial deficiency of lecithin-cholesterol acyl transferase (*LCAT*)		
Primary hypoalphalipoproteinemia	ABCA1 mutation (heterozygous)		
High HDL-C			
Cholesteryl ester storage disease	Lysosomal acid lipase (*LIPA*)		

(CHD: coronary heart disease; HDL-C: high-density lipoprotein cholesterol; LDL: low-density lipoprotein; TG: triglyceride; VLDL: very low-density lipoprotein)

MONOGENIC LIPID DISORDERS: PATHOGENETIC BASIS AND CLINICAL PRESENTATION

Lipid Disorder Associated With Elevated LDL-C and Normal Triglycerides

Familial Hypercholesterolemia

Familial hypercholesterolemia (FH) is characterized biochemically by elevated plasma levels of low-density lipoprotein cholesterol (LDL-C) but having normal TGs and clinically by tendon xanthomas, and premature coronary atherosclerosis. Inherited in an autosomal codominant pattern, FH is caused by a large number (>1,000) mutations in the LDL receptor gene.[2,3] The elevated levels of LDL-C in FH arise from a delayed removal of LDL from the blood as well as an increased production of LDL [since some of the intermediate-density lipoprotein (IDL) is cleared by LDL receptor-mediated endocytosis]. Individuals with both the LDL receptor alleles mutated (FH homozygotes) have much higher LDL-C levels than those having one mutant allele (FH heterozygotes).

Homozygous Familial Hypercholesterolemia

The prevalence of homozygous familial hypercholesterolemia (HoFH) is approximately one in a million persons. Based on the amount of LDL receptor (LDLR) activity, those with homozygous FH can be classified into either having <2% activity (receptor negative) or having 2–25% of activity (receptor defective). The typical LDL-C is between 500 to >1,000 mg/dL.[2] Without treatment, receptor-negative cases seldom live beyond 30 years. Receptor-defective cases have a slightly better prognosis but most develop atherosclerosis by the third decade.

Heterozygous Familial Hypercholesterolemia

One of the most common single gene disorder with a frequency of 1 in 250 individuals; it is characterized by LDL-C between 190 to 400 mg/dL. The risk of a myocardial infarction before 60 years of age is ~50% higher than the general population.

Familial Defective ApoB-100

The disease is rare outside Germany and the picture resembles heterozygous familial fypercholesterolemia (HeFH). Familial defective ApoB-100 (FDB) is caused by mutations in the LDL receptor-binding domain of ApoB-100, causing LDL to bind with reduced affinity.

Autosomal Dominant Hypercholesterolemia-3

It is caused by gain-of-function mutations in proprotein-convertase-subtilisin/kexin type 9 (PCSK9). PCSK9 binds to the LDL receptor so that after internalization of the LDL--LDL-R complex, the LDL-R is redirected to the lysosome rather than to the cell surface and the number of hepatic LDL-R is reduced.

Autosomal Recessive Hypercholesterolemia

Occurring due to mutations in LDLR adaptor protein, LDLRAP, autosomal recessive hypercholesterolemia (ARH) is a rare disorder (except in Sardinia, Italy). Due to the lack of LDLRAP, LDL binds to the LDL receptor but the receptor complex fails to be internalized.

Diagnosis of Familial Hypercholesterolemia

Three groups have developed diagnostic classification tools for FH patients and these include the Simon Broome Familial Hyperlipidemia Register (**Table 3**), the Dutch Lipid Clinic Network, and the Make Early diagnosis to Prevent Early Deaths (MEDPED). While the latter gives more objective cutoff points for diagnosing FH, the Simon Broome Register is easier to use in clinics (**Table 3**).[4]

TABLE 3: Simon Broome diagnostic criteria for familial hypercholesterolemia.	
Criteria	
A	Total cholesterol concentration >7.7 mmol/L in adults or >6.7 mmol/L in children <16 years of age; LDL-C concentration >4.9 mmol/L in adults or >4 mmol/L in children
B	Tendon xanthomas in the patient or a first degree relative
C	DNA-based evidence of mutation in the *LDLR* or *ApoB* gene
D	Family history of myocardial infarction before age 50 years in a second degree relative or before age 60 years in a first degree relative
E	Family history of raised total cholesterol >7.5 mmol/L in a first or second degree relative
Diagnosis	
Definite	Either criteria A and B or criterion C
Probable	Either criteria A and D or criteria A and C

Note: For total, HDL, and LDL cholesterol, values in mg/dL = values in mmol/L × 38.67
(LDLR: low-density lipoprotein receptor)

Sitosterolemia

Sitosterolemia occurs due to mutations in the members of the ATP-binding cassette (ABC) half transporter family, ABCG5, or ABCG8. These genes are expressed in enterocytes and hepatocytes. These heterodimerize to form a functional complex that pumps plant sterols such as sitosterol and campesterol, and animal sterols, predominantly cholesterol, into the lumen of the gut as well as into bile. Less than 5% of dietary plant sterols are absorbed by the proximal small intestine and delivered to the liver in normal individuals.[2] However, in sitosterolemia, the intestinal absorption of sterols is increased and biliary excretion reduced, resulting in increased plasma and tissue levels of both plant sterols and cholesterol. Incorporation of these plant sterols into the cell membranes results in misshapen red blood cells causing hemolysis and also megathrombocytes.[2]

Lysosomal-acid-lipase-deficiency (LALD): It is also known as cholesteryl ester storage disease (CESD) caused by loss-of-function variants in both alleles of the gene *LIPA*. Lysosomal acid lipase hydrolyses TG and cholesteryl esters after delivery to the lysosomes by the LDLR and is particularly important in the liver. Patients have elevated LDL-C with low HDL-C and variably raised TG. There is accumulation of neutral lipid in hepatocytes causing hepatic steatosis and progressive fibrosis. The most severe form of this disorder, Wolman's disease presents in infancy with failure to thrive, hepatosplenomegaly, jaundice, and adrenal calcification and is rapidly fatal within 1 year.[5]

Lipid Disorders Associated with Elevated Triglycerides

Familial Chylomicronemia Syndrome

The enzyme Lipoprotein-lipase (LPL) is necessary for hydrolyzing TGs in chylomicrons and very low density lipoprotein (VLDLs), and ApoC-II is a cofactor for LPL in this reaction. ApoA-V is required for the association of VLDL and chylomicrons with LPL. Therefore, the deficiency of any of these proteins results in profound elevations in plasma chylomicrons and elevated plasma levels of VLDL.

Familial chylomicronemia syndrome (FCS) can present in childhood or adulthood with recurrent bouts of acute pancreatitis. Eruptive xanthomas appearing occur on the buttocks, extensor surfaces of the arms and back. The retinal blood vessels may appear opalescent (lipemia retinalis). Premature coronary heart disease (CHD) is not seen. Extreme elevations of TGs above 10,000 mg/dL may rarely cause the hyperchylomicronemia syndrome manifesting as dyspnea and confusion.[3] A clinical diagnostic score has been proposed for its diagnosis (**Flowchart 1**).[6]

GPIHBP1 Deficiency

Lipoprotein-lipase is attached to the endothelial surface of capillaries by GPIHBP1. Homozygosity for mutations in GPIHBP1 can cause severe hypertriglyceridemia.

Lipid Disorders with Elevated Triglycerides and Cholesterol
Familial Dysbetalipoproteinemia

This disorder is also known as familial broad-β-disease and is characterized by a mixed hyperlipidemia due to accumulation of remnant lipoproteins. ApoE is present on chylomicron and VLDL remnants and mediates their removal via the hepatic lipoprotein receptors. Familial dysbetalipoproteinemia (FDBL) is due to genetic variations in ApoE. The *APOE* gene is polymorphic in sequence, and there are three common isoforms of ApoE: ApoE-3, which is the most common; and ApoE-2 and ApoE-4. ApoE-2 has low affinity for the LDLR; leading to slower clearance of chylomicron and VLDL remnants containing ApoE-2. Those having ApoE-4 are known to have an increased incidence of late onset Alzheimer's disease but not FDBL.

Familial dysbetalipoproteinemia is usually diagnosed in adulthood with moderate-to-severe hypertriglyceridemia and hypercholesterolemia, between 300 to 400 mg/dL. HDL-C levels are normal. Palmar xanthomas are pathognomonic while tendon xanthomas may also occur. Unlike FH where peripheral vascular disease is less common, in FDBL,

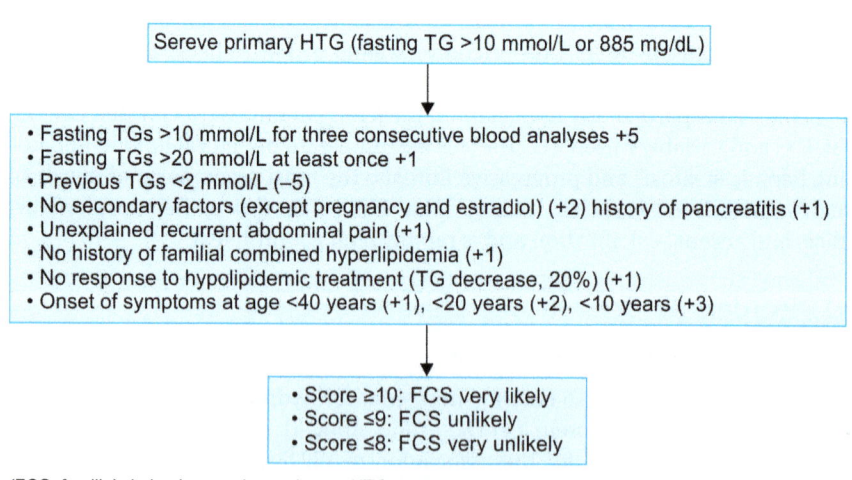

(FCS: familial chylomicronemia syndrome; HTG: hypertriglyceridemia; TG: triglyceride)

FLOWCHART 1: The familial chylomicronemia score.[6]

premature peripheral vascular disease is also common in addition to coronary artery disease (CAD).[3]

Hepatic Lipase Deficiency

Hepatic lipase (HL) hydrolyzes TGs and phospholipids in remnant lipoproteins and HDL. HL-deficiency manifests similar to FDB with elevated cholesterol and TGs with either a normal or elevated plasma level of HDL-C.

Polygenic Causes of Hypertriglyceridemia

Familial Hypertriglyceridemia

Familial hypertriglyceridemia (FHTG) is a relatively common (~1 in 500) autosomal dominant disorder characterized by moderately elevated TG, average to low LDL-C and HDL-C levels and a family history of hypertriglyceridemia. The risk for CAD is not significantly increased. However, in the presence of concomitant secondary causes of hypertriglyceridemia, the TG levels can rise to a level that could precipitate acute pancreatitis. A combination of genetic variants can cause this phenotype.[7]

Familial Combined Hyperlipidemia

In familial combined hyperlipidemia (FCHL), there is moderate elevation in plasma levels of TG (VLDL) and LDL-C and reduced plasma levels of HDL-C. It is estimated to occur in ~1 in 100 individuals. About one-fifth of individuals who develop CHD under the age of 60 have FCHL. There are three possible biochemical phenotypes: (1) Elevated plasma levels of LDL-C, (2) Elevated plasma levels of TG (VLDL), or (3) Elevated plasma levels of both. The lipoprotein profile in an individual can switch among these three over time depending on factors such as diet, exercise, and weight. Likely, defects in several different genes can cause FCHL. These patients have substantially elevated levels of ApoB, often disproportionately high relative to plasma LDL-C levels indicative of the presence of small, dense LDL particles.

Inherited Causes of Low Levels of ApoB-containing Lipoproteins

Abetalipoproteinemia

Abetalipoproteinemia (ABL) is caused by loss-of-function mutations in the gene which encodes for the microsomal triglyceride transfer protein (MTP), a protein responsible for transferring lipids to nascent chylomicrons and VLDL in the intestine and liver, respectively. These individuals have extremely low levels of cholesterol and TG while chylomicrons, VLDL, LDL, and ApoB are undetectable in plasma. This is an autosomal recessive disorder and the parents being obligate heterozygotes have normal plasma lipids and ApoB levels.

Most of the clinical manifestations arise from defects in the absorption of fat-soluble vitamins. It usually presents in early childhood with diarrhea and failure to thrive due to fat malabsorption. Initial neurologic manifestations include loss of deep tendon reflexes, decreased vibration and proprioception in lower extremities and ataxia progressing to a spastic gait by the third to fourth decade. There is pigmentary retinopathy ultimately progressing to near blindness.

Familial Hypobetalipoproteinemia

The generic term of familial hypobetalipoproteinemia (FHBL) is applied to all inherited disorders with low plasma levels of LDL-C (the "β-lipoprotein"). Most of the mutations causing FHBL interfere with the production of ApoB, resulting in a truncated ApoB protein with accelerated catabolism. Individuals heterozygous for these mutations usually have LDL-C levels <80 mg/dL and probably enjoy protection from atherosclerotic cardiovascular disease (ASCVD). Some heterozygotes have elevated liver fat with elevated transaminases. Mutations in both ApoB alleles resemble ABL though the neurologic features are less severe. However, the levels of LDL-C and ApoB are normal in the parents of patients with ABL and low in those with homozygous hypobetalipoproteinemia.

Proprotein-convertase-subtilisin/Kexin Type 9 Deficiency

Proprotein-convertase-subtilisin/kexin type 9 normally promotes the degradation of the LDLR. Mutations that interfere with the synthesis of PCSK9, common in Africans, result in increased LDLR activity and ~40% reduction in plasma level of LDL-C.[2] A sequence variation of higher frequency (R46L) is found predominantly in Europeans and is associated with a 15% reduction in LDL-C.[2,3] Individuals with inactivating mutations are protected from developing CHD relative to those without it.

Flowcharts 2A to C describe an algorithmic approach to the biochemical and genetic diagnoses of the three major dyslipidemia patterns in which monogenic lipoprotein disorders—severe hypertriglyceridemia, hypercholesterolemia and combined hyper-triglyceridemia with hypercholesterolemia.[8]

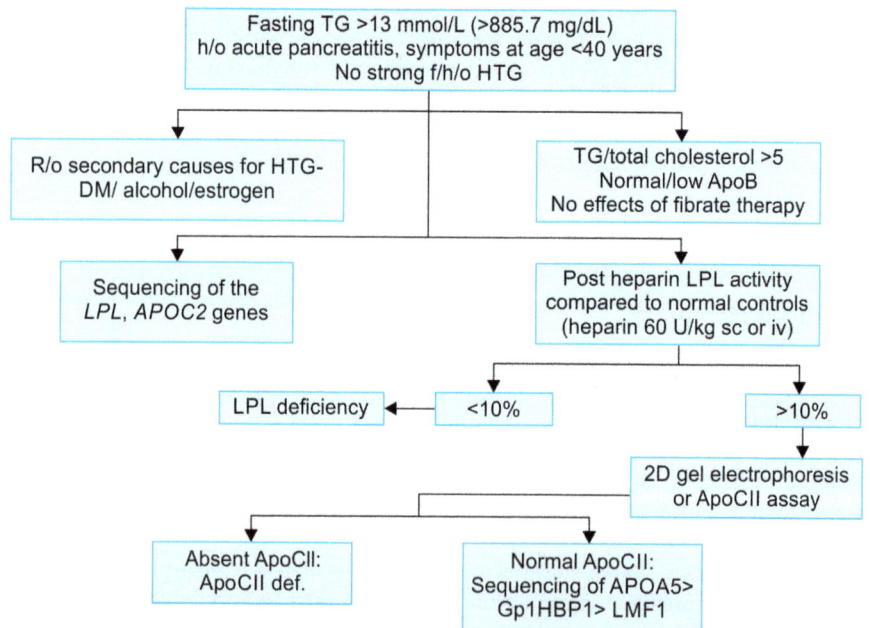

(DM: diabetes mellitus; f/h/o: family history of; h/o: history of; HTG: hypertriglyceridemia; LPL: lipoprotein lipase; R/o: rule out; TG: triglycerides)

FLOWCHART 2A: Approach to diagnosing monogenic lipoprotein disorders with isolated hypertriglyceridemia.

Monogenic Lipoprotein Disorders: From Phenotype to Genotype

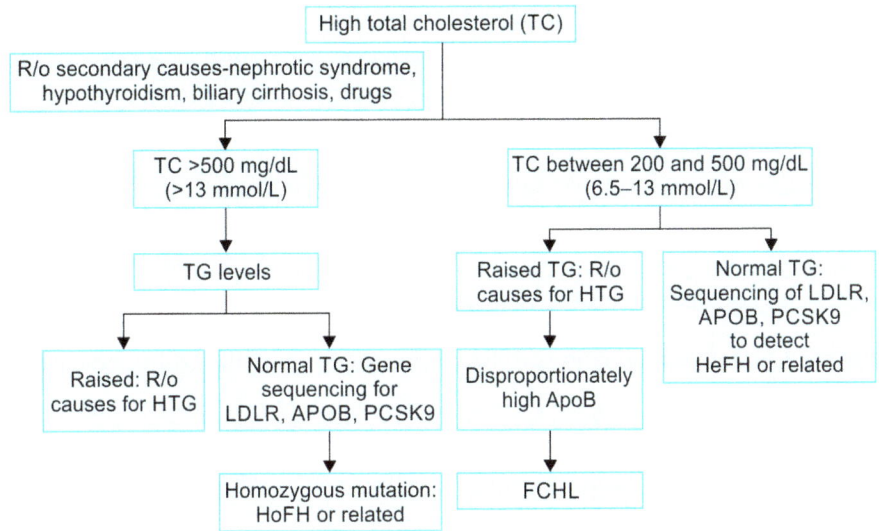

(R/o: rule out; FCHL: familial combined hyperlipidemia; HeFH: heterozygous familial hypercholesterolemia; HoFH: homozygous familial hypercholesterolemia; HTG: hypertriglyceridemia; TC: total cholesterol: TG: triglyceride)

FLOWCHART 2B: Approach to diagnosing monogenic lipoprotein disorders with hypercholesterolemia.

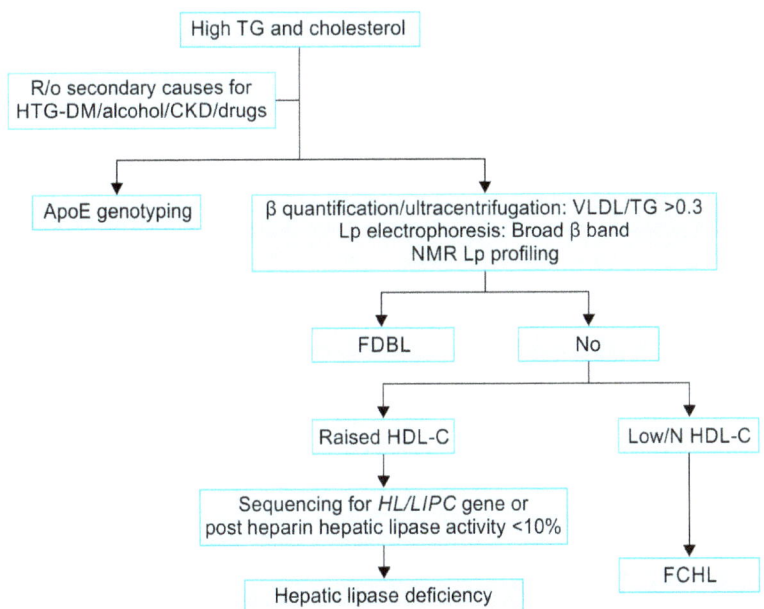

(CKD: chronic kidney disease; DM: diabetes mellitus; FCHL: familial combined hyperlipidemia; FDBL: familial dysbetalipoproteinemia; HDL-C: high-density lipoprotein cholesterol; HTG: hypertriglycerimidemia; Lp: lipoprotein; NMR: nuclear magnetic resonance; R/o: rule out; TG: triglyceride; VLDL: very low-density lipoprotein)

FLOWCHART 2C: Approach to diagnosing genetic lipoprotein disorders with combined hypercholesterolemia and hypertriglyceridemia.

Source: Adapted from Hegele RA, Tonstad S. Disorders of lipoprotein metabolism. In: Physician's Guide to the Diagnosis, Treatment, and Follow-Up of Inherited Metabolic Diseases. Berlin, Heidelberg: Springer; 2014. pp. 671-89.

Familial Combined Hypolipidemia

Nonsense mutations in both alleles of the gene angiopoietin-like 3 (*ANGPTL3*) lead to low plasma levels of TG, LDL-C, and HDL-C. ANGPTL3 inhibits LPL thus delaying the clearance of triglyceride-rich lipoprotein (TRL) from blood and therefore, its deficiency predominantly lowers TG. There is a reduced risk of CHD.

Genetic Disorders of High-density Lipoprotein Metabolism

Genetic forms of hypoalphalipoproteinemia (low HDL-C) are rare and not always associated with accelerated atherosclerosis.

Inherited Causes of Low Levels of High-density Lipoprotein Cholesterol

Gene Deletions in the ApoAV-AI-CIII-AIV Locus and Coding Mutations in ApoA-I

Complete deficiency of the *ApoA-I* gene results in the virtual absence of HDL from the plasma. The genes encoding *ApoA-I, ApoC-III, ApoA-IV,* and *ApoA-V* are present together in a clustered manner on the chromosome 11. Patients with absent ApoA-I often have genomic deletions that include other genes in this cluster. ApoA-I is required for LCAT activity. In the absence of LCAT, free cholesterol levels increase in both HDL and in tissues such as the cornea and in the skin, resulting in corneal opacities and planar xanthomas. Premature CHD is a common feature of ApoA-I deficiency, especially when additional genes in the complex are also deleted. Missense and nonsense mutations in the *ApoA-I* gene are very rare causes of low HDL-C levels. Patients heterozygous for an Arg173Cys substitution in ApoA-I (so-called ApoA-I Milano) have very low plasma levels of HDL and yet have no increased risk of premature CHD.

Tangier Disease (ABCA1 Deficiency)

Patients with Tangier disease have mutations in the gene encoding *ABCA1* which is a cellular transporter facilitating the efflux of unesterified cholesterol and phospholipids from cells to ApoA-I. In the absence of ABCA1, the nascent, poorly lipidated ApoA-I is immediately cleared from the circulation. Thus, patients with Tangier disease have extremely low circulating plasma levels of HDL-C (<5 mg/dL) and ApoA-I (<5 mg/dL). Cholesterol accumulates in the reticuloendothelial system of these patients, resulting in hepatosplenomegaly and pathognomonic enlarged, grayish yellow or orange tonsils.

Lecithin Cholesterol Acyl Transferase Deficiency

Lecithin Cholesterol Acyl Transferase (LCAT) is an enzyme activated by ApoA-I which mediates the esterification of cholesterol to form cholesteryl esters. Due to lack of normal cholesterol esterification, there is impaired formation of mature HDL particles. Proportion of free cholesterol in circulating lipoproteins is grossly increased from ~25% to >70% of total plasma cholesterol.

Primary Hypoalphalipoproteinemia (Isolated Low–HDL)

In these patients, the plasma HDL-C is below the tenth percentile though cholesterol and TG levels are relatively normal. There are no apparent secondary causes of low plasma HDL-C, and no clinical signs of LCAT deficiency or Tangier disease. The etiology of this is primarily an accelerated catabolism of HDL and its apolipoproteins.

Inherited Causes of High Levels of High-density Lipoprotein Cholesterol

Cholesterol Esteryl Transfer Protein deficiency

Cholesterol Esteryl Transfer Protein (CETP) facilitates the transfer of cholesteryl esters from HDL to ApoB-containing lipoproteins. The absence of this transfer results in an increase in the cholesteryl ester content of HDL which reduces their clearance rates to achieve very high HDL-C levels (usually >150 mg/dL), along with a reduction in plasma levels of LDL-C. CETP deficiency is rare outside of Japan. The relationship of CETP deficiency to ASCVD remains unresolved.

Familial Hyperalphalipoproteinemia

With plasma HDL-C level above the ninetieth percentile. this trait runs in families, but not all appear to have a reduced risk of CHD.

LABORATORY DIAGNOSIS AND GENETIC TESTING

The first step to diagnosing and managing lipoprotein disorders is to determine the particular class of lipoproteins that are increased or decreased. The next step involves ruling out possible secondary causes to the particular pattern of dyslipidemia. The basic panel for this must include a fasting blood sugar, thyroid profile, hepatic and renal profile to r/o hepatitis, CKD or nephrotic syndrome, history of smoking and alcohol, and a meticulous drug history. Once secondary causes are ruled out, the underlying genetic or primary lipid disorder should be categorized. Before proceeding to genetic tests, the particular clinical profile, biochemical pattern of dyslipidemia and family inheritance can give vital clues to the disorder. In most laboratories, the total cholesterol (TC) and TG levels in plasma are measured enzymatically. LDL-C is estimated using the Friedewald formula: LDL-C = TC − (TG/5) − HDL-C. However, LDL-C calculated from this formula becomes less accurate at TG level >200 mg/dL and it cannot be used at all for TG levels >400 mg/dL when LDL-C should also be directly measured.

Cases with low HDL (<19.3 mg/dL or <0.5 mmol/L) should undergo ApoA1 testing followed by assessment of LCAT activity.

Those with TC levels <19.3 mg/dL (<0.5 mmol/L) should undergo sequencing of the MTP followed by ApoB, PCSK9, and ANGPTL3 to detect homozygous mutations causing ABL or related phenotypes. Those with TC between 58 to 116 mg/dL should undergo ApoB testing and if it is below 0.5 mg/dL may be followed with sequencing of the same genes to detect heterozygous mutations causing hypobetalipoproteinemia and related phenotypes.

Importance of Genetic Testing

The increased availability and ease of DNA sequencing makes it a direct and cost-effective tool for diagnosing monogenic dyslipoproteinemias.[9] Apart from a confirmatory molecular diagnosis, identification of the relevant genetic defect can have other implications:

- *Genetic counseling*: Detection in the proband can dictate screening of other family members and timely intervention as well as provide a guide to family planning. Early pharmacologic therapy can prevent premature CHD in most patients with FH or FCH or FDBL.

- *Prognostic importance:* A diagnosis of FH should dictate the need for early screening for CAD. Certain disorders also predispose to characteristic complications such as peripheral vascular disease in FDBL and hepatic fibrosis in CESD or neurological complications of ABL.
- *Therapeutic implications:* As discussed under the "Treatment" section, genetic disorders with similar biochemical picture but different gene mutations may need moderately different treatment strategies.
- *Research implications*: Knowledge about the molecular defects in specific disorders can stimulate discovery of novel treatment strategies. The PCSK9 inhibitors were discovered following the observation that *PCSK9* gene deficient individuals have extremely low LDL-C levels. Similarly, therapies to antagonize ANGPTL3 are under development following discovery of ANGPTL3 mutations as the etiology behind familial combined hypolipidemia.

TREATMENT OF MONOGENIC LIPID DISORDERS

Hypercholesterolemia

Apart from dietary counseling to reduce saturated fats to <7% and exogenous cholesterol to <200 mg/day, patients with HeFH need statins, often in combination with other agents such as ezetimibe or preferably with PCSK9 inhibitors. The optimum age for initiation of treatment in children remains controversial but can be commenced at 10 years for all and <10 years for those with additional risk factors or obesity.[4] Patients with HoFH mostly have inadequate response to the highest tolerated doses of drugs. Additional agents approved for HoFH in the US include MTP inhibitor Lomitapide and antisense -oligonucleotide to ApoB, Mipomersen. Most patients would need serial plasma LDL apheresis. The indications for LDL apheresis include familial HoFH with LDL-C ≥500 mg/dL, familial HeFH with LDL ≥300 mg/dL or familial HeFH with documented ASCVD and LDL-C ≥200 mg/dL.[5] LDL apheresis acutely reduces LDL-C by 80% with long-term reductions between 20 and 40%.[5] Liver transplantation and gene therapy have been attempted, with variable results. In ARH, the LDLR function in cultured fibroblasts may be independent of LDLRAP and is often normal though the LDLR function in liver is negligible. Statins induce LDLR upregulation and enhances LDL-clearance by these to reduce plasma LDL-C by 30–60%.

Lysosomal acid lipase deficiency is often underdiagnosed though an effective enzyme replacement is available. However, the prognosis for Wolman disease is very unfavorable, in spite of some reports of remarkable results with bone marrow transplantation, enzyme therapy, and gene therapy. Treatment of sitosterolemia involves cessation of all plant sterols and reduction of dietary cholesterol. Statins have minimal response due to their selective reduction in plasma cholesterol but bile acid sequestrants and ezetimibe effectively lower both plasma cholesterol and plant sterol levels.

Hyperchylomicronemia

Familial hyperchylomicronemia arising from deficiency in LPL or ApoC-II is difficult to treat. Fibrates used to treat hypertriglyceridemia partially upregulates LPL activity, making them mostly unsuitable. Though niacin reduces VLDL secretion by hepatocytes, it is only marginally useful since chylomicrons have an intestinal origin. The most effective treatment modality is restriction of dietary TG to <25% of daily caloric intake. Fish oils may be effective. Plasma apheresis is almost never indicated. Expression of virus-recombinant human gain-of-function mutant LPLS447X in murine models has shown strongly favorable results.[8] Antisense oligonucleotide to suppress ApoC-3 is another promising approach.

Combined Hyperlipidemia

Combined hyperlipidemia such as FCHL and FDBL should maintain a balanced lifestyle with restriction of both simple carbohydrates and fats to as low as <10 % of total energy and statin therapy as for HeFH based on global cardiovascular risk. Patients with FDBL are particularly diet responsive whereas almost all with FHL require statin therapy.

Treatment for both ABL and homozygous FHBL include lifelong dietary supplementation with medium-chain-TGs and fat-soluble vitamins.

CONCLUSION

In clinical practice, the secondary lipoprotein disorders are more common than the monogenic disorders. Nevertheless, a detailed knowledge is necessary to treat the latter as they need individualized diagnostic and therapeutic approach.

KEY POINTS

- Monogenic lipoprotein disorders are rare, arise due to defects in enzymes or cofactors of lipoprotein metabolism pathways and have characteristic clinical pictures.
- The pattern of dyslipidemia can be grouped into categories such as hyperchylomicronemia, hypercholesterolemia, combined hyperlipidemia, hypobetalipoproteinemia, etc., and an algorithmic approach should be used for diagnosis and focused genetic testing.
- Genetic diagnosis is important before empirical treatment because individual disorders have unique therapeutic challenges such as nonresponse to statins or being extremely responsive to diet alone.
- Genetic counseling, screening of family members and timely intervention can prevent catastrophic complications.

REFERENCES

1. Fredrickson DS, Levy RI, Lees RS. Fat transport in lipoproteins—an integrated approach to mechanisms and disorders. N Engl J Med. 1967;276:32-44.
2. Rader DJ, Hobbs HH. Disorders of Lipoprotein Metabolism. In: Longo DL, Kasper DL, Jameson JL, Fauci AS, Hauser SL, Loscalzo J (Eds). Harrison's Principles of Internal Medicine, 20th edition. New York: McGraw Hill Education; 2018. pp 2889-902.
3. Semenkovich CF, Goldberg AC, Goldberg IJ. Disorders of lipid metabolism. In: Williams Textbook of Endocrinology; 2016. pp. 1660-700.
4. National Institute for Health and Clinical Excellence. Familial Hypercholesterolaemia: Identification and Management of Familial Hypercholesterolaemia. NICE; 2008.

5. Rahalkar AR, Hegele RA. Monogenic pediatric dyslipidemias: classification, genetics and clinical spectrum. Mol Genet Metab. 2008;93(3):282-94.
6. Moulin P, Dufour R, Averna M, Arca M, Cefalù AB, Noto D, et al. Identification and diagnosis of patients with familial chylomicronaemia syndrome (FCS): expert panel recommendations and proposal of an "FCS score". Atherosclerosis. 2018;275:265-72.
7. Soutar AK, Naoumova RP. Mechanisms of disease: genetic causes of familial hypercholesterolemia. Nat Clin Pract Cardiovasc Med. 2007;4(4):214-25.
8. Hegele RA, Tonstad S. Disorders of lipoprotein metabolism. In: Physician's Guide to the Diagnosis, Treatment, and Follow-Up of Inherited Metabolic Diseases. Berlin, Heidelberg: Springer; 2014. pp. 671-89.
9. Schaefer EJ, Tsunoda F, Diffenderfer M, Polisecki E, Thai N, Asztalos B. The Measurement of Lipids, Lipoproteins, Apolipoproteins, Fatty Acids, and Sterols, and Next Generation Sequencing for the Diagnosis and Treatment of Lipid Disorders. In: Feingold KR, Anawalt B, Boyce A, et al. (Eds). South Dartmouth (MA): MDText.com Inc.; 2000.

Section 2

Clinical Lipidology

CHAPTER 6

Dietary Fats and Cardiovascular Disease

Ajitesh Roy

ABSTRACT

Noncommunicable diseases (NCDs) comprise a vast group of diseases such as cardiovascular disease (CVD), cancer, diabetes, and chronic respiratory diseases contributing to 68% of all the deaths globally. Unhealthy nutrition, adverse lifestyle, and health-related behavior are recognized to be most important factors for prevention of CVD. Current meta-analysis suggested that industrial trans fatty acids (TFA) contributes to the genesis of heart disease, while n-3 polyunsaturated fatty acid (PUFA) and n-6 PUFA are protective. But it is uncertain whether total saturated and monounsaturated fats have important effects on risk of heart disease. There is also huge confusion regarding dietary n-6/n-3 ratios which is still not resolved. With these heterogeneous effects, it is possible to conclude that total fat and perhaps total saturated fat and monounsaturated fat in the diet appear to be less important than the types of foods consumed. In this article, we aim to analyze the recent evidences and opinions on the appropriate type of dietary fat for preventing CVD.

INTRODUCTION

Diabetes, cancer, chronic respiratory diseases, and cardiovascular diseases (CVDs) together comprise vast majority of noncommunicable diseases (NCDs), which are responsible for ~60% of mortality in India (World Health Organization-WHO, 2014). This recent prevalence highlighted the importance of reducing risk factors for respective diseases. Such risk factors for CVD include unhealthy diets, low physical activity, obesity, smoking, etc. Preventive treatment that reduces CVD by little changes can substantially reduce the number of people who develop CVD. Amongst all these risk factors unhealthy diet has immense effect on CVD and rationally the AHA (American Heart Association) has recommended reduction in dietary saturated fat to reduce the risk of CVD. The mechanisms underlying the effects of dietary fat on CVD are not straightforward. Several studies have shown the relationship between dietary saturated fatty acid (SFA) and trans fatty acids (TFA) to contribute to CVD by inflammatory and oxidative stress mechanisms.

The aim of this topic is to compile opinions concerning recommended intakes of various fatty acids and their roles in the CVD.

EVOLUTION OF DIETARY FAT GUIDANCE

A substantial change in Dietary Guidelines for Americans with regard to dietary fat was noticed with time since early 1960s. The persistent focus on reducing fats resulted in the decrease in this proportion to 37%E in 1988 and further down to 34%E by 1995.[1] In this time, contribution of carbohydrates (CHO) to the diet had increased significantly. By the turn of this century, it had become clear that the overwhelming focus on reducing fat intake was actually detrimental for the population, resulting in a dramatic increase in the incidence of overweight and obese population. Further studies have shown that fats, and even SFA, are not necessarily associated with risk for CVD.[2] In 1992, The US Food Pyramid put fats and oils, with sweets at the top. In the 2005, a redesigned version was introduced reflecting the evolution in dietary fat guideline by including a separate sector for liquid vegetable oils.

The most recent edition is focusing on dietary patterns and choosing "Fat- free or low-fat dairy including milk, yogurt, cheese, and/or fortified soy beverages"; "A variety of protein foods including seafood, lean meats and poultry, eggs, legumes, and nuts, seeds, and soy products"; and "Oils" and limiting "Saturated fats and *trans* fats, added sugars, and sodium."[3]

National Cholesterol Education Program and American Heart Association

From 1988 to 2001, National Cholesterol Education Program (NCEP) introduced the Adult Treatment Panel (ATP) I to ATP III recommendations with a change from low-fat to moderate-fat intake (25–35% of energy).[4]

In 1961, AHA first issued dietary guidelines recommending moderate-fat diet and replacing saturated fat with polyunsaturated fat. Further there is a change in concept with limit of total fat intake and a focus on replacing saturated fat with unsaturated fat. In 2013, the AHA and American College of Cardiology further updated this[5] which include consume "a dietary pattern that emphasizes vegetables, fruits, and whole-grains; including low-fat dairy products, poultry, fish, legumes, non-tropical vegetable oils, and nuts; and limiting intake of sweets, sugar-sweetened beverages, and red meats."

FATS AND FATTY ACIDS IN INDIAN DIETS: USE AND THE TRENDS

The main sources of fat in a typical Indian diet (which is predominantly either rice or wheat based) are vegetable oils, *ghee*, and dairy products. With the publication of the Indian recommended dietary allowance (RDA) and the hype against the use of SFA in the diet, the dietary advice given in the 1980s and 1990s was to reduce the intake of *ghee* and traditional oils such as coconut and groundnut oils and to use new polyunsaturated fatty acid (PUFA)-rich oils such as safflower and sunflower oils.

According to "National Nutrition Monitoring Bureau", between 1980 and 2002, there was an increase in the dietary intake of fat, both in urban and rural India.[6] But it was continued to be <15%E in many States. Pingali and Khwaja[7] have shown that contribution of animal food sources (that contain SFA) continues to be minimal. A trend was noticed in the favor of consumption of refined oils (such as sunflower, soybean, and other unsaturated vegetable oils), with decreased intakes of *ghee*/*vanaspati*.

Dietary fats and oils come from both animal and plant sources. In general, fats of animal origin tend to be relatively high in SFAs. Increasing evidence indicates that different SFAs and different food sources of saturated fat have divergent effects on cardiovascular (CV) and metabolic health. For example, SFAs with carbon chain lengths of 14 (myristic) and 16 (palmitic), chiefly found in dairy products and red meats, appear most potent in increasing both low-density lipoprotein cholesterol (LDL-C) and high-density lipoprotein cholesterol (HDL-C) and decreasing triglyceride-rich lipoproteins. Stearic acid (18 carbons), another component of beef has far smaller effects on LDL-C and HDL-C. Combining these effects, the consumption of myristic and palmitic acid has relatively little effect on the ratio of total cholesterol to HDL-C, while stearic acid and especially lauric acid (12 carbons) significantly lower this ratio. Also, a food source with higher percentage of processed meats imposes higher CV risk.[8]

Oils of plant origin tend to be relatively high in unsaturated fatty acids [monounsaturated fatty acid (MUFA) and PUFA]; however, contrary to that tropical oils, e.g., palm oil, coconut oils, etc., are high in SFAs but remain liquid because they contain a high proportion of short-chain fatty acids (SCFA). While partially hydrogenated vegetable oils are relatively high in TFA.

Polyunsaturated fatty acid can be categorized into two classes, the n-6 PUFA [e.g., linoleic acid (LA), arachidonic acid] and the n-3 family [e.g., alpha-linolenic acid (ALNA) eicosapentaenoic acid (EPA), docosahexaenoic acid (DHA)]. These distinct categories of fatty acids have been extensively studied in relation to CV risk factors and outcomes. The most abundant intake of n-6 PUFA in most plant products (soybean, safflower, and sunflower) is LA which lower serum LDL-C and triglycerides while also increasing HDL-C, leading to an overall reduction in the ratio of total to HDL-C. Dietary n-3 PUFA ALNA, is available in plant sources, e.g., from walnuts, soybean, and the long chain (LC) n-3 PUFA is present mainly in marine fish so its overall intake is low except in specific coastal regions.

A mechanism is postulated that high levels of LA can inhibit the conversion of ALNA.[9] It is, therefore, recommended that the dietary ratio of LA/ALNA should preferably be in the ratio of 5:1 though few researchers do not agree with this data because of lack of evidence. The increased consumption of n-3-PUFA-rich oils such as safflower and sunflower at the expense of more traditional oils such as coconut, groundnut, etc., has resulted in a very skewed ratio of n-6 and n-3 FA.

While n-6 PUFAs clearly have lipid-lowering effects, these are also proinflammatory in nature and have been shown to have adverse effects at higher intakes. Increased intake of LA during pregnancy has been found to be a postulated mechanism behind early life origin of obesity. Animal studies have shown that low ALNA intake can result in increased adiposity thus suggesting that an increased n-6/n-3 ratio could be an emerging risk factor for obesity.[10]

Monounsaturated fatty acids: In contrast to the diversity of individual saturated fatty acids, nearly all MUFA in foods is oleic acid, coming from red meats and dairy fats and a smaller amount coming from plant oils such as nuts and olive oil. Few studies showed that monounsaturated fats decrease LDL-C and triglycerides as well as maintain HDL-C.[11]

Trans fatty acids: Low levels of TFA occur naturally in some foods, especially dairy and meats. However, much higher levels of TFA consumption can occur as a result of the industrial partial hydrogenation.

These TFA raise blood LDL-C and lower HDL-C concentrations. By comparison, consumption of saturated fats, also raises LDL-C concentrations but concurrently lowers levels of triglyceride-rich lipoproteins and raises HDL-C. Thus, effects of saturated fats on the lipid profile are far less adverse than those of TFA. Globally TFA are being used lesser these days. However, TFA are still a component of many baked goods, such as cookies and deep-fried foods.

FATTY ACID INTAKES IN THE CONTEXT OF GLOBAL PATTERNS: THE CURRENT SITUATION IN INDIA

Global data on dietary intakes show significant differences in diet composition between the high-income and low-income countries (**Table 1**). Indian diets, particularly, contain lower fat as well as SFA, both in urban as well as rural surveys.[6]

In spite of having lower intakes of fat, SFA, and higher intakes of PUFA which could be an ideal diet for the population, risk of CVD is noticeably high in India. This clearly indicates that SFA may not be the main dietary culprit in increasing the risk of obesity and that it is important to look at other factors such as CHO and the PUFA, as more likely culprits.

DIETARY FAT AND CARDIOVASCULAR DISEASE: FACTS AND CONTROVERSIAL ASPECT

The age-old idea of high CV risk with total fat intake and high SFA had faced a challenge after few recent systematic reviews and meta-analysis that have questioned the relation between TFA and SFA intake and CVD events and mortality rates (**Fig. 1**).[12-15] For example, in Seven Countries Study, the locations with the lowest incidence of coronary heart disease (CHD) were Crete and Japan. However, the Japanese had a very low fat intake, while individuals in Crete typically had a high total fat intake, chiefly composed of MUFA from vegetable sources. Also in "Nurses' Health Study" involving 20-year follow-up,

TABLE 1: Average dietary intakes of macronutrients and fatty acids: A comparison between global intakes and intakes in India.

Mean intakes per day	RDA-India[21]	High-income economies	Low-/middle-income economies	Indian (rural-NNMB)	Indian (urban women)*	Indian-urban
Energy (kcal)		3,400	2,550	2,300	2,140	2,325
CHO (%E)	55–70	51	67	71	61	63
Protein (%E)	8–12	12.4	10.5	10	12.5	12.5
Fat (%E)	15–30	36	22.5	19	26	24
SFA (%E)	<10	12.3	13.1	4.5	8.8	7.2
LA (%E)	2.5–9	5.4	4.2	4.9	8.2	6.8
ALNA (%E)	>0.5	0.28	0.22	0.24	0.24	
LC n-3 PUFA (mg/day)	200	335	220	–	20	–

*Data on dietary intakes categorized into different economies based on the World Bank definitions (http://data.worldbank.org/country).

(ALNA: alpha-linolenic acid; CHO: carbohydrate; LA: linoleic acid; LC: long chain; NNMB: National Nutrition Monitoring Bureau; PUFA: polyunsaturated fatty acid; RDA: recommended dietary allowance; SFA: saturated fatty acid)

Dietary Fats and Cardiovascular Disease

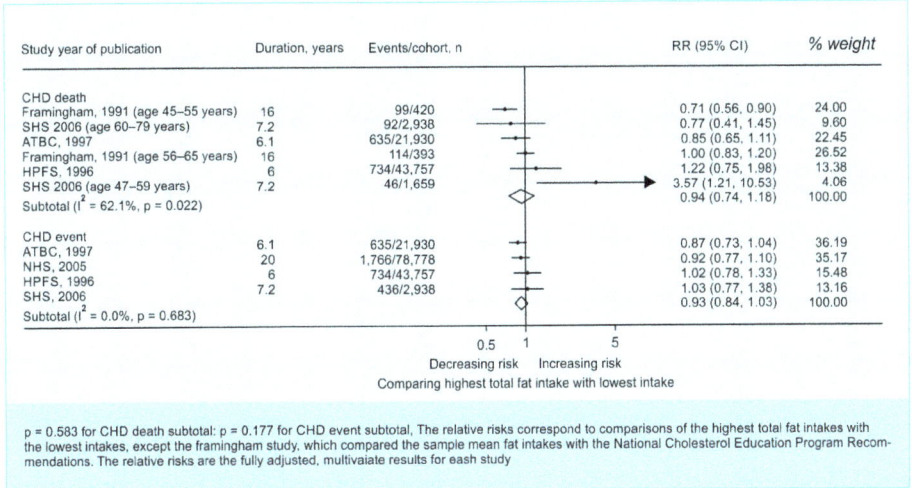

[ATBC: Alpha-Tocopherol, Beta-Carotene Cancer Prevention Study (Pietinen et al, 1997); Framinagham: Framingham study (Posner et al. 1991); HPFS: Health Professionals Follow-up Study (Ascherio et al. 1996); NHS: Nurses' Health Study (Oh et al. 2005); SHS: Strong Heart Study. (Xu et al. 2006)]

FIG. 1: Meta-analysis of total fat intake and coronary heart disease (CHD) prospective cohorts.
Source: Skeaff CM, Miller J. Dietary fat and coronary heart disease: summary of evidence from prospective cohort and randomised controlled trials. Ann Nutr Metab. 2009;55:173-201.

the multivariate adjusted relative risk (RR) of CHD for the highest quintile of dietary fat compared with the lowest was 0.92 (95% CI 0.77–1.09).[16]

Another landmark study, "Women's Health Initiative", randomly assigned women to an intensive behavior-modification group, with a goal of reducing total fat intake to 20% of calories and increasing daily intake of vegetables and grains, or to a control group.[17] After 6 years, intervention group had decreased total fat intake but the intervention had no effect on overall CHD (HR 0.94; 95% CI 0.86–1.02) or stroke (HR 1.02; CI 0.90–1.17).

In contrast to previous studies, few meta-analysis showed positive association between saturated fat and CVD.[18] When the data were analyzed on the basis of PUFA replacing SFA, there was a positive relation between dietary SFA and CVD (**Fig. 2**).[19] When carbohydrate replacing SFA, there was no definite relation.[16] These statistics were based on isocaloric substitution analyses.

Trans fatty acids and CV risk: In an analysis from the "Nurses' Health Study", for each increase in 2% of energy from trans fat, the relative risk for incident CHD was 1.93 (95% CI 1.43–2.61).[20] Meta-analysis showed intake of TFA was strongly associated with CHD mortality, with a RR of CHD death of 1.32 (95% CI 1.08–1.61; p = 0.006) for the highest compared with the lowest category (**Fig. 3**). An interesting fact is ruminant TFA were generally not associated with total CHD as compared to industrial TFA. This risk is also not applicable for transpalmitoleic acid, which is a TFA derived from naturally occurring dairy products.

In 2018, the Food and Drug Administration (FDA) implemented that partially hydrogenated vegetable oils are no longer "generally regarded as safe" (GRAS).

Monounsaturated fatty acids: Meta-analyses of prospective observational studies and randomized trials found no association between intake of monounsaturated fat and risk for CHD (**Fig. 4**). Mediterranean populations that consume high amounts of monounsaturated fats from extra-virgin olive and nuts appear to be protected against heart disease.[21]

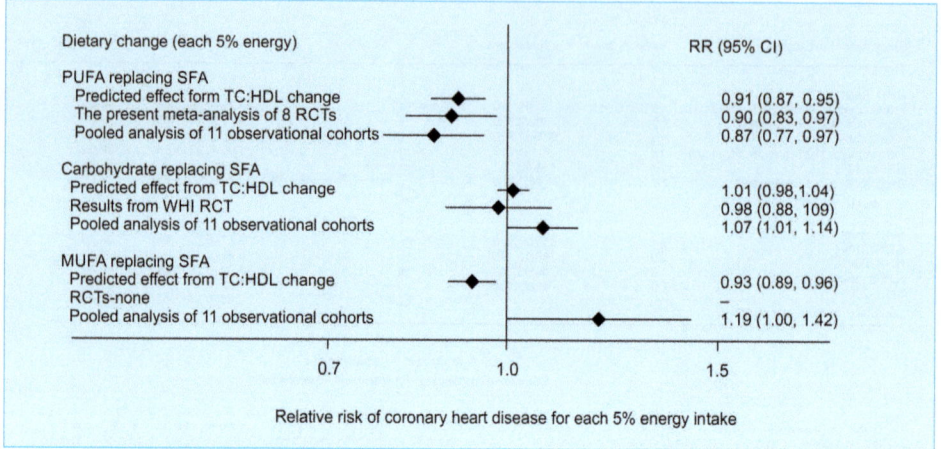

(CI: confidence interval; HDL: high-density lipoprotein; MUFA: monounsaturated fatty acids; PUFA: polyunsaturated fatty acid; RCT: randomized controlled trials; RR: relative risk; SFA: saturated fatty acid; TC: total cholesterol; WHI: women's health initiative)

FIG. 2: Effects on coronary heart disease risk of consuming PUFA, carbohydrate, or MUFA in place of saturated fatty acids.

Source: Reproduced from Mozaffarian D, Micha R, Wallace S. Effects on coronary heart disease of increasing polyunsaturated fat in place of saturated fat: a systematic review and meta-analysis of randomized controlled trials. PLoS Med. 2010;7:e1000252.

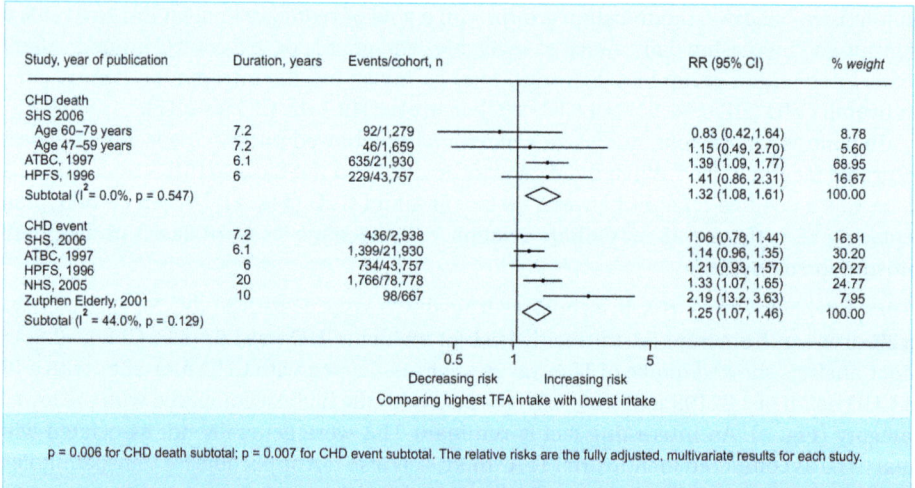

[ATBC: Alpha-Tocopherol, Beta-Carotene Cancer Prevention study (Pietinen et al, 1997); Framinagham: Framinagham study (Posner et al. 1991); HPFS: Health Professionals Follow-up Study (Ascherio et al. 1996); NHS: Nurses' Health Study (Oh et al. 2005); SHS: Strong Heart Study. (Xu et al. 2006)]

FIG. 3: Meta-analysis of prospective cohorts for trans fatty acids (TFA) and coronary heart disease (CHD).

Source: Skeaff CM, Miller J. Dietary fat and coronary heart disease: summary of evidence from prospective cohort and randomised controlled trials. Ann Nutr Metab. 2009;55:173-201.

Dietary Fats and Cardiovascular Disease

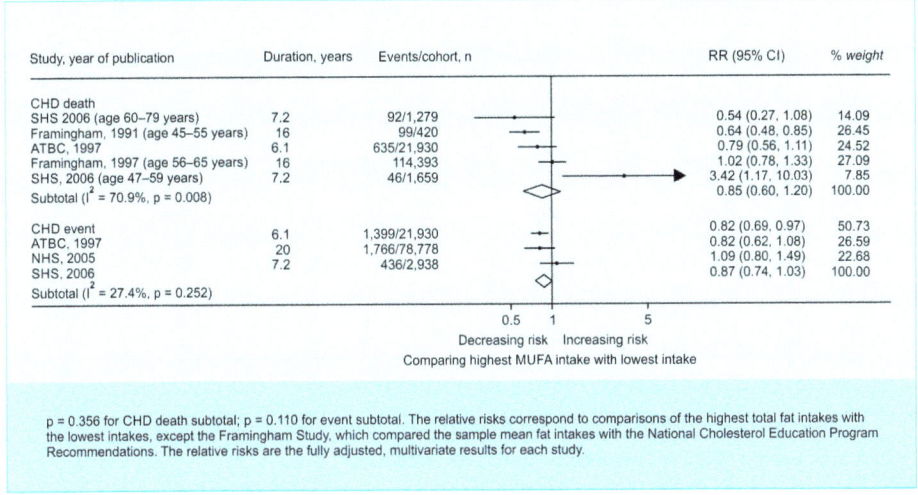

[ATBC: Alpha-Tocopherol, Beta-Carotene Cancer Prevention Study (Pietinen et al., 1197); Framingham: Framingham Study (Posner et al., 1991); HPFS: Health Professionals Follow-up Study (Ascherio et al.,1996); NHS: Nurses Health Study (Oh et al., 2005); SHS: Strong Heart Study (Xu et al., 2006)]

FIG. 4: Meta-analysis of prospective cohorts for monounsaturated fatty acids (MUFA) intake and coronary heart disease (CHD).
Source: Skeaff CM, Miller J. Dietary fat and coronary heart disease: summary of evidence from prospective cohort and randomised controlled trials. Ann Nutr Metab. 2009;55:173-201.

In sum, the evidence supports consuming plant foods rich in monounsaturated fat rather than trying to increase monounsaturated fats from all sources for being protective against CVD.

DIETARY FAT AND CARDIOVASCULAR DISEASE: IN A NUTSHELL

A Presidential Advisory entitled "Dietary Fats and Cardiovascular Disease"[22] was issued by AHA, at the end of 2017. They reviewed few "core" and "noncore" studies. "Core" studies met few criteria: The trials compared "saturated fat with high PUFA, did not include trans unsaturated fat as a major component, had at least a 2 years sustained dietary intervention, determined adherence using biomarkers (e.g., serum cholesterol, tissue concentration of PUFA, etc.), and validated information on CV events". Four "core" studies showed lowering of CHD by 29% with substitution of saturated fat with vegetable oil (**Fig. 5**).[23-26] The added benefits of modifying dietary fat might be the consequent reduction of dietary cholesterol in diet. Six additional "noncore" studies were reviewed.[27-32] These studies were categorized as "noncore" because either SFA was replaced with a combination of PUFA and carbohydrate and/or they had insufficient duration, or faulty study design. The effect of core studies becomes diluted when added with noncore trials in the meta-analysis. Two systematic reviews including noncore studies showed an overall risk reduction of ~20% by replacing SFA with PUFA.

American Heart Association also assessed few prospective observational studies.[22] A significant lower CHD risk (25%, 15%, and 9%, respectively) was observed with replacement of 5% of energy from SFA with PUFA, MUFA, or unrefined carbohydrate (**Fig. 6**).[33] In contrast, higher incidence of CHD was noticed when replacing SFA with refined carbohydrate (may be due to low fiber content and rapid absorption).

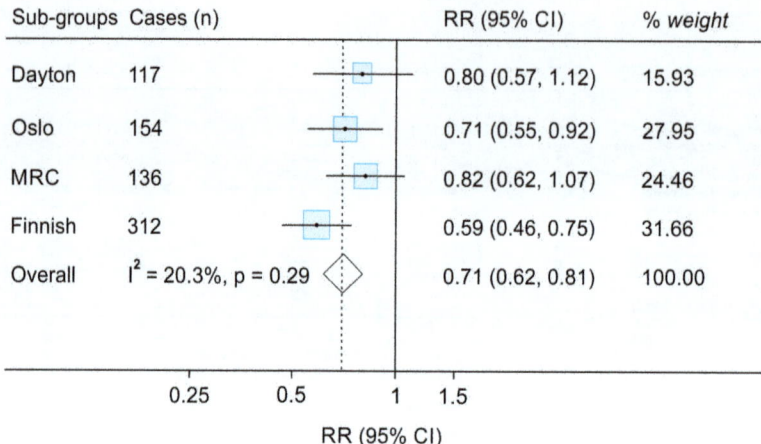

(CI: confidence interval; MRC: medical research council; RR: relative risk)

FIG. 5: Meta-analysis of core trials on replacing saturated with polyunsaturated fat.

Source: Reproduced from Sacks FM, Lichtenstein AH, Wu JHY, Appel LJ, Creager MA, Kris- Etherton PM, et al. Dietary fats and cardiovascular disease. A Presidential advisory from the American Heart Association. Circulation. 2017;136:e1–e23.

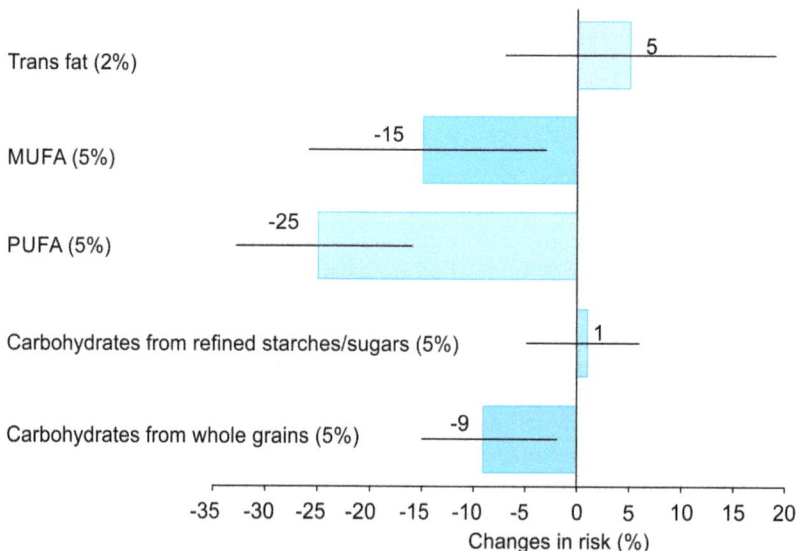

(PUFA: polyunsaturated fatty acid; SFA: saturated fatty acid)

FIG. 6: Placement of saturated fat with other types of fat or carbohydrate.

Source: Reproduced from Sacks FM, Lichtenstein AH, Wu JHY, Appel LJ, Creager MA, Kris- Etherton PM, et al. Dietary fats and cardiovascular disease. A Presidential advisory from the American Heart Association. Circulation. 2017;136:e1-e23.with permission from the AHA, Inc.

CONCLUSION

Given the rapid changes in concepts that have occurred in terms of advice, the subject of how much and what fats to eat is now very confusing. A number of issues need to be considered. Specifically, the daily requirement of n-3 PUFA needs to be reexamined. A large majority of Indians do not consume LC n-3 PUFA, and even nonvegetarians consume negligible amounts of marine fish. To generate evidence on the requirement of n-3 PUFA or LC n-3 PUFA and the n-6/n-3 PUFA ratio in diets, well-conducted randomized controlled trials need to be carried out.

Finally following points are worthy which concluded from evidences:
- PUFA from vegetable oils replacing saturated fats (<10%) from dairy and meat lowers CVD.
- Keep TFA consumption as low as possible by limiting intake of partially hydrogenated oils.
- A principle of reducing intake of TFA including saturated fat, and replacing the fats mainly with unspecified CHO does not prevent CHD.
- Lower intake of saturated fat coupled with higher intake of PUFA and MUFA is associated with lower rates of CVD and all-cause mortality.
- Overall, evidence supports the conclusion that PUFA from vegetable oils (mainly n-6, LA) reduces CVD somewhat more than MUFA (mainly oleic acid) when replacing saturated fat.
- Diet should include seafood, fruits, vegetables, whole grains and nuts, seeds, and soy products, as well as lean meats, poultry, eggs, and low-fat dairy products.

India is creating its own developmental milestones not to be confused with the West, where diets are increasing and becoming more accessible and diversified. So it is critical that a well-informed unique policy should be laid out, before lessons that have been learnt over the years in the West are forgotten.

KEY POINTS
- Types of fat consumed is much important than total fat intake in respect to CVD.
- Risk of CVD is reduced by lower intake of saturated fat coupled with higher intake of PUFA and MUFA but not when refined starches and added sugars are the replacement nutrients.
- Keep TFA consumption as low as possible.
- Diet should include seafood, fruits, vegetables, whole grains, as well as lean meats, eggs, and low-fat dairy products. So the goal of dietary advice should be on overall dietary patterns, not on single nutrient.

REFERENCES

1. Gifford KD. Dietary fats, eating guides, and public policy: history, critique, and recommendations. Am J Med. 2002;113(Suppl 9B):89S-106S.
2. Micha R, Mozaffarian D. Saturated fat and cardiometabolic risk factors, coronary heart disease, stroke, and diabetes: a fresh look at the evidence. Lipids. 2010;45:893-905.
3. Jahns L, Davis-Shaw W, Lichtenstein AH, Murphy SP, Conrad Z, Nielsen F. The history and future of dietary guidance in America. Adv Nutr. 2018;9:136-47.

4. Expert Panel on Detection, Evaluation, and Treatment of High Blood Cholesterol in Adults. Executive summary of the third report of the National Cholesterol Education Program (NCEP) expert panel on detection, evaluation, and treatment of high blood cholesterol in adults (Adult Treatment Panel III). JAMA. 2001;285(19):2486-97.
5. Eckel RH, Jakicic JM, Ard JD, de Jesus JM, Houston Miller N, Hubbard VS, et al. 2013 AHA/ACC guideline on lifestyle management to reduce cardiovascular risk: a report of the American College of Cardiology/American Heart Association Task Force on Practice Guidelines. J Am Coll Cardiol. 2014;63:2889-934.
6. National Nutrition Monitoring Bureau (NNMB) Technical Report 21: Diet and Nutritional status of rural population, Annexure 7.2. Hyderabad: National Institute of Nutrition; 2002.
7. Pingali P, Khwaja Y. Globalisation of Indian diets and the transformation of food supply systems. ESA Working Paper No. 04-05. 2004 [online] Available from http://www.fao.org/3/a-ae060t.pdf. [Last accessed July 2020].
8. Micha R, Peñalvo JL, Cudhea F, Imamura F, Rehm CD, Mozaffarian D. Association Between Dietary Factors and Mortality From Heart Disease, Stroke, and Type 2 Diabetes in the United States. JAMA. 2017;317(9):912.
9. Emken EA, Adlof RO, Gulley RM. Dietary linoleic acid influences desaturation and acylation of deuterium-labeled linoleic and linolenic acids in young adult males. Biochim Biophys Acta. 1994;1213:277-88.
10. Ailhaud G, Guesnet P, Cunnane SC. An emerging risk factor for obesity: does disequilibrium of polyunsaturated fatty acid metabolism contribute to excessive adipose tissue development? Br J Nutr. 2008;100:461-70.
11. Mata P, Alvarez-Sala LA, Rubio MJ, Nuño J, De Oya M. Effects of long-term monounsaturated- vs polyunsaturated-enriched diets on lipoproteins in healthy men and women. Am J Clin Nutr. 1992;55(4):846.
12. Mente A, de Koning L, Shannon HS, Anand SS. A systematic review of the evidence supporting a causal link between dietary factors and coronary heart disease. Arch Intern Med. 2009;169(7):659-69.
13. Jakobsen MU, O'Reilly EJ, Heitmann BL, Pereira MA, Bälter K, Fraser GE, et al. Major types of dietary fat and risk of coronary heart disease: a pooled analysis of 11 cohort studies. Am J Clin Nutr. 2009;89(5):1425-32.
14. Skeaff CM, Miller J. Dietary fat and coronary heart disease: summary of evidence from prospective cohort and randomised controlled trials. Ann NutrMetab. 2009;55:173-201.
15. Siri-Tarino PW, Sun Q, Hu FB, Krauss RM. Meta-analysis of prospective cohort studies evaluating the association of saturated fat with cardiovascular disease. Am J Clin Nutr. 2010;91:535-46.
16. Oh K, Hu FB, Manson JE, Stampfer MJ, Willett WC Dietary fat intake and risk of coronary heart disease in women: 20 years of follow-up of the nurses' health study. Am J Epidemiol. 2005;161(7):672-9.
17. Howard BV, Van Horn L, Hsia J, Manson JE, Stefanick ML, Wassertheil-Smoller S, et al. Low-fat dietary pattern and risk of cardiovascular disease: the Women's Health Initiative Randomized Controlled Dietary Modification Trial. JAMA. 2006;295(6):655.
18. Clifton PM, Keogh JB. A systematic review of the effect of dietary saturated and polyunsaturated fat on heart disease. Nutr Metab Cardiovasc Dis. 2017;27(12):1060-80.
19. Mozaffarian D, Micha R, Wallace S. Effects on coronary heart disease of increasing polyunsaturated fat in place of saturated fat: a systematic review and meta-analysis of randomized controlled trials. PLoS Med. 2010;7:e1000252.
20. Hu FB, Stampfer MJ, Manson JE, Rimm E, Colditz GA, Rosner BA, et al. Dietary fat intake and the risk of coronary heart disease in women.N Engl J Med. 1997;337(21):1491.
21. The diet and all-causes death rate in the Seven Countries Study. Lancet. 1981;2(8237):58.
22. Sacks FM, Lichtenstein AH, Wu JHY, Appel LJ, Creager MA, Kris-Etherton PM, et al. Dietary fats and cardiovascular disease. A Presidential advisory from the American Heart Association. Circulation. 2017;136:e1-e23.
23. Controlled trial of soya-bean oil in myocardial infarction. Lancet 1968;2:693-9.
24. Dayton S, Pearce ML. Prevention of coronary heart disease and other complications of arteriosclerosis by modified diet. Am J Med. 1969;46(5):751-62.
25. Leren P. The Oslo Diet-Heart study: eleven-year report. Circulation. 1970;42(5):935-42.
26. Turpeinen O, Karvonen MJ, Pekkarinen M, Miettinen M, Elosuo R, Paavilainen E. Dietary prevention of coronary heart disease: the Finnish Mental Hospital Study. Int J Epidemiol. 1979;8:99-118.
27. Watts GF, Lewis B, Brunt JN, Lewis ES, Coltart DJ, Smith LD, et al. Effects on coronary artery disease of lipid-lowering diet, or diet plus cholestyramine, in the St Thomas' Atherosclerosis Regression Study (STARS). Lancet. 1992;339:563-9.
28. Burr ML, Fehily AM, Gilbert JF, Rogers S, Holliday RM, Sweetnam PM, et al. Effects of changes in fat, fish, and fibre intakes on death and myocardial reinfarction: diet and reinfarction trial (DART). Lancet. 1989;2:757-61.

29. Houtsmuller AJ, van Hal-Ferwerda J, Zahn KJ, Henkes HE. Favorable influences of linoleic acid on the progression of diabetic micro- and macroangiopathy in adult onset diabetes mellitus. Prog Lipid Res. 1981;20:377-86.
30. Rose GA, Thomson WB, Williams RT. Corn oil in treatment of ischaemic heart disease. BMJ. 1965;1:1531-3.
31. Frantz ID, Jr, Dawson EA, Ashman PL, Gatewood LC, Bartsch GE, Kuba K, et al. Test of effect of lipid lowering by diet on cardiovascular risk: the Minnesota coronary survey. Arteriosclerosis. 1989;9(1):129-35.
32. Ramsden CE, Zamora D, Leelarthaepin B, Majchrzak-Hong SF, Faurot KR, Suchindran CM, et al. Use of dietary linoleic acid for secondary prevention of coronary heart disease and death: evaluation of recovered data from the Sydney Diet Heart Study and updated meta-analysis. BMJ. 2013;346:e8707.
33. Li Y, Hruby A, Bernstein AM, Ley SH, Wang DD, Chiuve SE, et al. Saturated fats compared with unsaturated fats and sources of carbohydrates in relation to risk of coronary heart disease: a prospective cohort study. J Am Coll Cardiol. 2015;66(14):1538-48.

CHAPTER 7

Dyslipidemia: Relationship with Insulin Resistance, Fatty Liver, and Subclinical Atherosclerosis

Kaushik Sen

ABSTRACT

Dyslipidemia is an established risk factor for atherosclerotic cardiovascular disease and is known to significantly contribute to ischemic heart disease and stroke. The typical pattern of dyslipidemia found in insulin resistance is termed as "atherogenic dyslipidemia" characterized by high triglycerides, low HDL and high small, dense LDL. A similar pattern of dyslipidemia is also seen in fatty liver disease which is believed to be a part in the broader definition of metabolic syndrome (MetS) these days because nonalcoholic fatty liver disease (NAFLD) is closely related to insulin resistance. Subclinical atherosclerosis is typified by a different pattern of dyslipidemia with high LDL and altered LDL:HDL ratio. Research during the last decade have shown difference in the risk profile of patients having similar defects in the standard lipid parameters which can be better explained by assessing the qualitative aspects of lipids. Despite the lack of concrete recommendations, statins remain the drugs of choice in all the three conditions.

INTRODUCTION

High cholesterol levels are believed to play a role in about 56% ischemic heart disease events and 18% of strokes worldwide. Insulin resistance, both hepatic and peripheral, as is well known, is the main driving force for the lipoprotein derangements. The three major interconnected metabolic pathways of lipoproteins in the body are: (1) Transport of exogenous or dietary lipoproteins, (2) Transport of hepatic or endogenous lipids, and (3) Reverse cholesterol transport. In this chapter, we will discuss different patterns of dyslipidemia in three closely linked conditions: Insulin resistance (hereinafter mentioned as IR), fatty liver disease and subclinical atherosclerosis (hereinafter mentioned as SA) and their relationship.

INSULIN RESISTANCE

Insulin resistance (IR) is established as central to the causation of the constellation of metabolic derangements, jointly known as metabolic syndrome (MetS) or IR syndrome

Dyslipidemia: Relationship with Insulin Resistance, Fatty Liver, and Subclinical Atherosclerosis

which confers increased risk of cardiovascular disease (CVD). The first definition of metabolic syndrome was given by the World Health Organization in 1998, since then there has been continuous evolution of the same and proposal of definitions by different authorities. The major features of MetS are central obesity, high triglyceride (TG), low high-density lipoproteins (HDLs), hypertension, and hyperglycemia (**Fig. 1**). The official definition of MetS given by NCEP/ATPIII is shown in **Table 1**.

The most common lipid abnormality seen in IR is the so called "Dyslipidemic triad" characterized by high TGs, low HDL, and elevated levels of catabolic products

TABLE 1: The NCEP-ATPIII 2001 definition of metabolic syndrome.	
Three or more of the following	Central obesity: Waist circumference >102 cm (M), >88 cm (F)
	Hypertriglyceridemia: Triglyceride level >150 mg/dL or on specific medication
	Low HDL cholesterol: HDL level <40 mg/dL (M), <50 mg/dL (F) or on specific medication
	Hypertension: Systolic BP ≥130 mm Hg or diastolic BP ≥85 mg/dL or on specific medication
	Fasting plasma glucose ≥100 mg/dL or on specific medication or previously diagnosed type 2 diabetes mellitus

(BP: blood pressure; HDL: high density lipoprotein; NCEP-ATPIII: National Cholesterol Education Programme and Adult Treatment Panel III)

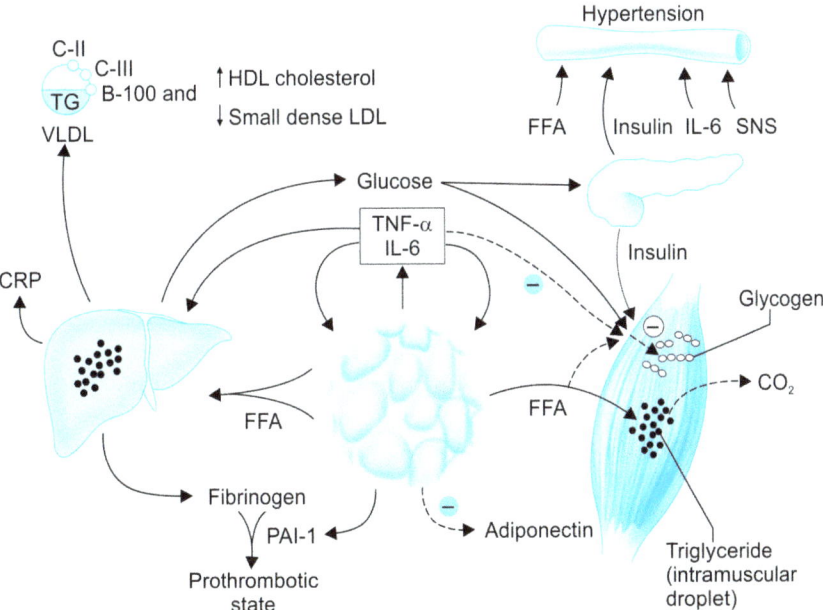

(CRP: C-reactive protein; FFA: FFA: free fatty acid; HDL: high-density lipoprotein; LDL: low-density lipoprotein; PAI-1: plasminogen activator inhibitor 1; TG: triglyceride; TNF-α: tumor necrosis factor-alpha; VLDL: very low-density lipoprotein)

FIG. 1: Schematic representation of metabolic syndrome.
Source: Modified from Jameson JL, Fauci AS, Kasper DL, Hauser SL, Longo DL, Loscalzo J (Eds). Harrison's Principles of Internal Medicine, 20th edition. New York: McGraw Hill; 2018.

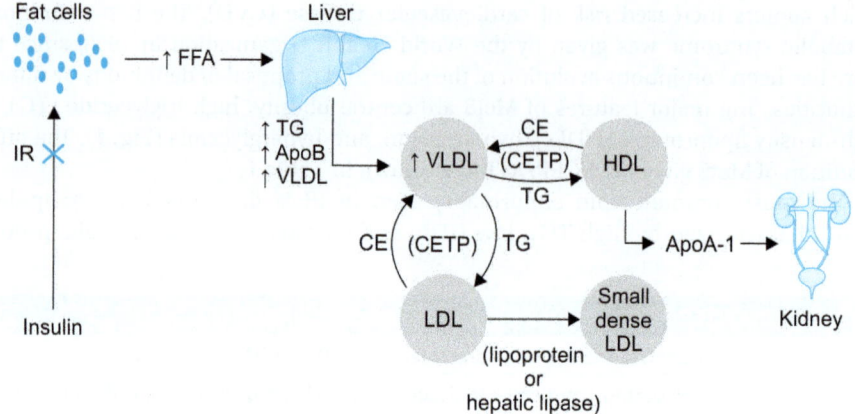

(CE: cholesteryl esters; CETP: cholesteryl ester transfer protein; FFA: free fatty acid; HDL: high-density lipoprotein; IR: insulin resistance; LDL: low-density lipoproteins; TG: triglyceride; VLDL: very low-density lipoprotein)

FIG. 2: The mechanism of generation of small-dense LDL particles and low HDL in MetS.
Source: Modified from Kwiterovich PO Jr. The metabolic pathways of high-density lipoprotein, low-density lipoprotein and triglycerides: A Current Review. Am J Cardiol. 2000;86(12A):5L-10L.

of triglyceride rich lipoproteins (TRLs), e.g., very low-density lipoprotein (VLDL) remnants, intermediate density lipoproteins (IDL), and low-density lipoproteins (LDLs).[1] The pattern of lipid abnormality is called as type B LDL profile (older term: hyperapobetalipoproteinemia) which is typified by normal or slightly elevated LDL level but an increase in small, dense LDL particles which are devoid of cholesteryl esters (CE). The two proposed mechanisms of such a lipoprotein patterns are: First, adipose tissue IR causes increased lipolysis leading to free fatty acid generation and secondly, decreased uptake and incorporation of fatty acids in TG in adipose tissue because of IR. The excess fatty acids are efficiently taken care of by the liver which synthesizes and secretes excess of VLDL-TG. In the blood stream, excess TG in the VLDL exchanges for CE in the LDL producing a CE-depleted and TG rich LDL. This TG-rich LDL is hydrolyzed by hepatic lipase so that a small-dense LDL is produced. The CE in HDL also gets exchanged with the TG in VLDL producing a TG-rich HDL. And this TG-rich HDL is catabolized by the kidneys giving rise to the low HDL in MetS (**Fig. 2**).

De-novo lipogenesis is also triggered by sterol regulatory element-binding protein-1c (SREBP-1c), a major lipogenic transcription factor.[2] Recent studies show that mTOR signaling is the central mediator for this "selective" insulin response to lipogenesis in an otherwise insulin resistant liver.[3,4]

Important to remember: Approximately 25% of individual with small, dense HDL "inherit" this defect and so, in them this is not accompanied by high TG. Measurement of ApoB will identify this population.

Another important aspect of dyslipidemia in IR is reduced activity of lipoprotein lipase (LPL) which is an insulin dependent endothelial enzyme mainly found in the capillaries. This results in decreased catabolism of chylomicron, VLDL, and IDL. The result is abnormally high postprandial lipemia which is documented to be linked to coronary artery disease (CAD).

Dyslipidemia: Relationship with Insulin Resistance, Fatty Liver, and Subclinical Atherosclerosis

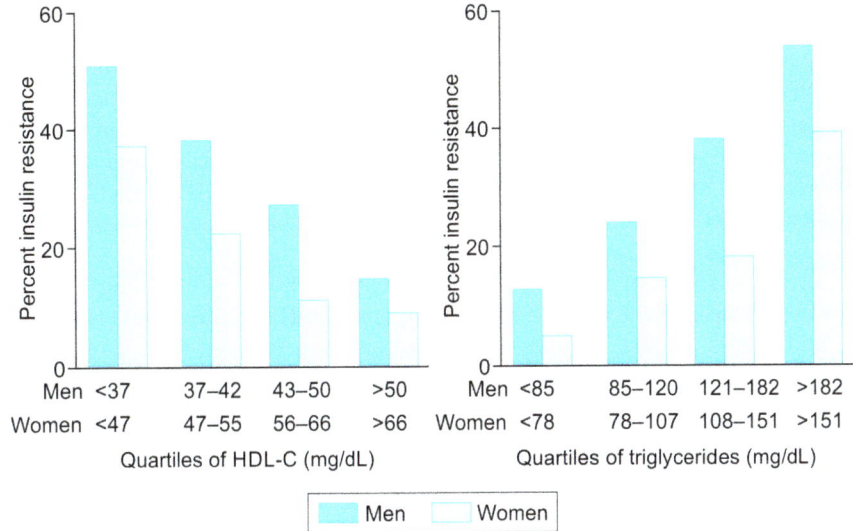

(HDL-C: high-density lipoprotein cholesterol; TG: triglycerides)

FIG. 3: Prevalence of insulin resistance according to quartiles of HDL and TG in Framingham offspring cohort.
Source: Modified from Robins SJ, Lyass A, Zachariah JP, Massaro Joseph M, Ramachandran S. Vasan insulin resistance and the relation of a dyslipidemia to coronary heart disease. The Framingham Heart Study. Arterioscler Thromb Vasc Biol. 2011;31(5):1208-14.

Causes of low LPL in obesity and IR: (1) Reduced transcription of LPL in adipose tissue and muscle due to tissue IR, and (2) Increased production of ApoC-3 by liver which is an inhibitor of LPL.[5]

Although, hypertriglyceridemia can be an independent genetic disorder, it is a well-accepted marker of IR. Even a slightly elevated level of TG (>150 mg/dL) may identify individuals who are at-risk of IR syndrome.

One of the initial documentations of dyslipidemia in IR was the IRAS (Insulin Resistance Atherosclerotic Study) which showed progressively more unfavorable lipoprotein profile with normal glucose tolerance, impaired glucose tolerance, and type 2 diabetes mellitus respectively.[6] Similar defects in lipid profile are also observed in pediatric population with IR. In a study on 1,175 school children between 7 and 12 years of age, significant correlation was found in different lipid parameters with IR. And type IIb and IV were found to have more IR compared to children with type IIa and normolipidemia.[7]

An interesting aspect of the relationship of altered lipid parameters and coronary heart disease (CHD) was found in a follow-up of Framingham Heart Study which followed up offspring of the original cohort of Framingham Heart Study participants and their spouses for incident myocardial infarction and CHD. After a mean follow-up of 14 years, it was found that the combination of high TG and low HDL was associated with increased incidence of CHD only when associated with increased IR and not when it was present without IR (**Fig. 3**).[8]

FATTY LIVER

Nonalcoholic fatty liver disease (NAFLD) has now become the most common cause of chronic liver disease with a spectrum from simple steatosis to nonalcoholic steatohepatitis (NASH) and may progress to cirrhosis and even to hepatocellular carcinoma. NAFLD is closely associated with abdominal obesity, dyslipidemia, hypertension, and IR which are all features of the MetS. Approximately 90% of patients with NAFLD have, at least, one of the features of MetS and about 33% present the complete diagnosis.[9] And it is well known that CVD is the most common cause of morbidity and mortality in NAFLD. An interesting aspect of this association is that a genetic variant in the gene encoding the protein patatin-like phospholipase domain containing 3 protein (PNPLA3), I148M (rs738409), a strong and specific cause of high liver fat content, is not associated with increased risk of ischemic heart disease.[10] This fact is a nice proof that the increased fat content of the liver *only* is not sufficient to increase cardiovascular (CV) risk, it is the metabolic derangements that is responsible.

Importantly, dyslipidemia in NAFLD is closely related to the presence and degree of IR. A recent cohort study that used magnetic resonance spectroscopy (MRS) to detect hepatic TG content and advanced lipoprotein profile found that the degree of lipoprotein abnormalities in NAFLD was dependent on liver fat content and IR, but not related to body mass index (BMI) or steatohepatitis.[11] It follows, therefore, that the pattern of lipid abnormality found in NAFLD would also be similar to that found in IR.

Obesity stimulates hepatocyte lipid accumulation by increasing intestinal permeability, thereby causing more exposure of the liver to gut-derived factors that promotes generation of inflammatory mediators to increase IR. Excess adipose tissue also increases IR by producing adipokines. Hyperinsulinemia, in-turn, promotes lipogenesis and storage. A meta-analysis published recently, showed that the presence of NAFLD significantly increases the risk of both subclinical atherosclerosis (SA), detected by increased carotid intima medial thickness (CIMT) and carotid plaque and CAD detected by coronary angiography.[9] The patterns of dyslipidemia in NAFLD are similar in pediatric age group also. In a retrospective analysis of 308 pediatric NAFLD patients, elevated TGs, low HDL and high non-HDL cholesterol were found in 88%, 77%, and 56% children, respectively. Among these, elevated TG and elevated non-HDL cholesterol showed a significant correlation with serum alanine aminotransferase (ALT) levels and persisted after correction of age and BMI.[12]

A relatively newer concept of lipoprotein abnormality in NAFLD is the qualitative aspects of lipoprotein profile, an issue that stems from the difference of CV risk in persons with seemingly similar quantitative lipoprotein profiles. Among the subfractions of HDL, HDL2, which is claimed to be the most anti-atherogenic, was found to be lower in comparison with total HDL in patients with NAFLD.[13] Subsequently, NASH patients were shown to have lower HDL2 levels compared to controls in another study.[14] Both studies indicate the presence of dysfunctional HDL particles present in NAFLD patients. On the other side, a multi-ethnic study of NAFLD patients with nuclear magnetic resonance (NMR) showed elevated TG and low HDL in patients with NAFLD than those in controls as expected, but failed to show significant difference in LDL between the two groups (p = 0.23). NMR study, however, showed increased LDL concentration and smaller particle size of LDL in the NAFLD patients.[15] Another study with the Mediterranean population showed similar results in the lipoprotein profile, but also showed a trend of increased TG content of all the lipoproteins with increasing markers of fatty liver.[16]

Regarding functionality of the HDL particles, two studies done on NAFLD patients, one with combined ultrasonography and elastography, and the other with fatty liver index (FLI) ≥60, showed impaired cholesterol efflux capacity (CEC) of HDL even after adjustment of HDL concentration.[17,18]

Fatty Liver Index (FLI) is calculated by a complex equation with four parameters: Triglyceride, BMI, GGT and waist circumference, for the details the reader is advised to check standard textbooks of Hepatology.

Important to remember: Triglyceride, *per se*, is not hepatotoxic, but its precursors (e.g., Fatty acids, glycerol) and metabolic by-products (e.g., ROS) are.[19]

Current recommendation: Unfortunately, there is real dearth of proper guideline adequately addressing the issue of either lipid management or overall CV risk management in NAFLD.[20,21] Like in most of other high CV-risk situations, statins are generally considered drugs of-choice in NAFLD. Randomized control trials on this issue are lacking. However, post-hoc analysis in related study showed more CV death reduction in those patients having elevated liver enzymes than those who did not have.[22] Statins are known to be safe even in the face of elevated liver enzymes in NAFLD and may even improve it. Ezetimibe and PCSK9 inhibitors also have shown some promise.

SUBCLINICAL ATHEROSCLEROSIS

Atherosclerosis, as is well known, is a chronic and progressive inflammatory disease that causes significant cardiovascular morbidity and mortality. The concept of SA stems from the fact that it develops silently, over decades, before being clinically manifest. And sudden, unexpected deaths are still common manifestations of major CVD. This gives us a unique opportunity to detect the disease earlier and to intervene. A study of young men, mean age 22.1 years, killed during the Vietnam conflict with the US revealed that about 45% had some evidence of coronary atherosclerosis and 5% even had grossly severe disease.[23] This was an irrefutable proof that the disease is present long before its clinical manifestation. Children and adolescents with heterozygous familial cholesterolemia are particularly prone to develop SA as evidenced from studies involving CIMT.[24] It has also been proved that among the different lipid parameters, LDL and the ratio of LDL and HDL are better predictors of increased CIMT indicating presence of SA.[25,26] As have been discussed in the previous section of this chapter, the functionality of lipoproteins have also been assessed in relation to SA. In a recent study done in newly diagnosed patients of NAFLD, it was found that impaired HDL-CEC was negatively correlated with CIMT and the presence of atherosclerotic plaque.[17] In an effort to look further into the details of dyslipidemia and relationship with SA, a recent study had recruited 422 subjects with primary dyslipidemia prior to therapy and studied role of mutant APOA5. It was found that compared to common homozygotes, carriers of rare alleles for -1131C and 19W had 24% ($p = 0.014$), and 71% ($p = 0.002$) higher TG levels respectively, and -1131C carriers had borderline 8% ($p = 0.032$) lower HDL-C levels.[27] NMR studies revealed larger VLDL and HDL particles and smaller LDL particle size in carriers of the rare allele. This change toward atherogenic dyslipidemia in 19W allele was also significantly associated with increased CIMT ($p = 0.028$).

Newer aspects: "Adropin" is a 76 amino acid peptide hormone and recently has been found to have important contribution to metabolic diseases. Serum adropin levels are negatively correlated with hyperlipidemia. A recent study from China has shown that serum adropin level is an independent predictor of SA.[28]

CONCLUSION

Patterns of dyslipidemia are similar in Insulin resistance and in fatty liver disease, but is somewhat different in subclinical atherosclerosis. Qualitative changes in the lipids e.g. particle size, TG content, cholesterol efflux capacity are newer parameters that can give valuable information over and above those obtained from standard quantitative "lipid profile." Statins remain the drugs of choice despite lack of specific disease-related recommendations.

KEY POINTS

- Dyslipidemia is an established contributor to atherosclerotic CVD and consequent morbidity and mortality.
- Although, dyslipidemia may also result from a primary genetic abnormality, insulin resistance is by far, the much more common driving factor for dyslipidemia as part of the spectrum of metabolic syndrome.
- Recent understanding incorporates NAFLD in the broader definition of metabolic syndrome and thus the pattern of dyslipidemia present in NAFLD is essentially similar to that seen in IR and characterized by high TG, low HDL and high small, dense LDL, the combination classically called as "atherogenic dyslipidemia."
- High LDL and LDL:HDL ratio are better predictors of increased CIMT indicating the presence of SA.
- The "qualitative" changes of lipoprotein profile, including particle size, TG content, subfraction pattern, cholesterol efflux capacity, etc., can give valuable information beyond that obtained from conventional "quantitative" lipid profile.

REFERENCES

1. Powers AC, Niswender KD, Evans-Molina CE. Diabetes Mellitus: Diagnosis, Classification and Pathophysiology. In: Jameson JL, Fauci AS, Kasper DL, Hauser SL, Longo DL, Loscalzo J. Harrison's Principles of Internal Medicine, 20th edition. New York: McGraw Hill; 2018. pp. 2850-9.
2. Soffer GR, Ginsberg HN. Lipid and lipoprotein metabolism, hypolipidemic agents and therapeutic goals. In: DeFronzo RA, Ferrannini E, Zimmet P, Alberti G (Ends). International Textbook of Diabetes Mellitus, 4th edition. New Jersey: Wiley Blackwell; 2015. pp. 262-74.
3. Lamming DEW, Sabatini DM. A central role for mTOR in lipid homeostasis. Cell Metal. 2013;18(4):465-9.
4. Kwiterovich PO. The metabolic pathways of high-density lipoprotein, low-density lipoprotein and triglycerides: A Current Review. Am J Cardiol. 2000;86(suppl):5L-10.
5. Rader DJ, Kathiresan S. Disorders of lipoprotein metabolism. In: Jameson JL, Fauci AS, Kasper DL, Hauser SL, Longo DL, Loscalzo J (Eds). Harrison's Principles of Internal Medicine, 20th edition. New York: McGraw Hill; 2018. pp. 2889-902.
6. Festa A, D'Agostino R Jr, Howard G, Mykkänen L, Tracy RP, Haffner SM. Chronic subclinical inflammation as part of the insulin resistance syndrome: The Insulin Resistance Atherosclerosis Study (IRAS): Circulation. 2000;102:42-7.
7. Asato Y, Katsuren K, Ohshiro T, Kikawa K, Shimabukuro T, Ohta T. Hyperlipidemia and Insulin Resistance in School Children. Arterioscler Thromb Vasc Biol. 2006;26:2781-6.

8. Robins SJ, Lyass A, Zachariah JP, Massaro Joseph M, Ramachandran S. Vasan insulin resistance and the relation of a dyslipidemia to coronary heart disease. The Framingham Heart Study. Arterioscler Thromb Vasc Biol. 2011;31(5):1208-14.
9. Ampuero J, Gallego-Durán R, Romero-Gómez M. Association of NAFLD with subclinical atherosclerosis and coronary-artery disease: Meta-analysis. Rev Esp Enferm Dig. 2015;107(1):10-6.
10. Lauridsen BK, Stender S, Kristensen TS, Kofoed KF, Køber L, Nordestgaard BG. Liver fat content, non-alcoholic fatty liver disease, and ischaemic heart disease: Mendelian randomization and meta-analysis of 279 013 individuals. Eur Heart J. 2018;39(5):385-93.
11. Bril F, Sninsky JJ, Baca AM, Superko HR, Portillo Sanchez P, Biernacki D, et al. Hepatic steatosis and insulin resistance, but not steatohepatitis, promote atherogenic dyslipidemia in NAFLD. J Clin Endocrinol Metab. 2016;101:644-52.
12. Dowla S, Aslibekyan S, Goss A, Fontaine K, Ashraf AP. Dyslipidemia is associated with pediatric nonalcoholic fatty liver disease. J Clin Lipidol. 2018;12:981-7.
13. Kantartzis K, Rittig K, Cegan A, Machann J, Schick F, Balletshofer B, et al. Fatty liver is independently associated with alterations in circulating HDL2 and HDL3 subfractions. Diabetes Care. 2008;31(2):366-8.
14. Corey KE, Misdraji J, Gelrud L, Zheng H, Chung RT, Krauss RM. Nonalcoholic steatohepatitis is associated with an atherogenic lipoprotein subfraction profile. Lipids Health Dis. 2014;13:100.
15. DeFilippis AP, Blaha MJ, Martin SS, Reed RM, Jones SR, Nasir K, et al. Nonalcoholic fatty liver disease and serum lipoproteins: the multi-ethnic study of atherosclerosis. Atherosclerosis. 2013;227:429-36.
16. Amor AJ, Pinyol M, Sola` E, Catalan M, Cofán M, Herreras Z, et al. Relationship between noninvasive scores of nonalcoholic fatty liver disease and nuclear magnetic resonance lipoprotein abnormalities: A focus on atherogenic dyslipidemia. J Clin Lipidol. 2017;11:551-61.e7.
17. Fadaei R, Poustchi H, Meshkani R, Moradi N, Golmohammadi T, Merat S, et al. Impaired HDL-cholesterol efflux capacity in patients with nonalcoholic fatty liver disease is associated with subclinical atherosclerosis. Sci Rep. 2018;8:11691.
18. Van den Berg EH, Gruppen EG, Ebtehaj S, Bakker SJL, Tietge UJF, Dullaart RPF. Cholesterol efflux capacity is impaired in subjects with an elevated Fatty Liver Index, a proxy of nonalcoholic fatty liver disease. Atherosclerosis. 2018;277:21-7.
19. Abdelmalek MF, Diehl AM. Nonalcoholic fatty liver diseases and nonalcoholic steatohepatitis. In: Jameson JL, Fauci AS, Kasper DL, Hauser SL, Longo DL, Loscalzo J (Eds). Harrison's Principles of Internal Medicine, 20th edition. New York: McGraw Hill; 2018. pp. 2401-5
20. Chalasani N, Younossi Z, Lavine JE, Charlton M, Cusi K, Rinella M, et al. The diagnosis and management of nonalcoholic fatty liver disease: practice guidance from the American Association for the Study of Liver Diseases. Hepatology. 2018;67:328-57.
21. Grundy SM, Stone NJ, Bailey AL, Beam C, Birtcher KK, Blumenthal RS, et al. AHA/ACC/AACVPR/AAPA/ABC/ACPM/ADA/AGS/APhA/ASPC/NLA/PCNA Guideline on the Management of Blood Cholesterol: executive Summary. J Am Coll Cardiol. 2019:3168-209.
22. Tikkanen MJ, Fayyad R, Faergeman O, Olsson AG, Wun CC, Laskey R, et al. Effect of intensive lipid lowering with atorvastatin on cardiovascular outcomes in coronary heart disease patients with mild-to-moderate baseline elevations in alanine aminotransferase levels. Int J Cardiol. 2013;168:3846-52.
23. McNamara JJ, Molot MA, Stremple JF. Coronary artery disease in combat casualties in Vietnam. JAMA. 1971;216:1185-7.
24. Braamskamp M, Langslet G, McCrindle BW, Cassiman D, Francis GA, Gagne C, et al. Effect of rosuvastatin on carotid intima-media thickness in children with heterozygous familial hypercholesterolemia: the CHARON study (Hypercholesterolemia in Children and Adolescents Taking Rosuvastatin Open Label). Circulation. 2017;136:359-66.
25. Aoki T, Yagi H, Sumino H, Tsunekawa K, Araki O, Kimura T, et al. Relationship between carotid artery intima-media thickness and small dense low-density lipoprotein cholesterol concentrations measured by homogenous assay in Japanese subjects. Clin Chim Acta. 2015;442:110-4.
26. Yang C, Sun Z, Li Y, Ai J, Sun Q, Tian Y. The correlation between serum lipid profile with carotid intima-media thickness and plaque. BMC Cardiovasc Disord. 2014;14:181.
27. Guardiola M, Cofán M, de Castro-Oros I, Cenarro A, Plana N, Talmud PJ, et al. APOA5 variants predispose hyperlipidemic patients to atherogenic dyslipidemia and subclinical atherosclerosis. Atherosclerosis. 2015;240:98-104.
28. Jian-Fang C, Mao-Song W, Cong-Yan B, Yi-Fei T, Tao Y, Wei-Ting X. Association between serum adropin and subclinical atherosclerosis in patients with hyperlipidemia. Int J Clin Exp Med. 2019;12(7):9351-8.

CHAPTER 8

Risk Assessment of Different Lipid Markers: Clinical Appraisal and Goal of Therapy

Amitabh Sur

ABSTRACT

Routine assessment of different lipid markers and keeping them within the prescribed target range is very vital for both primary and secondary prevention of cardiovascular (CV) disease. Increased low-density lipoprotein cholesterol (LDL-C), non-high-density lipoprotein (HDL), and ApoB are most predictive of worse CV outcomes and increased high-density lipoprotein cholesterol (HDL-C) is protective. Newer lipid markers such as ApoA-1, lipoprotein(a) [Lp(a)], and Lp-PLA2, seems promising but need further extensive data before they are used in routine clinical practice for therapeutic decision making.

INTRODUCTION

Plasma lipoproteins play a role in transporting lipids to tissue for lipid deposition, steroid hormone production, energy utilization, and bile acid formation. Lipoproteins contain esterified and unesterified cholesterol, phospholipids, triglycerides (Tg), and apolipoproteins which serves as structural components, enzyme activators and inhibitors, and ligands for cellular receptor binding. The endothelial barrier can be crossed by all ApoB containing lipoproteins, inclusive of smaller Tg rich lipoproteins and their remnant particles, and after crossing, they get trapped and start interacting with extracellular structures such as proteoglycan.

This process is expedited in the presence of endothelial dysfunction. Continuous exposure to ApoB containing lipoproteins in the arterial wall leads to the accumulation of additional particles which lead to the growth and progression of atherosclerotic plaques.[1,2] Therefore, those with higher concentrations of plasma ApoB containing lipoproteins tend to retain more particles and so have more rapid progression of atherosclerotic plaque volume. So, the total atherosclerotic plaque burden is proportional to cumulative exposure to these lipoproteins.[3] Six major lipoproteins in the blood are chylomicron, very low-density lipoprotein (VLDL), intermediate density lipoprotein (IDL), LDL, HDL, and Lp(a).

TABLE 1: Chemical and physical characteristics of human plasma lipoproteins.						
	Density (g/mL)	Diameter (nm)	Tg (%)	Cholesterol esters (%)	PLs (%)	Cholesterol (%)
Chylomicrons	<0.95	80–100	90–95	2–4	2–6	1
VLDL	0.95–1.006	30–80	50–65	8–14	12–16	4–7
IDL	1.006–1.019	25–30	25–40	20–35	16–24	7–11
HDL	1.063–1.210	8–13	7	10–20	55	5
LDL	1.019–1.063	20–25	4–6	34-35	22–26	6–15
Lp(a)	1.006–1.125	25–30	4–8	35–46	17–24	6–9

[HDL: high-density lipoprotein; IDL: intermediate density lipoprotein, LDL: low-density lipoprotein, Lp(a): lipoprotein A; VLDL: very low-density lipoprotein]
Source: ESC 2019 lipid guideline.

CHEMICAL AND PHYSICAL CHARACTERISTICS OF HUMAN PLASMA LIPOPROTEINS

The chemical and physical characteristics of human plasma lipoproteins are outlined in **Table 1**.

Atherosclerotic cardiovascular disease (ASCVD) events such as unstable angina and myocardial infarction (MI) are generally precipitated by disruption of plaque and formation of thrombus which acutely obstructs blood flow so their risk increases proportional to an increase in atherosclerotic plaque volume, and hence to cumulative exposure (quantum and length of exposure) to ApoB containing lipoproteins. Therefore, maintaining a healthy lifestyle, throughout life, to decrease cumulative exposure to ApoB containing lipoproteins is suggested to reduce ASCVD. Also, this forms the rationale of pharmacotherapy to reduce ApoB containing lipoproteins including LDL, for both primary and secondary prevention of ASCVD events.[3]

VARIOUS APOLIPOPROTEINS ASSOCIATED WITH HUMAN PLASMA LIPOPROTEINS

Various apolipoproteins associated with human plasma lipoproteins are given in **Table 2**.

RISK FACTORS OF ATHEROSCLEROTIC CARDIOVASCULAR DISEASE

Major: Advancing age, increased (Total cholesterol, LDL-C, and non-HDL-C), low HDL-C, smoking, diabetes mellitus (DM), chronic kidney disease (CKD), hypertension, and family history of premature ASCVD.

Additional risk factors: Abdominal obesity, increased small dense-LDL, increased ApoB, hypertriglyceridemia, and family history of dyslipidemia.

TABLE 2: Various apolipoproteins associated with human plasma lipoproteins.

	Major apolipoprotein	Others apolipoproteins
Chylomicrons	ApoB-48	ApoA-I, A-II, A-IV, A-V
VLDL	ApoB-100	ApoA-1, C-11, C-111, E, A-V
IDL	ApoB-100	ApoC-11, C-111, E
HDL	ApoA-1	ApoA-11, C-111, E, M
LDL	ApoB-100	
Lp(a)	ApoA	ApoB-100

[HDL: high-density lipoprotein; IDL: intermediate density lipoprotein, LDL: low-density lipoprotein, Lp(a): lipoprotein A; VLDL: very low-density lipoprotein]
Source: ESC 2019 lipid guideline.

Nontraditional risk factors: Increased Lp(a), high sensitivity C-reactive protein (hsCRP), Lp-PLA2, homocysteine, uric acid, ApoE-4, and Tg remnants.

(*Source*: AACE 2017, lipid guideline)

Based on the AACE 2017 guideline for managing dyslipidemias, the atherosclerotic cardiovascular disease risk can be categorized as follows:

Extreme risk:
- Progressive ASCVD including unstable angina, even after achieving LDL <70 mg/dL.
- Clinical CV disease in the patient of DM, CKD stage 3,4 or hereditary heterozygous hypercholesterolemia.
- History of premature ASCVD (<55 year male and <65 year female).

Very high risk:
- Established or recent hospitalization due to ACS, coronary, carotid or peripheral vascular disease, 10-year risk (of having MI or death due to CVD >20% based on Framingham risk score).
- Diabetes or CKD stage 3,4 with at least one more risk factor.
- Hereditary heterozygous hypercholesterolemia.

High risk:
- Diabetes or CKD stage 3,4 with no additional risk factor.
- Two or more risk factors, with 10 year risk 10–20%.

Moderate risk:
- Two or fewer risk factors and 10 year risk <10%.

Low risk:
- No risk factors.

LOW-DENSITY LIPOPROTEIN CHOLESTEROL AND ATHEROSCLEROSIS

Low-density lipoprotein is the most abundant of ApoB containing lipoproteins. Plasma LDL-C is an estimate of the concentration of circulating LDL. It is a measure of cholesterol mass carried by LDL particles. Various epidemiological studies, RCTs, and Mendelian randomization studies have proven log-linear relationship between LDL-C and risk of ASCVD, and also lowering LDL-C has been shown to lower the risk of ASCVD, proportional to an absolute reduction in the level of LDL-C.[4-8] Mendelian randomization studies have also shown that long-term reduced exposure to LDL-C is associated with a lower risk of ASCVD than relatively short-term exposure to low LDL-C.[5,9] So, the effect of LDL-C on ASCVD risk depends both on absolute magnitude and length of exposure to LDL-C. Also, atherosclerosis regression, characterized by a reduction in both size and inflammatory status of established atherosclerotic lesions is documented when LDL-C is <70–80 mg/dL. Direct measurement of LDL by enzymatic technique or preparative ultracentrifugation is more accurate than using Friedewald formula, as it underestimates LDL-C, especially when Tg >177 mg/dL. The formula becomes all the more inaccurate when Tg >400 mg/dL.

(Friedewald formula—LDL-C = Total Cholesterol – HDL-C – Tg/5).

TRIGLYCERIDE RICH LIPOPROTEINS AND ATHEROSCLEROSIS

Most of the circulating Tgs are carried by Tg rich VLDL particles, chylomicrons and chylomicron remnants, therefore circulating plasma Tg measurement is reflective of the concentration of ApoB containing Tg rich lipoproteins. Elevated Tg is associated with increased risk of ASCVD, but this association is nullified if adjusted for non-HDL-C, which reflects the total concentration of all ApoB containing lipoproteins.[4]

Therefore, the causal effect of Tg rich lipoproteins on ASCVD depends on the total circulating concentration of ApoB containing particles and not on Tg content itself. Increased Tg is associated with insulin resistance, hypertension, and a procoagulant state. Ideally fasting Tg should be <150 mg/dL. Postprandial Tg is gradually emerging as a more potent risk predictor of ASCVD.[10,11] Elevated postprandial Tg is the only variable, which independently predicts CV events in women with normal HDL-C >50 mg/dL.[11]

Postprandial highly atherogenic Tg remnants may be causal for the association of postprandial hypertriglyceridemia and CV events. Increased Tg >2,000 mg/dL can also lead to pancreatitis.

HIGH-DENSITY LIPOPROTEIN CHOLESTEROL AND ATHEROSCLEROSIS

The inverse relationship between HDL-C and risk of atherosclerosis is well established in observational epidemiology.[4,12] But no RCT or genetic study to date has shown the benefit of therapeutically raising HDL-C on ASCVD risk. Whether the altering function of HDL particles can alter ASCVD risk or not, is a matter of future investigations.[13-18]

HDL-C >60 mg/dL is an independent negative risk factor for ASCVD in both sexes.

LIPOPROTEIN(A) AND ATHEROSCLEROSIS

Lipoprotein(a), is a <70 nm diameter LDL particle where an ApoA moiety is covalently bonded to an ApoB component. It can freely flux across endothelial barrier where if retained within the arterial wall can increase risk of atherosclerosis.[19] Lp(a) is structurally related to procoagulant plasminogen and also carries oxidized phospholipid, and these may contribute to the proatherogenic potential of Lp(a).[20] Though Mendelian randomization studies have noted a strong association between lifetime exposure of Lp(a) and ASCVD, the risk prediction from Lp(a), is much weaker compared to LDL-C.[21-23] Very high Lp(a) levels >180 mg/dL, may be associated with high lifetime risk of ASCVD. More than 90% of Lp(a) is hereditary controlled and so very high Lp(a) may suggest an inherited lipid disorder similar but more frequent to hereditary heterozygous hypercholesterolemia. A clinically meaningful reduction in ASCVD requires a large reduction of Lp(a) levels from its baseline values.[24] Reduction of up to 20% by Niacin or CETP inhibitors have not seen to change ASCVD risk, but the use of PCSK 9 inhibitors have suggested a possible role of Lp(a) reduction in ASCVD risk.[23]

ApoB AND ATHEROSCLEROSIS

Direct measurement of the circulating concentration of atherogenic ApoB gives an accurate estimate of the number of atherogenic particles in the plasma. Fasting is not mandatory, as ApoB-48 containing chylomicrons form <1% of the total concentration of circulating ApoB containing lipoproteins. Analytical performance of test measuring ApoB is better than the test used to measure or calculate LDL-C or non-HDL-C (Total cholesterol – HDL-C).[25] ApoB is identified as an independent additional risk factor for ASCVD by AACE dyslipidemia guideline 2017.

TOTAL CHOLESTEROL AND ATHEROSCLEROSIS

Total cholesterol as a major risk factor for ASCVD has been identified in the Framingham study, MRFIT study and guidelines, and position statements of AACE. Plasma lipids (including cholesterol and Tg) are highly dependent on diet and lifestyle apart from genetics.[26] Total cholesterol in humans is distributed in VLDL, LDL, and HDL, along with minor amounts in IDL and Lp(a). Plasma total cholesterol is associated with the risk of ASCVD over a broad range of values, therefore normal level is defined not based on population average but based on value, below which risk of ASCVD is minimal, i.e., 150 mg/dL. Genetic and epidemiological studies have shown cholesterol deposition in the arterial wall is a causal factor for ASCVD and therapeutic intervention with a statin to reduce cholesterol has proven to reduce the risk of ASCVD.

NON-HDL CHOLESTEROL AND ATHEROSCLEROSIS

Useful for assessment of total atherogenic lipid burden in patients of established ASCVD.

$$Non\text{-}HDL\text{-}C = Total\ cholesterol - HDL\text{-}C$$

OTHER POSSIBLE LIPID MARKERS

Small Dense LDL-C

Hyperglycemia, hyperinsulinemia, and hypertriglyceridemia including increased flux of Tg loaded VLDL particles, and genetics all play role in the formation of small dense LDL particles. They are more atherogenic perhaps due to high oxidative susceptibility and can easily pass through the endothelial barrier. Increased small dense LDL, along with high Tg and low HDL, is seen in diabetic dyslipidemia. ApoB and elevated non-HDL-C, give an estimate of small dense LDL.

Total cholesterol/HDL cholesterol ratio: Ratio <3.5 is considered normal and >6 is suggestive of the high risk of ASCVD.

LDL cholesterol/HDL cholesterol ratio: More than 3.5 in male and >3.0 in female indicates increased risk of ASCVD.

ApoB/ApoA-1 ratio: More than 1.0 in male and >0.9 in female indicates increased risk of ASCVD.

HsCRP: Especially useful to quantify the risk of ASCVD in otherwise healthy postmenopausal women with LDL <130 mg/dL. It is an independent predictor of CV events in men aged 40–84 years after adjusting for other risk factors. HsCRP <1 mg/L is considered normal, 1–3 mg/L is intermediate, and >3 mg/L is high.

Lipoprotein-associated Phospholipase A2

It is an enzyme that hydrolyzes oxidized phospholipids causing atherogenic vascular inflammation. It is a strong and independent risk factor of CV events and CVA in both symptomatic and asymptomatic ASCVD. It is more specific than hsCRP. Lp PLA2 levels <200 ng/mL—low, 201–223 ng/mL—intermediate and >223 ng/mL—high. It acts synergistically to hsCRP, and risk is profound if both are elevated.

ApoA-1: The presence of normal ApoA-1 in a person having low HDL-C signifies less risk of ASCVD as it indicates adequate number of HDL particles that contain less cholesterol.

ApoE-4 isoform: Apolipoprotein E is a major ligand of the LDL receptor. It has three isoforms: E2, E3, and E4, which differ structurally by 1–2 amino acids. ApoE-2 and E-4 increase risk of ASCVD by increasing LDL-C levels. ApoE-2 binds poorly to LDL receptors and E4 causes downregulation of LDL receptors by preferentially binding to Tg rich VLDL particles.

Fibrinogen: It is a clotting factor that can causes a prothrombotic state. Adding fibrinogen to routine lipid estimation may improve ASCVD risk prediction.

Homocysteine: Elevated homocysteine level may stratify people at intermediate risk of ASCVD to high risk, but routine screening is not recommended. Therapeutically lowering homocysteine does not offer any CV benefit.

Uric acid: Increased serum uric acid is associated with insulin resistance, dyslipidemia, and hypertension. Few epidemiological studies did indicate uric acid >6.99 mg/dL in male and >5.6 mg/dL in females to be associated with a significant increase in ASCVD mortality, but lowering uric acid therapeutically has not shown a positive effect on CV outcome.[27,28,]

GOALS OF THERAPY FOR ADULT PATIENTS AT RISK FOR ASCVD

The goals of therapy for adult patients at risk for ASCVD are given in **Table 3**.

IDEAL LIPID LEVEL FOR CHILDREN AND ADOLESCENTS WITH DYSLIPIDEMIA

Ideal lipid level for children and adolescents with dyslipidemia is given in **Table 4**.

TABLE 3: Goals of therapy for adult patients at risk for atherosclerotic cardiovascular disease (ASCVD).

Lipid parameter	Goal (mg/dL)
TC	<200 mg/dL
LDL-C	<55 mg/dL (extreme risk) <70 mg/dL (very high risk) <100 mg/dL (high risk) <100 mg/dL (moderate risk) <130 mg/dL (low risk)
Non-HDL-C	Up to 25 mg/dL above LDL (extreme risk) Up to 30 mg/dL above LDL (for all other)
Triglyceride	<150 mg/dL
ApoB	<70 mg/dL (extreme risk) <80 mg/dL (very high risk) <90 mg/dL (high risk, moderate risk)

(HDL-C: high-density lipoprotein cholesterol; LDL: low-density lipoprotein; TC: total cholesterol)
Source: AACE 2017 guideline.

TABLE 4: Ideal lipid level for children and adolescents with dyslipidemia.

Category	LDL (mg/dL)
Acceptable	<110
Borderline	110–129
High	130 or higher
Category	Total cholesterol (mg/dL)
Acceptable	<170
Borderline high	170–199
High	200 or more
Category	HDL-C (mg/dL)
Acceptable	>45
Borderline low	40–45
Low	<40
Category (10–19 year age group)	Triglyceride (mg/dL)
Acceptable	<90
Borderline high	90–129
High	130 or more

Source: Endocrine society clinical practice guideline 2017.

CONCLUSION

Routine assessment of traditional and newer lipid markers and evidence based therapeutic interventions to keep them in desired ranges, are pivotal for both primary and secondary prevention of cardiovascular diseases.

> **KEY POINTS**
>
> - Routine assessment of various lipid markers forms back bone of strategy for both primary and secondary prevention of adverse cardiovascular outcome.
> - Although role of measurement of ApoB containing lipid markers like ApoB, LDL and Non-HDL-C cannot be overemphasized, the emerging role of newer markers like Lipoprotein A, Small dense LDL, ApoA-1, LpPLA2, hsCRP are increasingly being recognized.
> - Pharmacotherapy to keep ApoB containing lipid particles in desired target range to reduce total volume of atherosclerotic particles, forms cornerstone of management of dyslipidemia.

REFERENCES

1. Tabas I, Williams KJ, Boren J. Subendothelial lipoprotein retention as the initiating process in atherosclerosis: update and therapeutic implications. Circulation. 2007;116:1832-44.
2. Boren J, Williams KJ. The central role of arterial retention of cholesterol-rich apolipoprotein-B-containing lipoproteins in the pathogenesis of atherosclerosis: a triumph of simplicity. Curr Opin Lipidol. 2016;27:473-83.
3. Ference BA, Graham I, Tokgozoglu L, Catapano AL. Impact of lipids on cardiovascular health: JACC Health Promotion Series. J Am Coll Cardiol. 2018;72:1141-56.
4. Emerging Risk Factors Collaboration, Di Angelantonio E, Gao P, Pennells L, Kaptoge S, Caslake M, Thompson A, et al. Lipid-related markers and cardiovascular disease prediction. JAMA. 2012;307:2499-506.
5. Ference BA, Yoo W, Alesh I, Mahajan N, Mirowska KK, Mewada A, et al. Effect of long-term exposure to lower low-density lipoprotein cholesterol beginning early in life on the risk of coronary heart disease: a Mendelian randomization analysis. J Am Coll Cardiol. 2012;60:2631-9.
6. Holmes MV, Asselbergs FW, Palmer TM, Drenos F, Lanktree MB, Nelson CP, et al. Mendelian randomization of blood lipids for coronary heart disease. Eur Heart J. 2015;36:539-50.
7. Silverman MG, Ference BA, Im K, Wiviott SD, Giugliano RP, Grundy SM, et al. Association between lowering LDL-C and cardiovascular risk reduction among different therapeutic interventions: a systematic review and meta-analysis. JAMA. 2016;316:1289-97.
8. Baigent C, Keech A, Kearney PM, Blackwell L, Buck G, Pollicino C, et al; Cholesterol Treatment Trialists' (CTT) Collaborators. Efficacy and safety of cholesterol-lowering treatment: prospective meta-analysis of data from 90,056 participants in 14 randomised trials of statins. Lancet. 2005;366:1267-78.
9. Cohen JC, Boerwinkle E, Mosley TH Jr, Hobbs HH. Sequence variations in PCSK9, low LDL, and protection against coronary heart disease. N Engl J Med. 2006;354:1264-72.
10. Ridker PM. Fasting compared with nonfasting triglycerides and risk of cardiovascular events in women. JAMA. 2007;298:309-16.
11. Nordestgaard BG, Benn M, Schnohr P, TybjaergHansen A. Nonfasting triglycerides and risk of myocardial infarction, ischemic heart disease, and death in men and women. JAMA. 2007;298:299-308.
12. Prospective Studies Collaboration, Lewington S, Whitlock G, Clarke R, Sherliker P, Emberson J, Halsey J, et al. Blood cholesterol and vascular mortality by age, sex, and blood pressure: a meta-analysis of individual data from 61 prospective studies with 55,000 vascular deaths. Lancet. 2007;370:1829-39.
13. Lincoff AM, Nicholls SJ, Riesmeyer JS, Barter PJ, Brewer HB, Fox KAA, et al; ACCELERATE Investigators. Evacetrapib and cardiovascular outcomes in high-risk vascular disease. N Engl J Med. 2017;376:1933-42.
14. HPS/TIMI/REVEAL Collaborative Group; Bowman L, Hopewell JC, Chen F, Wallendszus K, Stevens W, Collins R, et al. Effects of anacetrapib in patients with atherosclerotic vascular disease. N Engl J Med. 2017;377:1217-27.

15. Schwartz GG, Olsson AG, Abt M, Ballantyne CM, Barter PJ, Brumm J, et al; dal-OUTCOMES Investigators. Effects of dalcetrapib in patients with a recent acute coronary syndrome. N Engl J Med. 2012;367:2089-99.
16. Aim-High Investigators; Boden WE, Probstfield JL, Anderson T, Chaitman BR, Desvignes-Nickens P, Koprowicz K, et al. Niacin in patients with low HDL cholesterol levels receiving intensive statin therapy. N Engl J Med. 2011;365:2255-67.
17. Group HTC, Landray MJ, Haynes R, Hopewell JC, Parish S, Aung T, et al. Effects of extended-release niacin with laropiprant in high-risk patients. N Engl J Med. 2014;371:203-12.
18. Andrews J, Janssan A, Nguyen T, Pisaniello AD, Scherer DJ, Kastelein JJ, et al. Effect of serial infusions of reconstituted high-density lipoprotein (CER-001) on coronary atherosclerosis: rationale and design of the CARAT study. Cardiovasc Diagn Ther. 2017;7:4551.
19. Nordestgaard BG, Langsted A. Lipoprotein(a) as a cause of cardiovascular disease: insights from epidemiology, genetics, and biology. J Lipid Res. 2016;57:1953-75.
20. van der Valk FM, Bekkering S, Kroon J, Yeang C, Van den Bossche J, van Buul JD, et al. Oxidized phospholipids on lipoprotein(a) elicit arterial wall inflammation and an inflammatory monocyte response in humans. Circulation. 2016;134:611-24.
21. Nordestgaard BG, Chapman MJ, Ray K, Boreń J, Andreotti F, Watts GF, et al; European Atherosclerosis Society Consensus Panel. Lipoprotein(a) as a cardiovascular risk factor: current status. Eur Heart J. 2010;31:2844-53.
22. Emerging Risk Factors Collaboration; Erqou S, Kaptoge S, Perry PL, Di Angelantonio E, Thompson A, White IR, et al. Lipoprotein(a) concentration and the risk of coronary heart disease, stroke, and nonvascular mortality. JAMA. 2009;302:412-23.
23. Kamstrup PR, Tybjaerg-Hansen A, Steffensen R, Nordestgaard BG. Genetically elevated lipoprotein(a) and increased risk of myocardial infarction. JAMA. 2009;301:2331-9.
24. Parish S, Hopewell JC, Hill MR, Marcovina S, Valdes-Marquez E, Haynes R, et al; HPS2-THRIVE Collaborative Group. Impact of apolipoprotein(a) isoform size on lipoprotein(a) lowering in the HPS2-THRIVE Study. Circ Genom Precis Med. 2018;11:e001696.
25. Langlois MR, Chapman MJ, Cobbaert C, Mora S, Remaley AT, Ros E, et al; European Atherosclerosis Society (EAS) and the European Federation of Clinical Chemistry and Laboratory Medicine (EFLM) Joint Consensus Initiative. Quantifying atherogenic lipoproteins: current and future challenges in the era of personalized medicine and very low concentrations of LDL cholesterol. A consensus statement from EAS and EFLM. Clin Chem. 2018;64:10061033.
26. Kato H, Tillotso J, Nichaman MZ, Rhoads GG, Hamilton HB. Epidemiologic studies of coronary heart-disease and stroke in Japanese men living in Japan, Hawaii and California—serum-lipids and diet. Am J Epidemiol. 1973;97:372-85.
27. Fang J, Alderman MH. Serum uric acid and cardiovascular mortality the NHANES I epidemiologic followup study, 1971-1992. National Health and Nutrition Examination Survey. JAMA. 2000;283:2404-10.
28. Culleton BF, Larson MG, Kannel WB, Levy D. Serum uric acid and risk for cardiovascular disease and death: the Framingham Heart Study. Ann Intern Med. 1999;131:7-13.

CHAPTER 9

Rethinking "Good" Cholesterol: Understanding the Role of High-density Lipoprotein

Kaushik Biswas, Satyam Chakraborty

ABSTRACT

Atherosclerotic vascular disease is one of the leading causes of death throughout the world. High-density lipoprotein cholesterol (HDL-C) is traditionally considered as "good" cholesterol over decades since the discovery of its negative association with cardiovascular disease (CVD). Genetic studies have failed to show any strong association between HDL and CVDs. Multiple therapeutic intervention trials raising the levels of HDL also remain futile in terms of cardiovascular protection. These findings instigated researches, revealing many complex roles and functioning of HDL in health and disease. There seems to be a disease specific, "patterned" structure-composition-function of HDL with multiple subclasses. In this review, we explore novel and paradigm-shifting understanding of HDL's biology in health and disease leaving behind the "good-bad" binary.

INTRODUCTION

The Framingham Heart Study in 1960, reported first the inverse relationship between cardiovascular (CV) risk and plasma high-density lipoprotein cholesterol (HDL-C).[1] This landmark discovery encouraged many mechanistic investigations on HDL's atheroprotection which ultimately helped in identification of the reverse cholesterol transport (RCT) pathway.[2] RCT is defined as the process by which cholesterol moves out of cells in peripheral tissues (i.e., foam cells in atherosclerotic plaques), enters the circulation, and is excreted in the feces. HDL's CV protection has been conferred to its multifaceted ability to accept cholesterol from cells, carrying the cholesterol in the RCT pathway and finally delivering the cholesterol into the liver. Observational studies indicated that CV risk decreases by ≈2–3% per 1 mg/dL increase in HDL-C.[3] The level of HDL-C in the plasma was conceptualized as a faithful marker of the ability of the HDL to mediate RCT. In 2020, there appears to be a paradigm shift in the whole HDL saga in that HDL-C measurements is no longer considered to the honest reflector of the overall abundance of HDL particles, the distribution of HDL subspecies,[4] or RCT efficiency.[5] Moreover, several human genetic studies coupled with many negative HDL-raising trials have generated much controversy over the HDL hypothesis.[6]

UNFOLDING MYTH OF HDL-C: THE "GOOD" CHOLESTEROL

Evidence from Genetic Analysis

As evident from the study of rare genetic conditions of HDL, not all mutations that lead to increased HDL-C should be predicted to result in reduced CV risk. Indeed, deficiency of endothelial lipase (LIPC) or scavenger receptor B1 (SCARB1) results in elevated HDL-C but it does not translates into CV protection (**Table 1**). This observation is consistent with the interpretation of Mendelian randomization studies of HDL-C in humans.

Evidence from HDL-raising Therapeutics

Fibrates

It is not clear whether the clinical effects of fibrates are due to the increase in HDL-C only or due to the decrease in triglycerides (TGs) and slight reduction in low-density lipoprotein (LDL). In patients with type 2 diabetes mellitus (T2DM) already on statin therapy, fibrates have not been shown to provide any added CV benefits.[7]

Niacin

Niacin reduces TG, very low-density lipoprotein (VLDL) cholesterol, LDL cholesterol (LDL-C) and lipoprotein(a) [Lp(a)], and they increase HDL-C. The AIMHIGH, ACCORD, and HPS2-THRIVE studies indicate that there is no sufficient evidence for the reduction of CV events through the use of fibrates and niacin[8] especially on top of with statin therapy (**Table 2**).

TABLE 1: Genes–HDL–CV outcome relationship.		
Gene	Effect on HDL-C	Effect on CVD
APOA1	↓	↑
ABCA1	↓	↑/-
LCAT	↓	↑/-
CETP	↑	↓/-
LPL	↓	↑/-
LPIC	↑	↑
LPIG	↑	↓/-
APOC2	↓	ND
APOC3	↑	↓
PLTP	↓	ND
SRAB1	↑	↑?

(–: no effect; ?: hypothesized effect based on data from mice; ABCA1: ATP-binding cassete transporter A1; APOA1: apolipoprotein A1; APOC2: apolipoprotein C2; APOC3: apolipoprotein C3; CETP: cholesterol ester transfer protein CVD: cardiovascular disease; HDL-C: high-density lipoprotein cholesterol; LCAT: lecithin–cholesterol acyltransferase; LPIC: endothelial lipase; LPIG: hepatic lipase; LPL: lipoprotein lipase ND: no data available; PLTP: phospholipid transfer protein; SCARB1: scavenger receptor B1)

Source: Adapted from Brunham LR, Hayden MR. Human genetics of HDL: Insight into particle metabolism and function. Progress in Lipid Research. 2015;58:14-25.

TABLE 2: Summary of HDL-targeted therapy trials.

	Phase	Intervention	Condition	n	Primary endpoint	Status of result
HDL-infusion therapies						
CSL112 (AEGIS-II)	3	CSL112 6 g apolipoprotein A1 per dose (~80 mg/kg) infusion vs. placebo; one dose per week for 4 weeks	ACS	17,400	Cardiovascular death, myocardial infarction, or stroke	Ongoing; estimated completion 2021
CSL112 (AEGIS-I)	2b	CSL112 low-dose (2 g apolipoprotein A1 per dose) or high-dose (6 g apolipoprotein A1 per dose) vs. placebo; one dose per week for 4 weeks	ACS	1,258	Drug-induced liver injury or change in renal status	Well tolerated, no safety concerns
CER-001 (CARAT)	2	CER-001 3 mg/kg infusion vs. placebo; one dose per week for 10 weeks	ACS	301	Coronary artery percentage atheroma volume	No effect on primary endpoint
MDCO-216 (MILANO-PILOT)	2	MDCO-216 (recombinant apolipoprotein A1 Milano) 20 mg/kg infusion vs. placebo; one dose per week for 5 weeks	ACS	126	Coronary artery percentage atheroma volume	No effect on primary endpoint
Oral HDL-C-raising drugs						
Anacetrapib (REVEAL)	3	Anacetrapib (CETP inhibitor) 100 mg/day vs. placebo	Cardiovascular disease	30,449	Cardiovascular death, myocardial infarction, or coronary revascularization	Increased HDL-C concentrations and positive effect on primary endpoint (development program terminated because of small effect size)

Continued

TABLE 2: Summary of HDL-targeted therapy trials.

	Phase	Intervention	Condition	n	Primary endpoint	Status of result
Evacetrapib (ACCELERATE)	3	Evacetrapib (CETP inhibitor) 130 mg/day vs. placebo	Cardiovascular disease, cerebrovascular disease, peripheral vascular disease	12,092	Cardiovascular death, myocardial infarction, stroke, revascularization, or admission to hospital for unstable angina	Terminated (lack of efficacy); reduced LDL-C and increased HDL-C concentrations
Dalcetrapib (dal-OUTCOMES)	3	Dalcetrapib (CETP inhibitor) 600 mg/day vs. placebo	Cardiovascular disease	15,865	Cardiovascular death or morbidity	Increased HDL-C but had no effect on primary endpoint
Torcetrapib (ILLUMINATE)	3	Torcetrapib (CETP inhibitor) 60 mg/day plus atorvastatin vs. placebo plus atorvastatin	Cardiovascular disease, type 2 diabetes mellitus	15,067	Major cardiovascular disease event	Terminated (safety); increased risk of the primary endpoint
Niacin (AIM-HIGH)	3	Niacin 1,500–2,000 mg/day vs. low-dose niacin (up to 200 mg/day)	Cardiovascular disease, cerebrovascular disease	3,414	Cardiovascular death, myocardial infarction, stroke, or coronary or cerebral revascularization	Terminated (lack of efficacy); marginally increased HDL-C and reduced triglyceride concentrations
Niacin and laropiprant (HPS2-THRIVE)	3	Niacin 2,000 mg/day and laropiprant 40 mg/day plus simvastatin vs. placebo plus simvastatin (plus ezetimibe 10 mg/day if LDL-C not adequately controlled)	Cardiovascular disease, cerebrovascular disease, peripheral vascular disease	25,673	Cardiovascular death, myocardial infarction, stroke, or coronary or cerebral revascularization	Did not reduce risk of the primary endpoint, but increased risk of serious adverse events; marginally increased HDL-C

(ACS: acute coronary syndrome; CETP: cholesteryl ester transfer protein; HDL-C: high-density lipoprotein cholesterol; LDL-C: low-density lipoprotein cholesterol)

Source: Adapted from Xiang AS, Kingwell BA, Rethinking good cholesterol: a clinicians' guide to understanding HDL. Lancet Diabetes Endocrinol. 2019;7(7):575-82.

Continued

Cholesteryl Ester Transfer Protein Inhibition

Cholesteryl ester transfer protein (CETP) mediates the transfer of cholesterol and TG among lipoproteins (**Fig. 1**). Therapeutic CETP-inhibition in humans is associated with a significant increase in HDL-C and a drop in LDL-C. However, three out of four randomized control trials with CETP inhibitors had to be terminated prematurely due to increased mortality in the treatment group or due to futility (**Table 2**).[9]

HDL STRUCTURE-COMPOSITION-FUNCTION

High-density Lipoproteins' Lifecycle

The building block of HDL is apolipoprotein A1 (ApoA-1), synthesized via forkhead box protein A3[10] in the liver and intestine. ATP-binding Cassete Transporter A1 (ABCA1) mediates cholesterol efflux and lapidates ApoA-1 to form nascent discoidal pre-β- HDLs. Lipidation and storage of free cholesterol as cholesterol esters (CE) by the enzyme, lecithin–cholesterol acyltransferase (LCAT) form mature spherical α-HDL.[11] Having passed through constant dynamic remodeling due to interactions with multiple other lipoproteins and variety of enzymes such as hepatic and endothelial lipase for 4–5 days, mature HDL converts to smaller subspecies (e.g., pre-β HDL).[10] In addition, HDLs collects antioxidants such as carotenoids from the intestine and transport in the systemic circulation and[10] HDL structure and lipidome can be modified postprandially.[10] It has been suggested that nonfasting HDL concentrations may be more relevant predictors of CV events than fasting levels.[10] Each mature HDL particle contains ~100 phospholipids (PLs) (such as phosphatidylcholines, sphingolipids, etc.), up to 30–90 CE molecules, 10–50 free cholesterol molecules, 10–20 TG, up to 80 proteins (e.g., ApoA-I and ApoA-II)

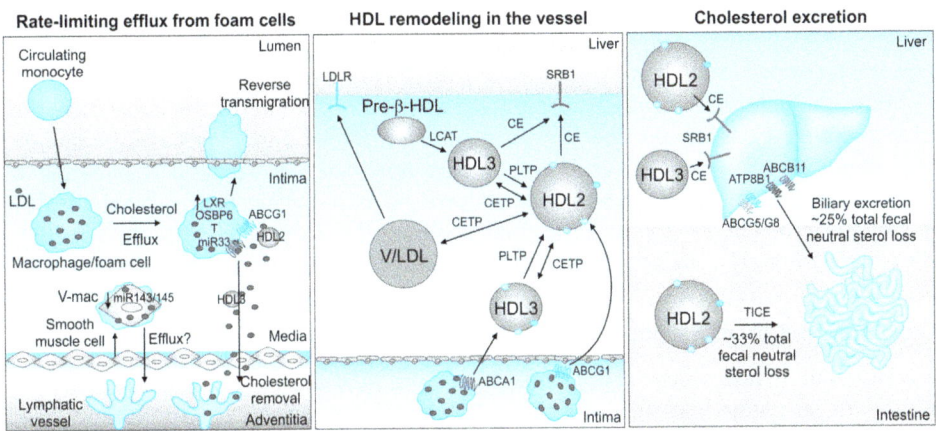

(ABCA1: ATP-binding cassete transporter A1; ABCG1: ATP binding cassette subfamily G member 1; ABCG5/G8: ATP binding cassette subfamily G member 5/8; CE: cholesterol esters; CETP: cholesterol ester transfer protein; LCAT: lecithin–cholesterol acyltransferase; LDL: low-density lipoprotein; LDLR: low-density lipoprotein receptor; LXR: liver X receptor; PLTP: phospholipid transfer protein; SRB1: scavenger receptor class B type 1; V/LDL: very low-density lipoprotein)

FIG. 1: Cholesterol efflux capacity (CEC), reverse cholesterol transport (RCT), and elimination of high-density lipoprotein (HDL).

Source: Adapted from Ouimet M, Barrett T J, Fisher A. HDL and reverse cholesterol transport basic mechanisms and their roles in vascular health and disease. Circ Res. 2019;124:1505-18.

and enzymes.[12,13] The CE and TG content in the hydrophobic core modify the size of HDL.[12] However, the biological properties of HDL and its beneficial functions are dependent on the type of transported proteins, lipids, and nucleic acids. CE in the HDL core can be relocated to TG-rich lipoproteins by CETP and is eliminated via the hepatic LDL receptor (LDL-R) or selectively taken up via SR-B1 acting as a hepatic receptor for HDL. HDLs deliver cholesterol to the liver and steroidogenic tissues mainly through binding to SR-B1.[14] De-lipidated ApoA-1 are also cleared through SR-B1, preferentially in the kidneys.[11]

REVERSE CHOLESTEROL TRANSPORT (FIGURE 1)

Cholesterols are present in high concentrations in cellular and subcellular membranes of mammalian cells. The amount of free cholesterol is checked within a narrow range to prevent toxicity. Any surplus cholesterol is effluxed from the cells to extracellular acceptors or esterified to CE and stored in cytosolic lipid droplets (LDs). The Liver X receptor (LXR) in liver regulates expression of numerous genes involved in efflux pathway including ABCA1/G1.[15] A recent report by the Fernandez–Hernando group demonstrates that repression of ABCA1 is the primary mechanism by which miR-33 regulates macrophage cholesterol efflux and atherogenesis.[16] Foam cells are traditionally thought to be cholesterol-laden macrophages originating from monocytes, but they can also be macrophage-like cells originating from vascular smooth muscle cells (VSMCs). VSMC plasticity in atherosclerosis is well recognized.[17] ATP binding cassette subfamily G member 1 (ABCG1) preferentially lipidates small HDL particles, specifically ApoA-I to form nascent HDL.[17] ABCG1 stimulates net cholesterol efflux to larger HDL but not to lipid-poor ApoA-I.[17] Finally, RCT to the liver of cholesterol involves two routes: (1) Direct (HDL > liver SR-B1), and (2) Indirect (HDL > LDL/VLDL > liver LDL-R). In the liver, the CE is hydrolyzed and the free cholesterol is either converted to bile acids or transported by ABCG5 and ABCG8 into the bile for excretion into the feces.

Thus, the term RCT currently encompasses all potential routes of net cholesterol flux from peripheral tissues into the feces[17] including artificial ones that have therapeutic potential. They are artificial cholesterol acceptors such as non-HDL particles of 2-hydroxypropyl-β-CD (cyclodextrin) and have been shown in vivo to mediate RCT and atheroprotection.[18] In transintestinal cholesterol transport (TICE), cholesterol is transported directly from blood, through the enterocytes, into the lumen of the intestine.[19]

Quantification of Reverse Cholesterol Transport

The HDL-controversy does not undervalue its role in RCT. The HDL-C measurement may not be the surrogate marker for CV protection but the HDL particle's cholesterol efflux capacity (CEC) could be. Indeed, an independent inverse association between HDL CEC and incident CV events has been shown both in the Dallas Heart Study and in the European Prospective Investigation of Cancer-Norfolk study.[20,21] Cuchel et al.[22] adapted the conventional RCT method to allow for quantification of RCT in humans. This method involves intravenous delivery of 3H-cholesterol nanoparticles, followed by blood and sample collection to quantify tracer counts in plasma, non-HDL, and HDL fractions, as well as fecal fractions. This is an exciting advancement in the field, providing a feasible approach to quantify RCT in vivo in humans.

High-density Lipoproteins Structural Diversity

High-density lipoproteins are complex particles with the potential to be classified into several subclasses depending on their differing physicochemical properties but this unique identification system is still in its infancy. Experts have devised a five-part sub classification for HDLs, which tries to include all aforementioned properties. It has been classified as very large HDL, large HDL, medium HDL, small HDL, and very small HDL. The function and metabolism of HDLs can be influenced by the subclass it belongs to,[23-25] and the ability to distinguish between HDL-subclasses may be clinically relevant.[26] By ultracentrifugation, two HDL subclasses can be obtained: HDL2 (1.063–1.125 g/mL) and HDL3 (1.125–1.21 g/mL). On the other hand, agarose gel electrophoresis separates HDL based on surface charge and shape into α- or pre-β-migrating particles (α-HDL or pre-β-HDL). Further clinical studies are required to establish reference values for nuclear magnetic resonance (NMR)[27] and should assess whether integration of HDL subclasses measurements and parameters of HDL functionality help in therapeutic decision making and alter the outcome of an intervention.

FUNCTIONAL PROPERTIES OF HDL IN CEC, RCT, AND PLEIOTROPISM

The focus has been shifted to consider the functionality of HDL particle keeping aside traditional cholesterol content estimation and to investigate its pleiotropic actions besides cholesterol efflux. The heterogeneity of HDL particles is of clinical relevance. Independent of its ability to mediate RCT, HDL is also known to exert potent antioxidant and anti-inflammatory effects that can improve RCT, retard atherosclerotic plaque progression, and promote plaque regression.[17] Overexpression of paraoxonase-1 (PON-1), an antioxidant enzyme, improves the efflux capacity of HDL and drives RCT in mice.[28]

Proteomic and lipidomic analyses have added complexities in HDL structure and function over time. In addition to protein and lipid cargo, HDL can transport functional noncoding ribonucleic acids (RNAs), such as miRNAs, and this pool of lipoprotein-associated RNA can be altered in disease.[17] It is now appreciated that how specific HDL functions (in CEC/RCT, thrombosis, inflammation, etc.) are related to HDL compositional heterogeneity in humans and how HDL subspecies altered during coronary artery disease (CAD) could lead to the identification of new diagnostic tools and therapies.[17]

HDL FUNCTION IN HEALTHY CONDITIONS

The variable protein and lipid composition of HDL may underlie its atheroprotective function (**Fig. 2**). One of the most important properties of HDL is its ability to induce NO-production in endothelial cells, through activation of surface receptors, such as SR-B1[29] HDL. In atherosclerotic cardiovascular diseases (ASCVD), larger HDL particles have lesser antioxidative capacity than smaller, denser ones,[30] which can be explained by a modified proteome. Larger HDL particles correlate to apolipoprotein A2,which reduces the association between HDL and PON-1.[31] Small, dense HDL3 have a more potent anti-inflammatory effect than larger HDL2. HDL3 is highly effective in inhibiting tumor necrosis factor alpha (TNF-α) induced vascular cell adhesion molecule-1 (VCAM-1) expression in an in-vitro endothelial cell model. Evidences are mounting on a disease-specific HDL size function relationship. While small HDLs are traditionally regarded as

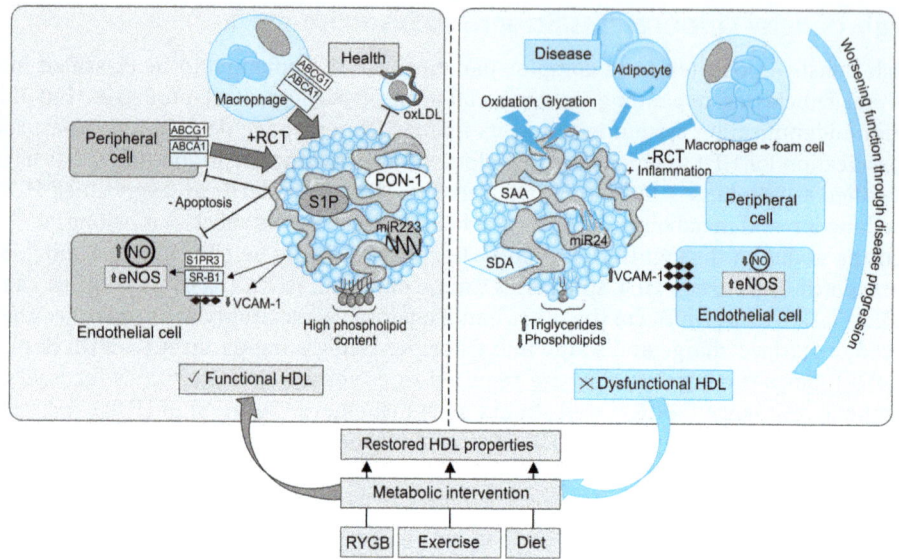

(ABCA1: ATP-binding cassete transporter A1; ABCG1: ATP binding cassette subfamily G member 1; eNOS: endothelial nitric oxide synthase; LDL: low density lipoprotein; PON-1: paraoxonase-1; RCT: reverse cholesterol transport; SAA: serum amyloid A; VCAM-1: vascular cell adhesion molecule-1; SDA: symmetric dimethylarginine)

FIG. 2: High-density lipoprotein (HDL) in health and disease.

Source: Adapted from Rayner KJ, Hennessy EJ. Extracellular communication via microRNA: lipid particles have a new message. J Lipid Res. 2013;54:1174-81.

antiatherosclerotic, in T2DM, larger HDLs seem beneficial,[31] potentially due to a different molecular composition. Normally PLs are the dominant HDL associated lipids (up to 50% of HDL-lipids) which stabilize the particle.[31] An increase in phosphatidylcholine promotes cholesterol efflux and an increase in sphingomyelin decreases influx of cholesterol via scavenger receptor class B type 1 (SRB1).[32] Nowadays, the sphingosine-1-phosphate receptors (S1PR) is in the limelight as an HDL target receptor. About 50–70% of S1P is carried by HDL particles in circulation. Finally, HDLs are an effective carrier of circulating microRNA (miRNA) to target cells.[33] The miRNA function of miR-223 and miR-24 are best understood, with miR-223 providing a beneficial anti-inflammatory profile in contrast to miR-24 which could be atherogenic[31] and in pathological conditions the nature or balance of miRNA in HDL particles get deranged.

HDL DYSFUNCTION IN A PATHOPHYSIOLOGICAL STATE

Chronic Kidney Disease

In chronic kidney disease (CKD), patient's HDL levels are usually reduced in relation to decreased levels of apolipoproteins and LCATs and its abnormal post-translational modifications.[34.] The diminished levels of HDL in CKD is associated with the decline in the catabolism of TG-rich lipoproteins, the slow rate of ApoA-I and A-II synthesis by the liver, and the increased activity of CETP.[34] Modifications observed in HDL particles in pathological states impair cholesterol efflux,[34] antioxidative functions, vasorelaxation, the ability to suppress TNF-α induced nuclear factor kappa B (NF-κB) activation, and adhesion molecule expression.[34] Growing evidence suggests that in patients with chronic inflammatory disorders, cardiovascular disease (CVD), CKD, and T2DM, HDL may lose important antiatherosclerotic properties and become dysfunctional (**Fig. 2**). The differences in HDL properties might be attributable to altered HDL-associated proteome and lipidome. Such modifications involve alterations in the amount and type of proteins and lipids bound to the HDL particle as well as post-translational modifications. Low ApoA-I levels in HDL in inflammatory states are associated with accelerated HDL catabolism and the serum amyloid A (SAA) concentrations are elevated in HDL particles.[34] Honda et al.[35] showed decreasing levels of HDL3 and ApoA-I in HDL3 subfraction along with the worsening of CKD severity; however, in their study, the level of HDL2 and ApoA-I in HDL2 subfraction was stable. Tolle et al.[36] showed that HDL in CKD was enriched in SAA and exhibited decreased anti-inflammatory capacity. The loss of PON-1 activity in HDL fails to prevent oxidation of its own lipids and proteins. It appears that HDL disturbances in CKD play a key important role in retarded VLDL catabolism and the development of hypertriglyceridemia. Decreased concentration and function of HDL coupled with hypertriglyceridemia in CKD creates a "vicious circle" where these factors significantly influence one another, perpetuating lipid disturbances and driving atherosclerosis development in this group of population.[34] Meta-analysis found that asymmetric dimethylarginine was strongly associated with CVD outcomes in patients with kidney diseases.[37]

Cardiovascular Disease

As mentioned before (**Table 1**) more recent results suggest that persons with high HDL levels are not always protected from CVD and therapeutic HDL-rising medications do not reduce the risk of CV events.[38] In a large study of >33,000 hemodialysis patients, U-shaped association between HDL concentrations of <30 mg/dL and >60 mg/dL and increased risk for total and CVD mortality was observed.[39,40] Similar to the state of CKD many dysfunctional changes in HDL are apparent in CVD too. Myeloperoxidase, the enzyme found in granulocyte is increased in ASCVD which can catalyze deleterious changes to HDL associated proteins, namely ApoA-1. This can cause an impaired CEC and increase in inflammatory pathways. SAA is a causal factor of HDL dysfunction that includes a loss of anti-inflammatory and RCT function and a decreased ability of HDLs to interact with the plasma membrane of adipocytes.[41]

Type 2 Diabetes Mellitus

Lipid composition, size, and structure of HDLs are closely linked. In T2DM patients, several studies have shown that there is a preponderance of smaller HDL particles, and an increase in TG content in HDL, which may render them more hydrophobic and therefore, challenging the idea that small HDL is always protective. On the other hand, HDLs can directly influence whole-body glucose homeostasis.[42] HDL promotes AMPK-dependent glucose uptake, increases insulin secretion, and protect pancreatic ß-cells from apoptosis. So, there is a bidirectional relationship between HDL and dysglycemia.

THE PARADOX OF EXTREMELY HIGH HDL

Recent studies have indicated that a high level of HDL-C could be potentially deleterious to CV health. The cutoff for pathological elevation of HDL has tentatively been placed as HDL-C levels ranging from 60 to 80 mg/dL. A cross-sectional study determined that in men, and after adjusting for CV risk factors, extremely high HDL-C was associated to endothelial dysfunction, as measured in-vivo by flow mediated vasodilation (FMD).[43] Extremely high HDL-C levels are more frequent in type 1 diabetes mellitus (T1DM) in comparison to general population.[44]

HIGH-DENSITY LIPOPROTEINS: THERAPEUTIC AVENUES

Recovery from metabolic and CVD is associated with restoration of HDL functionality and concentration. Roux-en-Y gastric bypass leads to an early improvement of HDL function such as enhanced CEC, antiapoptotic, antioxidant and anti-inflammatory activity, and it also increased capacity to produce nitric oxide (NO).[45]

Exercise and diet can improve HDL function too. In chronic heart failure patients, a 15-week exercise intervention significantly improved the ability of HDLs to activate endothelial nitric oxide synthase (eNOS) and produce NO.[46] One study demonstrated that HDLs isolated before and after an exercise-based weight loss program demonstrated significant correlation between RCT and amount of weight lost.[47] The effects of diet and exercise have been summarized in **Table 3**.

HDL-BASED THERAPIES

Manipulation of HDL components has been shown to have beneficial effects.[9] Enrichment of S1P to reconstituted HDL (rHDL) leads to greater vasorelaxation than control rHDL.[48,49] In humans, a small trial found that short-term infusion (4 weeks, 1 infusion per week) of rHDL was able to significantly diminish endothelial progenitor cell (EPC) apoptosis in patients with acute coronary syndrome (ACS), and increase the circulating chemokine levels which are involved in EPC recruitment such as stromal cell-derived factor-1 (SDF-1) or vascular endothelial growth factor.[33] Increase in ApoA-1 levels alone, either through genetic manipulation in animal models or through exogenous infusion in animals and humans was shown to provide beneficial effects on atherosclerotic plaque regression[33] (**Table 1**). Larger trials added more complexity, indicating no detectable effect after supplementation of an engineered pre-β HDL mimetic on atherosclerotic plaque composition or regression compared to placebo.[33]

TABLE 3: Influence of lifestyle modifications on HDL-C.		
Intervention	HDL-increase in %	Effect
Physical activity	5–10	Increased LPL, pre-β-HDL, reverse cholesterol transport, protective lipoproteins
Smoking secession	5–10	Increased LCAT, reverse cholesterol transport, inhibits CETP
Weight loss	5–20	Increased LCAT, LPL and reverse cholesterol transport
Alcohol consumption	5–15	Increased ABCA1, APOA1 and paraoxonase, reduced CETP
Mediterranean diet	0–5	Increase in atheroprotective lipoproteins

(APOA1: apolipoprotein A1; ABCA 1: ATP-binding cassete transporter A1; CETP: cholesterol ester transfer protein; HDL-C: high-density lipoprotein cholesterol; LCAT: lecithin–cholesterol acyltransferase; LPL: lipoprotein lipase)

Source: März W, Kleber ME, Scharnag H, Speer T, Zewinger S, Ritsch A, et al. HDL cholesterol: reappraisal of its clinical relevance. Clin Res Cardiol. 2017;106:663-75.

HIGH-DENSITY LIPOPROTEINS AS A DRUG DELIVERY VEHICLE

Encapsulating vaso-protective compounds within rHDL could be a promising therapeutic venture. A study has shown that the infusion of rHDL containing a potent LXR agonist enabled atherosclerotic plaque regression in the ApoE-knock out mouse model. There was significant accumulation of the synthetic HDL in the atherosclerotic lesions.[50] Likewise, rHDL encapsulating statins (S-rHDL) are shown to be more effective at reducing atherosclerosis-induced inflammation than statin or rHDL alone in mice.[51] The combined therapeutic and diagnostic (containing drugs and imaging tracers respectively) rHDL can be developed as "theranostics."

CONCLUSION

High-density lipoprotein is a complex and dynamic particle, the full biology and therapeutic potentials of which are incompletely understood. Today, HDLs are considered as multifaceted entities beyond their cholesterol-carrying action. In this review, we tried to analyze the versatility of the HDL's functions and the responsible mechanisms behind them. We are living behind the dipole model of "good" and "bad" cholesterol and are representing a more complex and realistic image of HDLs. The growing evidence for HDL dysfunction in pathologic conditions instigates the investigations for disease-specific structure-composition-function of HDL. This includes identifying diverse subclasses of HDLs using new techniques, and typifying the proteome and lipidome and micro-RNA profiles of different HDL subclasses in health, disease. Though the clinical results on HDL-based therapies remain controversial, more clinically relevant insights are waiting from further research. Finally, the good cholesterol label should be recognized as outdated and the prognostic, diagnostic, and therapeutic potentials of HDL have to be seen from a new perspective.

> **KEY POINTS**
>
> - Atherosclerotic vascular disease is one of the leading causes of death throughout the world. HDL-C is traditionally considered as "good" cholesterol over decades since the discovery of its negative association with CVD. However, Mendelian randomization studies and multiple therapeutic interventions associated with raised HDL have failed to show any strong association between HDL and CVDs. The structure, function, and composition of HDLs change dynamically during health and disease paralleling differential contribution from diverse lipidome, proteome, and other molecules. The good cholesterol label should be recognized as outdated and the prognostic, diagnostic and therapeutic potentials of HDL have to be seen from a new perspective.

REFERENCES

1. Wilson PW, Garrison RJ, Castelli WP, Feinleib M, McNamara PM, Kannel WB. Prevalence of coronary heart disease in the Framingham offspring study: role of lipoprotein cholesterols. Am J Cardiol. 1980;46:649-54.
2. Glomset JA, Janssen ET, Kennedy R, Dobbins J. Role of plasma lecithin: cholesterol acyltransferase in the metabolism of high density lipoproteins. J Lipid Res. 1966;7:638-48.
3. Gordon DJ, Probstfield JL, Garrison RJ, Neaton JD, Castelli WP, Knoke JD, et al. High-density lipoprotein cholesterol and cardiovascular disease. Four prospective American studies. Circulation. 1989;79:8-15.
4. Hutchins PM, Ronsein GE, Monette JS, Pamir N, Wimberger J, He Y, et al. Quantification of HDL particle concentration by calibrated ion mobility analysis. Clin Chem. 2014;60:1393-1401.

5. de la Llera-Moya M, Drazul-Schrader D, Asztalos BF, Cuchel M, Rader DJ, Rothblat GH. The ability to promote efflux via ABCA1 determines the capacity of serum specimens with similar high-density lipoprotein cholesterol to remove cholesterol from macrophages. Arterioscler Thromb Vasc Biol. 2010;30:796-801.
6. Brunham LR, Hayden MR. Human genetics of HDL: Insight into particle metabolism and function. Progress in Lipid Research. 2015;58:14-25.
7. Keech A, Simes RJ, Barter P, Best J, Scott R, Taskinen MR, et al. Effects of long-term fenofibrate therapy on cardiovascular events in 9795 people with type 2 diabetes mellitus (the FIELD study): randomised controlled trial. Lancet. 2005;366(9500):1849-61.
8. Administration FaD. Withdrawal of approval of indications related to the coadministration with statins in applications for niacin extended-release tablets and fenofibric acid delayedrelease capsules. Fed Reg. 2016;81(74).
9. Xiang AS, Kingwell BA, Rethinking good cholesterol: a clinicians' guide to understanding HDL. Lancet Diabetes Endocrinol. 2019;7(7):575-82.
10. Jomard A, Osto E. High Density Lipoproteins: Metabolism, Function, and Therapeutic Potential. Front Cardiovasc Med. 2020;7:39.
11. Zannis VI, Fotakis P, Koukos G, Kardassis D, Ehnholm C, Jauhiainen M, et al. HDL biogenesis, remodeling, and catabolism. In: von Eckardstein A, Kardassis D (Eds). Handbook of Experimental Pharmacology. Cham: Springer International Publishing; 2015. pp. 53-111.
12. Linton MF, Yancey PG, Davies SS, Jerome WG, Linton EF, Song, WL, et al. The Role of Lipids and Lipoproteins in Atherosclerosis. South Dartmouth, MA: MDText.com Inc.; 2019.
13. Karathanasis SK, Freeman LA, Gordon SM, Remaley AT. The changing face of HDL and the bestway to measure it. Clin Chem. 2017;63:196-210.
14. Marques PE, Nyegaard S, Collins RF, Troise F, Freeman SA, Trimble WS, et al. Multimerization and retention of the scavenger receptor SR-B1 in the plasma membrane. Dev Cell. 2019;50:283-95.e5.
15. Rayner KJ, Suárez Y, Dávalos A, Parathath S, Fitzgerald ML, Tamehiro N, et al. MiR-33 contributes to the regulation of cholesterol homeostasis. Science. 2010;328:1570-3.
16. Price NL, Rotllan N, Zhang X, Canfrán-Duque A, Nottoli T, Suarez Y, et al. Specific disruption of abca1 targeting largely mimics the effects of miR-33 knockout on macrophage cholesterol efflux and atherosclerotic plaque development. Circ Res. 2019;124:874-80.
17. Ouimet M, Barrett T J, Fisher A. HDL and reverse cholesterol transport basic mechanisms and their roles in vascular health and disease. Circ Res. 2019;124:1505-18.
18. Mendelsohn AR, Larrick JW. Preclinical reversal of atherosclerosis by FDA-Approved compound that transforms cholesterol into an anti- inflammatory "Prodrug". Rejuvenation Res. 2016;19:252-5.
19. Paalvast Y, de Boer JF, Groen AK. Developments in intestinal cholesterol transport and triglyceride absorption. Curr Opin Lipidol. 2017;28:248-54.
20. Rohatgi A, Khera A, Berry JD, Givens EG, Ayers CR, Wedin KE, et al. HDL cholesterol efflux capacity and incident cardiovascular events. N Engl J Med. 2014;371:2383-93.
21. Saleheen D, Scott R, Javad S, Zhao W, Rodrigues A, Picataggi A, et al. Association of HDL cholesterol efflux capacity with incident coronary heart disease events: a prospective case-control study. Lancet Diabetes Endocrinol. 2015;3:507-13.
22. Cuchel M, Raper AC, Conlon DM, Pryma DA, Freifelder RH, Poria R, et al. A novel approach to measuring macrophage-specific reverse cholesterol transport in vivo in humans. J Lipid Res. 2017;58:752-62.
23. Kontush A, Lindahl M, Lhomme M, Calabresi L, Chapman MJ, Davidson WS. Structure of HDL: particle subclasses and molecular components. In: von Eckardstein A, Kardassis D, (Eds). Handbook of Experimental Pharmacology. Cham: Springer International Publishing; 2015. pp. 3-51.
24. Rosenson RS, Brewer HBJ, Chapman MJ, Fazio S, Hussain MM, Kontush A, et al. HDL measures, particle heterogeneity, proposed nomenclature, and relation to atherosclerotic cardiovascular events. Clin Chem. 2011;57:392-410.
25. Kontush A. HDL particle number and size as predictors of cardiovascular disease. Front Pharmacol. 2015;6:218.
26. Martin SS, Khokhar AA, May HT, Kulkarni KR, Blaha MJ, Joshi PH, et al. HDL cholesterol subclasses, myocardial infarction, and mortality in secondary prevention: the Lipoprotein Investigators Collaborative. Eur Heart J. 2015;36:22-30.
27. Otvos JD, Jeyarajah EJ, Bennett DW, Krauss RM. Development of a proton nuclear magnetic resonance spectroscopic method for determining plasma lipoprotein concentrations and subspecies distributions from a single, rapid measurement. Clin Chem. 1992;38:1632-8.

28. Ikhlef S, Berrougui H, Kamtchueng Simo O, Zerif E, Khalil A. Human paraoxonase 1 overexpression in mice stimulates HDL cholesterol efflux and reverse cholesterol transport. PLoS One. 2017;12:e0173385.
29. Yuhanna IS, Zhu Y, Cox BE, Hahner LD, Osborne-Lawrence S, Lu P, et al. High-density lipoprotein binding to scavenger receptor-BI activates endothelial nitric oxide synthase. Nat Med. 2001;7:853-7.
30. Karlsson H, Kontush A, James RW. Functionality of HDL: antioxidation and detoxifying effects. In: von Eckardstein A, Kardassis D (Eds). Handbook of Experimental Pharmacology. Cham: Springer International Publishing; 2015. pp. 207-28.
31. Jomard A, Osto E. High Density Lipoproteins: Metabolism, Function, and Therapeutic Potential. Front Cardiovasc Med. 2020;7:39.
32. Yancey PG, de la Llera-Moya M, Swarnakar S, Monzo P, Klein SM, Connelly MA, et al. High density lipoprotein phospholipid composition is a major determinant of the bi-directional flux net movement of cellular free cholesterol mediated by scavenger receptor BI. J Biol Chem. 2000;275:36596-604.
33. Rayner KJ, Hennessy EJ. Extracellular communication via microRNA: lipid particles have a new message. J Lipid Res. 2013;54:1174-81.
34. Rysz J, Brzozka AG, Gorzynska M, Franczyk B. The Role and Function of HDL in Patients with Chronic Kidney Disease and the Risk of Cardiovascular Disease. Int J Mol Sci. 2020;21:601.
35. Honda H, Hirano T, Ueda M, Kojima S, Mashiba S, Hayase Y, et al. High-density lipoprotein subfractions and their oxidized subfraction particles in patients with chronic kidney disease. J Atheroscler Thromb. 2016;23:81-94.
36. Tolle M, Huang T, Schuchardt M, Jankowski V, Prufer N, Jankowski J, et al. High-density lipoprotein loses its anti-inflammatory capacity by accumulation of pro-inflammatory-serum amyloid A. Cardiovasc. Res. 2012;94:154-62.
37. Willeit P, Freitag DF, Laukkanen JA, Chowdhury S, Gobin, R, Mayr M, et al. Asymmetric dimethylarginine and cardiovascular risk: Systematic review and meta-analysis of 22 prospective studies. J Am Heart Assoc. 2015;4:e001833.
38. Hassan M, Philip P. CANHEART: Is HDL cholesterol a cardiovascular specific risk factor? Glob Cardiol Sci Pract. 2016;2016:e201634.
39. Kronenberg, F. HDL in CKD-The Devil Is in the Detail. J Am Soc Nephrol. 2018;29:1356-71.
40. Moradi H, Streja E, Kashyap ML, Vaziri ND, Fonarow GC, Kalantar-Zadeh K. Elevated high-density lipoprotein cholesterol and cardiovascular mortality in maintenance hemodialysis patients. Nephrol Dial Transplant. 2014;29:1554-62.
41. Han CY, Tang C, Guevara ME, Wei H, Wietecha T, Shao B, et al. Serum amyloid A impairs the antiinflammatory properties of HDL. J Clin Invest. 2016;126:266-81.
42. Siebel AL, Heywood SE, Kingwell BA. HDL and glucose metabolism: current evidence and therapeutic potential. Front Pharmacol. 2015;6:258.
43. Takaeko Y, Matsui S, Kajikawa M, Maruhashi T, Kishimoto S, Hashimoto H, et al. Association of extremely high levels of high-density lipoprotein cholesterol with endothelial dysfunction in men. J Clin Lipidol. 2019;13:664-72.e1.
44. Chiesa ST, Charakida M, McLoughlin E, Nguyen HC, Georgiopoulos G, Motran L, et al. Elevated high-density lipoprotein in adolescents with type 1 diabetes is associated with endothelial dysfunction in the presence of systemic inflammation. Eur Heart J. 2019;40:3559-66.
45. Osto E, Doytcheva P, Corteville C, Bueter M, Dorig C, Stivala S, et al. Rapid and body weight-independent improvement of endothelial and high-density lipoprotein function after Roux-en-Y gastric bypass: role of glucagon-like peptide-1. Circulation. 2015;131:871-81.
46. Adams V, Besler C, Fischer T, Riwanto M, Noack F, Hollriegel R, et al. Exercise training in patients with chronic heart failure promotes restoration of high-density lipoprotein functional properties. Circ Res. 2013;113:1345-55.
47. Kralova Lesna I, Suchanek P, Kovar J, Poledne R. Life style change and reverse cholesterol transport in obese women. Physiol Res. 2009;58(Suppl 1):S33-8.
48. März W, Kleber ME, Scharnag H, Speer T, Zewinger S, Ritsch A, et al. HDL cholesterol: reappraisal of its clinical relevance. Clin Res Cardiol. 2017;106:663-75.
49. Sattler K, Graler M, Keul P, Weske S, Reimann CM, Jindrova H, et al. Defects of high-density lipoproteins in coronary artery disease caused by low sphingosine-1-phosphate content: correction by sphingosine-1-phosphate-loading. J Am Coll Cardiol. 2015;66:1470-85.
50. Guo Y, Yuan W, Yu B, Kuai R, Hu W, Morin EE, et al. Synthetic high-density lipoprotein-mediated targeted delivery of liver x receptors agonist promotes atherosclerosis regression. EBioMedicine. 2018;28:225-33.
51. Duivenvoorden R, Tang J, Cormode DP, Mieszawska AJ, Izquierdo-Garcia D, Ozcan C, et al. A statin-loaded reconstituted high-density lipoprotein nanoparticle inhibits atherosclerotic plaque inflammation. Nat Commun. 2014;5:3065.

CHAPTER

10

Triglycerides, Remnant Cholesterol, and Atherosclerotic Cardiovascular Disease Risk

Amritava Ghosh

ABSTRACT

While low-density lipoprotein (LDL) is an established risk factor for atherosclerotic cardiovascular disease (ASCVD) there is increasing discussion on the contribution of hypertriglyceridemia to cardiovascular disease (CVD) risk. Lifestyle changes focused on dietary modification, increased physical activity, and weight loss are the mainstay of therapy for hypertriglyceridemia. High-risk patients with hypertriglyceridemia may benefit from additional drug treatment in addition to statin. In this chapter, we review the role of hypertriglyceridemia and associated lipoproteins in the pathogenesis of ASCVD, the cardiovascular outcomes of triglycerides (TGs)-lowering intervention trials, and the current guidelines for treating hypertriglyceridemia.

INTRODUCTION

Elevated triglycerides (TGs) are common among individuals with type 2 diabetes mellitus (T2DM). Mild-to-moderate hypertriglyceridemia is also commonly associated with lifestyle factors such as diet rich in simple carbohydrates and saturated fat, excessive alcohol consumption, sedentary behavior and obesity. Other secondary causes of hypertriglyceridemia include hypothyroidism, renal disease (nephrotic syndrome, glomerulonephritis, chronic kidney disease), acute hepatitis, Cushing syndrome, pregnancy and drugs (nonselective β blockers, corticosteroids, oral estrogens, thiazide diuretics, tamoxifen, raloxifene, androgens, bile acid sequestrants, cyclosporine, atypical antipsychotics, sirolimus, tacrolimus, isotretinoin, cyclophosphamide, protease inhibitors, interferon). Genetic causes of hypertriglyceridemia include familial hyperchylomicronemia (autosomal recessive mutations in *LPL*, *APOC2*, *APOA5*, *LMF1*, *GPIHBP1*, *GPD1*), familial hypertriglyceridemia (polygenic), familial combined hyperlipidemia and dysbetalipoproteinemia (autosomal recessive mutations in *APOE*). The increasing prevalence of obesity, metabolic syndrome and T2DM has been associated with a rising prevalence of hypertriglyceridemia.[1] While low-density lipoprotein (LDL) is an established risk factor and reduction of LDL with statins is the first-line therapy for cardiovascular risk reduction, "residual risk" of cardiovascular disease (CVD) persists even after achievement of LDL goal. While this "residual risk" may be related to a number of non-lipid risk factors (such as smoking, hypertension, inflammation, and prothrombotic milieu), a part of it may be attributable to raised TG and low high-density lipoprotein (HDL) cholesterol concentration. A meta-analysis of 14 trials proved that presence of low HDL and high TG limits the efficacy of statin therapy in reducing vascular events despite achievement of target LDL levels.[2] Similarly, a meta-analysis of 17 prospective studies showed TG to be an independent risk factor for CVD after adjusting for variables such as HDL, total cholesterol, and other risk factors.[3] In the Pravastatin or Atorvastatin Evaluation and Infection Therapy - Thrombolysis in Myocardial Infarction 22 (PROVE IT-TIMI 22) trial (2008), an on-treatment TG <150 mg/dL was found to be independently associated with a lower risk of recurrent coronary heart disease events.[4] However, in ACCORD (Action to Control Cardiovascular Risk in Diabetes) Lipid trial (2010), the addition of fenofibrate to simvastatin did not reduce the rate of major adverse cardiovascular events (MACE).[5] Ever since, there is increasing discussion about cardiovascular risk associated with moderate hypertriglyceridemia.[6] Major fibrate trials have yielded conflicting results regarding the cardiovascular benefit of lowering TG.[5,7-12] A systematic review and meta-analysis of these trials has shown that fibrates may be of benefit among patients with high TG and low HDL, a classic lipid phenotype seen in people with DM and metabolic syndrome.[13] In addition, clinical and epidemiologic studies[14-16] have also highlighted TG as a modifiable risk factor in CVD. Although it is possible that TG directly promotes atherosclerosis, there is growing consensus that TG serves as a marker for circulating TG-rich lipoproteins (TRL) which in turn are causally related to atherogenesis. Therefore, there has been renewed interest in targeting TGs to address the issue of "residual risk" of CVD discussed above.

TRIGLYCERIDE METABOLISM

Triglycerides consist of fatty acid(s) esterified to a molecule of glycerol. They form the bulk of esterified fatty acids in TRL which include chylomicrons, very low-density lipoproteins (VLDL), and their remnants. **Figure 1** summarizes the metabolism of TRL.[17]

Fatty acids derived from the diet are largely esterified to form TGs in intestinal mucosal cells which are subsequently packaged into chylomicrons. The chylomicrons are secreted into intestinal lymph and enter the systemic circulation through the thoracic duct, bypassing the liver. In the peripheral tissues, lipoprotein lipase (LPL) hydrolyzes the TG in chylomicrons to release free fatty acids (FFAs), in the process generating chylomicron remnants.

Nonesterified fatty acid (NEFA)s are released from adipocytes through the action of hormone sensitive lipase (HSL) and adipocyte triglyceride lipase, a process which is inhibited by insulin.[18] Conversely, insulin deficiency and/or resistance is associated with excessive release of NEFA from adipocytes, a major driver of dyslipidemia in DM and obesity.[19]

The liver derives fatty acids from additional lipolysis, uptake of remnant lipoproteins, de novo hepatic lipogenesis, uptake of albumin bound NEFA circulating in plasma. VLDL assembled and secreted from the liver undergo lipolysis by LPL in peripheral tissues resulting in release of fatty acids and production of VLDL remnant particles called intermediate-density lipoproteins (IDL). IDL return to the liver. The liver removes

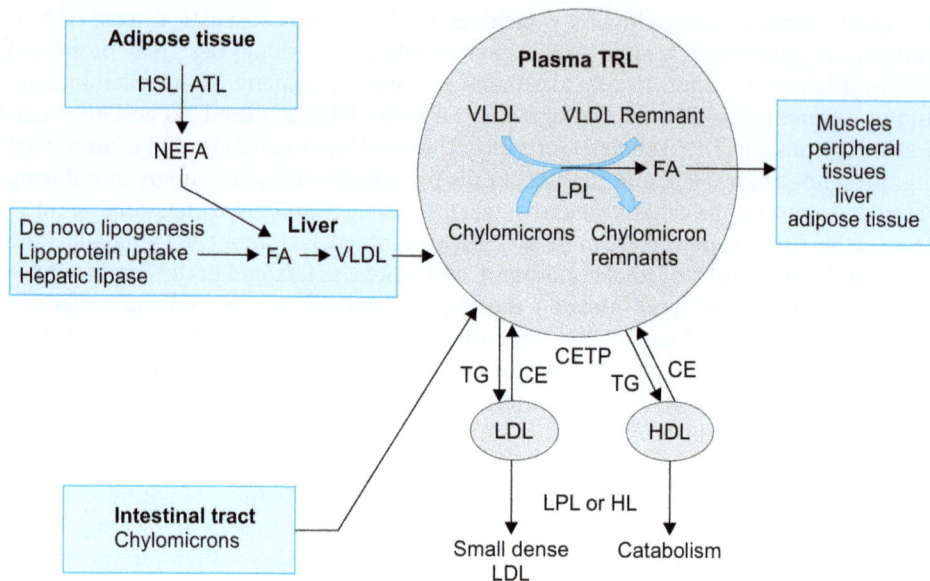

(ATL: adipocyte triglyceride lipase; CE: cholesterol esters; CETP: cholesteryl ester transfer protein; FA: fatty acids; HDL: high-density lipoprotein; HL: hepatic lipase; HSL: hormone sensitive lipase; LDL: low-density lipoprotein; LPL: lipoprotein lipase; NEFA: nonesterified fatty acids; TG: triglycerides; TRL: triglyceride-rich lipoproteins; VLDL: very low-density lipoprotein)

FIG. 1: Triglyceride-rich lipoprotein metabolism.

40–60% of IDL by receptor mediated endocytosis, while the remainder is acted upon by hepatic lipase (HL) at the cell surface resulting in formation of LDL. Insulin promotes production of FFAs and degradation of apolipoprotein B100 (ApoB-100). Apo B is the primary protein constituent of VLDL, IDL, and LDL. As insulin resistance develops and levels of insulin rise, the pathway for FFA production remains insulin sensitive resulting in increased hepatic production of FFA; while there is reduction of insulin stimulated intracellular degradation of ApoB. These changes at the level of hepatocytes, together with insulin resistance at the level of adipocytes, causing increased FFA efflux in bloodstream, is thought to be central to the pathogenesis of ADD. This results in increased production of VLDL cholesterol from the liver.[20] Insulin stimulates LPL activity. Insulin deficiency and/or resistance also results in suboptimal metabolism of VLDL particles. Thus hyperglycemia leads to excess VLDL production and/or inefficient lipolysis.

Triglyceride rich lipoproteins are involved in exchange of TG and cholesteryl esters with LDL and HDL facilitated by cholesteryl ester transfer protein (CETP). This leads to TG enrichment of LDL and HDL particles. Through subsequent action of HL, the LDL particles become small, dense, and more atherogenic. The TG rich HDL undergo similar hydrolysis by HL or LPL resulting in low HDL; ApoA-I dissociates from the reduced-size HDL, which due to its small size is filtered by the renal glomeruli and degraded in renal tubular cells.[21,22] This leads to the development of the classic triad of elevated TG, low HDL, and increased proportion of small dense LDL, the pattern of dyslipidemia associated with DM and insulin resistance,[19] known as atherogenic dyslipidemia. While poorly controlled

DM can be associated with extreme hypertriglyceridemia and risk of pancreatitis, even after modest to tight glycemic control is achieved, the dyslipidemic triad tends to persist.

The cholesterol content of partially lipolyzed TRLs is also called residual cholesterol (RC). Evidence from a number studies supports an association between residual cholesterol and ASCVD. RC levels can be estimated using the formula: RC = Total cholesterol − LDL − HDL. Additionally, RC can be measured directly by a number of methods including ultracentrifugation, nuclear magnetic spectroscopy and a direct automated assay. However, the assays available so far do not measure all residual cholesterol in plasma and have limited clinical applicability.[1]

GLUCOSE-LOWERING MEDICATIONS AND TRIGLYCERIDES

Various glucose-lowering medications reduce TG to varying degrees in addition to that caused by reduction of glucose levels.

Insulin lowers circulating TG through induction of LPL and suppression of HSL.

Metformin is an insulin sensitizer and reduces TG independent of its effects on weight or glucose levels. In a systematic review of 37 studies, metformin (at doses >1,700 mg/day) was found to produce an average 11.5 mg/dL reduction in serum TG.[23]

Pioglitazone, a peroxisome proliferator-activated receptor-γ (PPAR-γ) agonist is an insulin sensitizer with potent effect on TG. It has been shown to reduce TG by up to 50 mg/dL and raise HDL-C by 5 mg/dL.[24-27] Although, it tends to raise LDL-C, pioglitazone reduces dense atherogenic LDL particles[28] and the observed rise in LDL likely reflects an increase in LDL particle size rather than particle number. Thus, in spite of rise in LDL, pioglitazone use is not associated with any increased cardiovascular risk. Unlike pioglitazone, rosiglitazone has less favorable impact on lipids. While some studies report a neutral effect, others note increase in TG.[24-27]

Sulfonylureas have not demonstrated any consistent effect on lipids.[27,29]

Glucagon-like peptide-1 (GLP-1) receptor agonists have been shown to improve hypertriglyceridemia with mean reduction up to 27 mg/dL, however no consistent effect on HDL was noted.[24,26,30-32] DPP-4 inhibitors have modest effect on TG[24,27] with the exception of saxagliptin which appears lipid-neutral.[24,27] Aside from weight loss and glycemic control, incretin-based therapies may lower TG by delay in gastric emptying, and decreasing intestinal TG absorption resulting in decrease in chylomicron synthesis.[33]

Sodium glucose cotransporter-2 (SGLT-2) inhibitors are associated with 10% reduction in TG and increase in HDL. Although LDL rise, small, dense LDL particles are reduced with a consequent shift toward large, buoyant, less atherogenic LDL and SGLT-2 inhibitors have shown cardioprotective effects.[24,26,34]

ESTABLISHED METHODS OF LOWERING TRIGLYCERIDES

Diet and Lifestyle

Lifestyle changes focused on dietary modification, increased physical activity, and weight loss are the mainstay of therapy for hypertriglyceridemia. A combination of dietary regulation, exercise, and moderation of alcohol intake have been shown to reduce TG by up to 60%.[35] A weight loss of 5–10% of body weight can reduce TG by 25% and raise HDL by 8%.[35] The key dietary recommendations aimed at TG reduction include avoidance of high glycemic index foods and sugar-sweetened or naturally sweet beverages. Weight loss and low glycemic index diet can reduce TG via reduction in circulating NEFA and decreased de novo hepatic lipogenesis. Very low fat diet is generally not necessary and should be considered only in select cases such as in those with very high TG levels (e.g., >1,000 mg/dL) or patients with severely impaired LPL activity, e.g., familial chylomicronemia syndrome.[36]

Treatment of Secondary Causes

Evaluation should be made for other secondary causes of hypertriglyceridemia (including endocrine diseases, renal or liver diseases, and drugs) and treatment focused on same constitutes another key component of management of hypertriglyceridemia.

Statins

Statins are the cornerstone of treatments for LDL and cardiovascular risk reduction. Statins can reduce TG by 22–45%.[37]

Fibrates

Fibrates are often considered first-line adjuncts to manage hypertriglyceridemia if lifestyle modifications and statin therapy fail to achieve goals. They act by activating PPAR-α, which in turn enhances expression of LPL and other lipid-modifying genes.[38] Fibrates have been shown to reduce TG by 30–50%.[39] Fibrates are generally well tolerated, but can cause myopathy especially when combined with other lipid lowering agents (statin, niacin). While one should avoid gemfibrozil in statin-treated patients, fenofibrate is usually safe but should be combined with statins with caution.[40]

Studies of fibrates and cardiovascular outcomes have shown mixed results. In the HHS (Helsinki Heart Study), gemfibrozil reduced the rate of incident coronary heart disease by 32% compared to placebo.[7] VA HIT (Veterans Affairs High-Density Lipoprotein Cholesterol Intervention Trial) demonstrated efficacy of gemfibrozil in reducing cardiovascular events by 24% compared to placebo among participants with coronary disease and low HDL.[10] The FIELD (Fenofibrate Intervention and Event Lowering in Diabetes) study specifically examined participants with T2DM (who had poor representation in previous studies). In the trial, fenofibrate failed to demonstrate a reduction in the primary outcome of coronary events (coronary heart disease death or nonfatal myocardial infarction). However, fewer nonfatal myocardial infarctions and revascularizations were observed with fenofibrate compared to placebo.[8] ACCORD (Action to Control Cardiovascular Risk in Diabetes) Lipid was the first study to examine the cardiovascular benefit of fibrates as add-on therapy to statins. No additional cardiovascular benefit was seen with the combination of fenofibrate and simvastatin compared with simvastatin alone; however, subgroup analysis suggested that patients with high TG ($\geq$204 mg/dL) and low HDL ($\leq$34 mg/dL) may derive benefit.[5] Meta-analysis of several trials involving fibrates [ACCORD Lipid, FIELD, BIP (Bezafibrate Infarction Prevention) study, HHS, VA-HIT] by Sacks et al. showed that fibrates are very likely to be beneficial as an add on therapy in subgroup of participants characterized by high TG and low HDL-C despite statin usage.[41] Similar systematic review on usefulness of fibrate therapy by Jun et al. demonstrated that those participants with a high mean TG concentration (>204 mg/dL) obtained greater reduction in cardiovascular outcomes.[42] Additional cardiovascular outcome studies of fibrates are under way. New PPAR agonists are under investigation.[43] PROMINENT (Pemafibrate to Reduce Cardiovascular

OutcoMes by Reducing Triglycerides IN patiENts with diabeTes) is a multicenter phase III trial which will examine major cardiovascular outcomes with pemafibrate versus placebo among participants with DM having elevated TGs and low HDL-C. In short-term trials, pemafibrate has shown good lipid lowering efficacy with better tolerability compared to fenofibrate.[44,45]

Glitazars

Combined PPAR-α/γ agonists (glitazars) may be expected to be ideal agents to address the "residual risk" of CVD persisting even after achievement of LDL goal, since they would lower TG not only directly (PPAR-α activity), but also by improving insulin sensitivity and glycemic control (PPAR-γ activity). Several pharmaceutical agents with such action have been developed.

Depending on their molecular structure, these molecules exert dual action with varying degrees of PPAR-α and PPAR-γ agonism. Faglitazar, the first glitazar to be tested, was discontinued very early in development phase secondary to significant edema.[46] Similarly, ragaglitazar was also dropped early on due to its carcinogenic potential on urothelial cells in rodent models.[47] Muraglitazar, although proved efficacious in improving insulin sensitivity and treating diabetic dyslipidemia, was suspended in 2006 due to significant adverse cardiovascular effects.[48] Tesaglitazar showing similar promise was withdrawn secondary to its toxicity on bone marrow and kidney.[49] Aleglitazar was associated with significant increase in gastrointestinal bleeding and renal dysfunction.[50] Although these tested molecules resulted in adverse events, these have been compound specific and of diverse origin, that is, urothelial, renal, and cardiac raising hopes of a potential drug that could mitigate these side effects and still have a positive influence on insulin sensitivity and dyslipidemia dominated by high TG, increased proportion of small dense LDL, and low HDL.

Saroglitazar is a dual PPAR-α/γ agonist with predominant PPAR-α and moderate PPAR-γ agonistic activities developed indigenously in India.[51] Saroglitazar is the first approved agent in the Glitazar class for the management of diabetic dyslipidemia and has been approved in India.[52] In a phase III clinical trial of participants with DM on treatment with atorvastatin 10 mg and with residual dyslipidemia consisting of TG between 200 and 500 mg/dL, addition of saroglitazar showed significant reduction in TG concentration (percentage change from baseline −45.5 ± 3.03% and −46.7 ± 3.02% for 2 mg/day and 4 mg/day doses respectively) compared to placebo.[53] In a recent study on Indian patients with diabetic dyslipidemia, the add on therapy of saroglitazar with metformin reduced TG, glycosylated hemoglobin, and fasting and post prandial glucose levels to a greater extent compared to the add on therapy of fenofibrate with metformin.[54] Saroglitazar appears a promising agent for the management of diabetic dyslipidemia. Saroglitazar appears to be safe and well tolerated and has not shown any of the adverse effects that are commonly recognized with its class of molecules.[53] But given the historical concerns, proof of long-term safety is required. Also, a longer term clinical trial focusing on cardiovascular outcome, currently lacking, is much desired.

Omega-3 Fatty Acids

Omega-3 fatty acids decrease TG through suppression of hepatic VLDL-TG production.[55-57] Doses of 3–4 g/day of eicosapentaenoic acid (EPA) ± docosahexaenoic acid (DHA) reduce TG by up to 45%.[58,59] Ninety five percent icosapent ethyl is an ethyl ester of EPA[60-63] which has similar efficacy as EPA/DHA formulation in reducing TG. However, while EPA/DHA formulation may raise LDL modestly, icosapent ethyl does not have significant effect on LDL.[58,59] The clinical relevance of the LDL rise seen with EPA/DHA formulations is unclear.[46]

Until 2018, there was limited evidence to suggest added cardiovascular benefit of omega-3 fatty acids in patients who were already taking a statin. Only the JELIS trial (Japan EPA Lipid Intervention Study)[64] had shown a reduction in major coronary events with this class of drugs. However, doses used in initial studies (1 g) were much lower than those currently recommended.[65-67] Also, people with hypertriglyceridemia were not targeted.[65,66] The landmark study, JELIS, demonstrated a 19% reduction in major coronary events with a combination of EPA ethyl esters (1.8 g) and statin compared with statin alone.[64] The REDUCE IT trial (Reduction of Cardiovascular Events with Icosapent Ethyl Intervention Trial), published in 2018, involved participants who had either established atherosclerotic cardiovascular disease (ASCVD) or DM and other risk factors, with TG between 135 and 499 mg/dL and LDL between 41 and 100 mg/dL. In the trial, the addition of icosapent ethyl 4 g/day to statin was associated with a significant 25% reduction in MACE compared to statin alone.[68] Thus, omega-3 fatty acids may be particularly beneficial in high-risk patients with significant atherogenic dyslipidemia despite statin-optimized LDL. In addition to TG reduction, other potential pleiotropic effects which might contribute to the cardiovascular benefits of omega-3 fatty acids include- improvement in endothelial function, anti-inflammatory action, antioxidant effects, greater functionality of HDL particles, suppression of platelet aggregability through inhibition of thrombin generation, and antiarrhythmic properties.[69] Additional cardiovascular outcome studies of moderate- and high-dose omega-3 fatty acids are under way the RESPECT-EPA (Randomized trial for Evaluation in Secondary Prevention Efficacy of Combination Therapy) evaluating the effects of addition of EPA to standard therapy compared to standard therapy alone, EVAPORATE (Effect of Vascepa on Improving Coronary Atherosclerosis in People with High Triglycerides Taking Statin Therapy) assessing the effect of 4 g/day icosapent ethyl on coronary atherosclerosis in participants with TG 135-499 mg/dL on-statin therapy. Icosabutate is a synthetic omega-3 fatty acid under development.[70]

Niacin

Niacin, or nicotinic acid, reduces TG and increases HDL by up to 30% at doses of 1.5–3 g daily.[71] It also achieves a 15% reduction in LDL and 25% reduction in lipoprotein(a) levels.[71, 72] However, its use has been limited on account of its poor tolerability due to bothersome flushing. It has also been implicated in worsening of glycemic control.[73-75] Other less frequent adverse effects include eczema-like skin rash, acanthosis nigricans, and gastrointestinal distress.

Older trials of secondary cardiovascular prevention, done before the advent of statin therapy, had demonstrated absolute mortality reductions of up to 7.8% with statin therapy.[76] However, later studies—AIM-HIGH (Atherothrombosis Intervention in Metabolic Syndrome with low HDL/HIGH Triglycerides) trial and HPS2-THRIVE (Heart Protection Study 2-Treatment of HDL to Reduce the Incidence of Vascular Events) found no incremental cardiovascular benefit of adding niacin (1.5-2 g) among participants already on simvastatin ± ezetimibe.[77,78] These results argue against the use of niacin as first-line adjunct for lowering TG in patients already on-statin therapy.

Triglyceride-lowering Therapies in Development

Apolipoprotein CIII Inhibitors

Apolipoprotein (Apo) CIII increases TG by inhibiting LPL and reducing hepatic uptake of remnant lipoproteins.[79] Volanesorsen is an antisense oligonucleotide, inhibitor of ApoCIII mRNA, which decreases ApoCIII expression thereby reducing TG levels. In a phase II trial involving participants with T2DM and TG levels of 201–499 mg/dL, volanesorsen reduced TG by 69%, and raised HDL by 42% compared to placebo. It also improved insulin sensitivity and reduced glycosylated hemoglobin (−0.44%).[80] It appears as a promising agent; however, thrombocytopenia and serious bleeding are concerning side effects.[81]

Angiopoietin-like 3 Protein Inhibitors

Angiopoietin-like 3 protein (ANGPTL3) is an endogenous inhibitor of LPL. Evinacumab, a monoclonal antibody against ANGPTL3, has been shown to lower fasting TG by up to 76% and LDL by 23% in a dose-dependent manner.[82] An antisense oligonucleotide against ANGPTL3 has also demonstrated efficacy in reduction of atherogenic lipoproteins and appears to be promising.[83] So far, ANGPTL3 appears to be a promising therapeutic target.

APPROACH TO MODERATE HYPERTRIGLYCERIDEMIA

Many organizations have published guidelines for the diagnosis and categorization of hypertriglyceridemia. **Table 1** summarizes the classification of TG levels by the various organizations.

All guidelines recommend that adults should be screened for hypertriglyceridemia at least once every 5 years,[84-88] and American Association of Clinical Endocrinologists (AACE) further advises annual screening for dyslipidemia in people with DM.[84] In addition to lifestyle modifications, pharmacotherapy is generally recommended for patients with

TABLE 1: Categorization of triglyceride levels by lipid guidelines.[84-88]

Categorization*	American Association of Clinical Endocrinologists (AACE)	National Lipid Association (NLA)	National Cholesterol Education Program Expert Panel on Detection, Evaluation, and Treatment of High Blood Cholesterol in Adults (NCEP)	Endocrine Society
Normal	<150	<150	<150	<150
Borderline high (TES: Mild)	150–199	150–199	150–199	150–199
High (TES: Moderate)	200–499	200–499	200–499	200–999
Very high (TES: Severe)	≥500	≥500	≥500	1,000–1,999
Very severe	N/A	N/A	N/A	≥2,000

*Triglyceride levels are in mg/dL

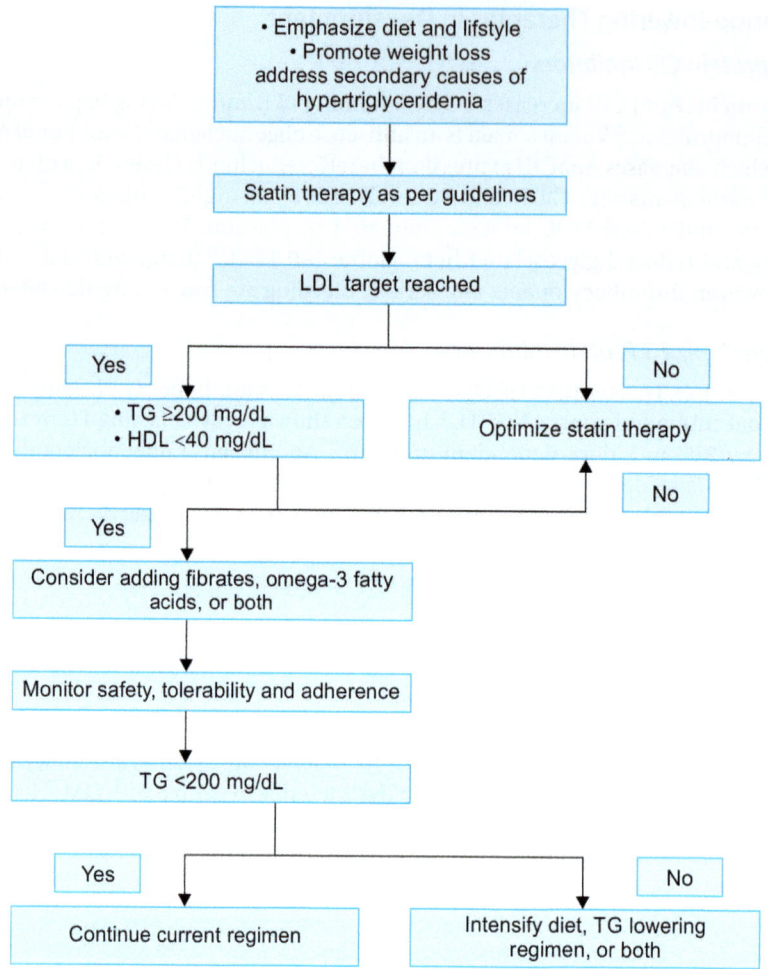

(HDL: high-density lipoprotein; LDL: low-density lipoprotein; TG: triglycerides)

FLOWCHART 1: Proposed approach to management of triglycerides in atherogenic dyslipidemia.

very high/severe hypertriglyceridemia in order to decrease pancreatitis risk.[84-88] While recent guidelines recognize the association of cardiovascular risk with moderate TG elevations (200–499 mg/dL),[86] there is a lack of agreement on management of sustained moderate hypertriglyceridemia despite appropriate lifestyle modifications and statin-optimized LDL. The American Diabetes Association (ADA) Standards of Medical Care in Diabetes—2020 has recommended consideration for addition of icosapent ethyl for DM patients with ASCVD or other cardiovascular risk factors on a statin with controlled LDL-C but elevated TGs (135–499 mg/dL). However, the guidelines actively discourage the use of combination therapy with niacin or fibrates to target HDL-C or TG.[89] In contrast, the American Association of Clinical Endocrinologists and American College of Endocrinology Guidelines for Management of Dyslipidemia and Prevention of Atherosclerosis has recommended the use of fibrates to improve ASCVD outcomes in primary and secondary prevention when TG concentrations are ≥200 mg/dL and HDL cholesterol concentration <40 mg/dL. Niacin has also been recommended as an adjunct for reducing TG. A TG goal of <150 mg/dL has been recommended by the guidelines.[84] The 2019 ESC (European Society of Cardiology)/EAS (European Atherosclerosis Society) Guidelines for the management of Dyslipidaemias also recommend drug therapy in order to reduce CVD risk among high-risk individuals with hypertriglyceridemia (>200 mg/dL). While statin has been recommended as the first drug of choice, omega-3 fatty acids (icosapent ethyl) or fibrates have been suggested as add on drugs.[90] The RSSDI-ESI Clinical Practice Recommendations for the Management of Type 2 Diabetes Mellitus 2020 have also suggested addition of fibrates if TGs remain >200 mg/dL in DM patients receiving statin therapy.[91] Based on recent evidence, a therapeutic approach as detailed in **Flowchart 1** may be adopted toward management of TG.[17]

CONCLUSION

Atherogenic dyslipidemia is characterized by the classic triad of elevated triglycerides (TGs), low high density lipoprotein (HDL), and increased proportion of small dense low density lipoprotein (LDL). While LDL is an established risk factor and reduction of LDL with statins is the first-line therapy for cardiovascular risk reduction, residual risk of cardiovascular disease (CVD) persists even after achievement of LDL goal. Several studies support the role of TG and residual cholesterol as modifiable risk factors for CVD. While lifestyle modification, optimization of glycemic control and addressing other secondary causes of hypertriglyceridemia form the cornerstone of management of hypertriglyceridemia, the role of pharmacotherapy (fibrates, omega-3 fatty acids) in the management of persistently elevated TG and reduced HDL is increasingly being recognized.

KEY POINTS

- Hypertriglyceridemia is very common in people with obesity, diabetes mellitus and a number of other conditions.
- Besides the critical role played by low density lipoprotein cholesterol, hypertriglyceridemia contributes significantly toward atherogenesis and increased cardiovascular risk.
- Optimization of glycemic control, lifestyle changes, and addressing other secondary causes of hypertriglyceridemia are integral to the management of hypertriglyceridemia.
- There is growing evidence supporting the use of additional triglyceride-lowering therapies (particularly omega-3 fatty acids) in selected patients with persistently elevated triglycerides and reduced high density lipoprotein despite statin therapy.

REFERENCES

1. Sandesara PB, Virani SS, Fazio S, Shapiro MD. The forgotten lipids: triglycerides, remnant cholesterol, and atherosclerotic cardiovascular disease risk. Endocr Rev. 2019;40(2):537-57.
2. Cholesterol Treatment Trialists' (CTT) Collaborators, Kearney PM, Blackwell L, Collins R, Keech A, Simes J, Peto R, et al. Efficacy of cholesterol lowering therapy in 18,686 people with diabetes in 14 randomised trials of statins: A meta-analysis. Lancet. 2008;371:117-25.

3. Hokansen JE, Austin MA. Plasma triglyceride level is a risk factor for cardiovascular disease independent of high density lipoproteincholesterol level: Meta analysis of population based prospective studies. J Cardiovasc Risk. 1996;3:213-9.
4. Miller M, Cannon CP, Murphy SA, Qin J, Ray KK, Braunwald E, et al. Impact of triglyceride levels beyond low-density lipoprotein cholesterol after acute coronary syndrome in the PROVE IT-TIMI 22 trial. J Am Coll Cardiol. 2008;51(7):724-30.
5. Ginsberg HN, Elam MB, Lovato LC, Crouse JR 3rd, Leiter LA, Linz P, et al. Effects of combination lipid therapy in type 2 diabetes mellitus. N Engl J Med. 2010;362(17):1563-74.
6. Nichols GA, Philip S, Reynolds K, Granowitz CB, Fazio S. Increased cardiovascular risk in hypertriglyceridemic patients with statin-controlled LDL cholesterol. J Clin Endocrinol Metab. 2018;103(8):3019-27.
7. Frick MH, Elo O, Haapa K, Heinonen OP, Heinsalmi P, Helo P, et al. Helsinki Heart Study: primary-prevention trial with gemfibrozil in middle-aged men with dyslipidemia. Safety of treatment, changes in risk factors, and incidence of coronary heart disease. N Engl J Med. 1987;317(20):1237-45.
8. Keech A, Simes RJ, Barter P, Best J, Scott R, Taskinen MR, et al. Effects of long-term fenofibrate therapy on cardiovascular events in 9795 people with type 2 diabetes mellitus (the FIELD study): randomised controlled trial. Lancet. 2005;366(9500):1849-61.
9. Koskinen P, Manttari M, Manninen V, Huttunen JK, Heinonen OP, Frick MH. Coronary heart disease incidence in NIDDM patients in the Helsinki Heart Study. Diabetes Care. 1992;15(7):820-5.
10. Rubins HB, Robins SJ, Collins D, Fye CL, Anderson JW, Elam MB, et al. Gemfibrozil for the secondary prevention of coronary heart disease in men with low levels of high-density lipoprotein cholesterol. Veterans Affairs high-Density Lipoprotein Cholesterol Intervention Trial Study Group. N Engl J Med. 1999;341(6):410-8.
11. Scott R, O'Brien R, Fulcher G, Pardy C, D'Emden M, Tse D, et al. Effects of fenofibrate treatment on cardiovascular disease risk in 9,795 individuals with type 2 diabetes and various components of the metabolic syndrome: the Fenofibrate Intervention and Event Lowering in Diabetes (FIELD) study. Diabetes Care. 2009;32(3):493-8.
12. Bezafibrate Infarction Prevention Study. Secondary prevention by raising HDL cholesterol and reducing triglycerides in patients with coronary artery disease. Circulation. 2000;102(1):21-7.
13. Bruckert E, Labreuche J, Deplanque D, Touboul PJ, Amarenco P. Fibrates effect on cardiovascular risk is greater in patients with high triglyceride levels or atherogenic dyslipidemia profile: a systematic review and meta-analysis. J Cardiovasc Pharmacol. 2011;57(2):267-72.
14. Toth PP, Granowitz C, Hull M, Liassou D, Anderson A, Philip S. High triglycerides are associated with increased cardiovascular events, medical costs, and resource use: a real-world administrative claims analysis of statin-treated patients with high residual cardiovascular risk. J Am Heart Assoc. 2018;7(15):e008740.
15. Kasai T, Miyauchi K, Yanagisawa N, Kajimoto K, Kubota N, Ogita M, et al. Mortality risk of triglyceride levels in patients with coronary artery disease. Heart. 2013;99(1):22-9.
16. Sarwar N, Danesh J, Eiriksdottir G, Sigurdsson G, Wareham N, Bingham S, et al. Triglycerides and the risk of coronary heart disease: 10,158 incident cases among 262,525 participants in 29 Western prospective studies. Circulation. 2007;115(4):450-8.
17. Alexopoulos A-S, Qamar A, Hutchins K, Crowley MJ, Batch BC, Guyton JR. Triglycerides: Emerging Targets in Diabetes Care? Review of Moderate Hypertriglyceridemia in Diabetes. Curr Diab Rep. 2019;19(4):13.
18. Lampidonis AD, Rogdakis E, Voutsinas GE, Stravopodis DJ. The resurgence of Hormone-Sensitive Lipase (HSL) in mammalian lipolysis. Gene. 2011;477(1–2):1-11.
19. Ginsberg HN. Insulin resistance and cardiovascular disease. J Clin Invest. 2000;106(4):453-8.
20. Lewis GF. Fatty acid regulation of very low density lipoproteins production. Curr Opin Lipidol. 1997;8:146-53.
21. Bruce C, Chouinard RA Jr, Tall AR. Plasma lipid transfer proteins, high-density lipoproteins and reverse cholesterol transport. Annu Rev Nutr. 1998;18:297-330.
22. Austin MA, King MC, Vranizan KM, Krauss RM. Atherogenic lipoprotein phenotype. A proposed genetic marker for coronary heart disease risk. Circulation. 1990;82:495-506.
23. Wulffele MG, Kooy A, de Zeeuw D, Stehouwer CD, Gansevoort RT. The effect of metformin on blood pressure, plasma cholesterol and triglycerides in type 2 diabetes mellitus: a systematic review. J Intern Med. 2004;256(1):1-14.
24. Rosenblit PD. Common medications used by patients with type 2 diabetes mellitus: what are their effects on the lipid profile? Cardiovasc Diabetol. 2016;15:95.
25. Ovalle F. Cardiovascular implications of antihyperglycemic therapies for type 2 diabetes. Clin Ther. 2011;33(4):393-407.

26. Szalat A, Durst R, Leitersdorf E. Managing dyslipidaemia in type 2 diabetes mellitus. Best Pract Res Clin Endocrinol Metab. 2016;30(3):431-44.
27. Chaudhuri A, Dandona P. Effects of insulin and other antihyperglycaemic agents on lipid profiles of patients with diabetes. Diabetes Obes Metab. 2011;13(10):869-79.
28. Winkler K, Konrad T, Fullert S, Friedrich I, Destani R, Baumstark MW, et al. Pioglitazone reduces atherogenic dense LDL particles in nondiabetic patients with arterial hypertension: a double-blind, placebo-controlled study. Diabetes Care. 2003;26(9):2588-94.
29. Chen YH, Du L, Geng XY, Peng YL, Shen JN, Zhang YG, et al. Effects of sulfonylureas on lipids in type 2 diabetes mellitus: a meta-analysis of randomized controlled trials. J Evid Based Med. 2015;8(3):134-48.
30. Rizzo M, Rizvi AA, Spinas GA, Rini GB, Berneis K. Glucose lowering and anti-atherogenic effects of incretin-based therapies: GLP-1 analogues and DPP-4-inhibitors. Expert Opin Investig Drugs. 2009;18(10):1495-503.
31. Hjerpsted JB, Flint A, Brooks A, Axelsen MB, Kvist T, Blundell J. Semaglutide improves postprandial glucose and lipid metabolism, and delays first-hour gastric emptying in subjects with obesity. Diabetes Obes Metab. 2018;20(3):610-9.
32. Edwards KL, Minze MG. Dulaglutide: an evidence-based review of its potential in the treatment of type 2 diabetes. Core Evid. 2015;10:11-21.
33. Farr S, Adeli K. Incretin-based therapies for treatment of postprandial dyslipidemia in insulin-resistant states. Curr Opin Lipidol. 2012;23(1):56-61.
34. Hayashi T, Fukui T, Nakanishi N, Yamamoto S, Tomoyasu M, Osamura A, et al. Dapagliflozin decreases small dense low-density lipoprotein-cholesterol and increases high-density lipoprotein 2-cholesterol in patients with type 2 diabetes: comparison with sitagliptin. Cardiovasc Diabetol. 2017;16(1):8.
35. Watts GF, Ooi EM, Chan DC. Demystifying the management of hypertriglyceridaemia. Nat Rev Cardiol. 2013;10(11):648-61.
36. Faeh D, Minehira K, Schwarz JM, Periasamy R, Park S, Tappy L. Effect of fructose overfeeding and fish oil administration on hepatic de novo lipogenesis and insulin sensitivity in healthy men. Diabetes. 2005;54(7):1907-13.
37. Stein EA, Lane M, Laskarzewski P. Comparison of statins in hypertriglyceridemia. Am J Cardiol. 1998;81(4A):66B–9B.
38. Rosenson RS, Davidson MH, Hirsh BJ, Kathiresan S, Gaudet D. Genetics and causality of triglyceride-rich lipoproteins in atherosclerotic cardiovascular disease. J Am Coll Cardiol. 2014;64(23):2525-40.
39. Shapiro MD, Fazio S. From lipids to inflammation: new approaches to reducing atherosclerotic risk. Circ Res. 2016;118(4):732-49.
40. Rosenson RS. Current overview of statin-induced myopathy. Am J Med. 2004;116(6):408-16.
41. Sacks FM, Carey VJ, Fruchart JC. Combination lipid therapy in type 2 diabetes. N Engl J Med. 2010;363:692-4.
42. Jun M, Foote C, LV J, Neal B, Patel A, Nicholls SJ, et al. Effects of fibrates on cardiovascular outcomes: A systematic review and meta-analysis. Lancet. 2010;375:1875-84.
43. Hong F, Xu P, Zhai Y. The opportunities and challenges of peroxisome proliferator-activated receptors ligands in clinical drug discovery and development. Int J Mol Sci. 2018;19(8):E2189.
44. Araki E, Yamashita S, Arai H, Yokote K, Satoh J, Inoguchi T, et al. Effects of pemafibrate, a novel selective PPAR alpha modulator, on lipid and glucose metabolism in patients with type 2 diabetes and hypertriglyceridemia: a randomized, double-blind, placebo-controlled, phase 3 trial. Diabetes Care. 2018;41(3):538-46.
45. Arai H, Yamashita S, Yokote K, Araki E, Suganami H, Ishibashi S, et al. Efficacy and safety of pemafibrate versus fenofibrate in patients with high triglyceride and low HDL cholesterol levels: a multicenter, placebo-controlled, double-blind, randomized trial. J Atheroscler Thromb. 2018;25(6):521-38.
46. Henke BR, Blanchard SG, Brackeen MF, Brown KK, Cobb JE, Collin JL, et al. N-(2-Benzoylphenyl)-L-tyrosine PPARgamma agonists. Discovery of a novel series of potent antihyperglycemic and antihyperlipidemic agents. J Med Chem. 1998;41:5020-36.
47. Egerod FL, Nielsen HS, Iversen L, Thorup I, Storgaard T, Oleksiewicz MB. Biomarkers for early effects of carcinogenic dual-acting PPAR agonists in rat urinary bladder urotheliumin vivo. Biomarkers. 2005;10:295-309.
48. Kendall DM, Rubin CJ, Mohideen P, Ledeine JM, Belderet R, Grossal J, et al. Improvement in glycemic control, triglycerides, and HDL cholesterol levels with muraglitazar, a dual (alpha/gamma) peroxisome proliferator-activated receptor activator, in patients with type 2 diabetes inadequately controlled with Metformin monotherapy: A double blind, randomized, pioglitazone-comparative study. Diabetes Care. 2006;29:1016-23.

49. Fagerberg B, Edwards S, Halmos T, Lopatynsky J, Schuster H, Stender S, et al. Tesaglitazar, a novel dual peroxisome proliferator-activated receptor alpha/gamma agonist, dose-dependently improves the metabolic abnormalities associated with insulin resistance in a non-diabetic population. Diabetologia. 2005;48:1716-25.
50. Lincoff AM, Tardif JC, Schwartz GG, Nicholls SJ, Ryden L, Neal B, et al. Effect of aleglitazar on cardiovascular outcomes after acute coronary syndrome in patients with type 2 diabetes mellitus: the AleCardio randomized clinical trial. JAMA. 2014;311(15):1515-25.
51. Jani RH, Kansagra K, Jain MR, Patel H. Pharmacokinetics, safety and tolerability of saroglitazar (ZYH1), a predominantly PPARα agonist with moderate PPARγ agonist activity in healthy human subjects. Clin Drug Investig. 2013;33:800-16.
52. Agrawal, R. The first approved agent in the Glitazar's Class: Saroglitazar. Curr Drug Targets. 2014;15:151-5.
53. Jani RH, Pai V, Jha P, Jariwala G, Mukhopadhyay S, BhansaliA, et al. A multicenter, prospective, randomised, double blind study to evaluate the safety and efficacy of saroglitazar 2 and 4 mg compared with placebo in type 2 diabetes mellitus patients having hypertriglyceridemia not controlled on Atorvastatin therapy (PRESS VI). Diabetes Technol Ther. 2014;16:63-71.
54. Ghosh A, Sahana PK, Das C, Mandal A, Sengupta N. Comparison of effectiveness and safety of add-on therapy of saroglitazar and fenofibrate with metformin in Indian patients with diabetic dyslipidaemia. J Clin Diagn Res. 2016;10:FC01-FC04.
55. Nestel PJ, Connor WE, Reardon MF, Connor S, Wong S, Boston R. Suppression by diets rich in fish oil of very low density lipoprotein production in man. J Clin Invest. 1984;74(1):82-9.
56. Harris WS, Connor WE, Illingworth DR, Rothrock DW, Foster DM. Effects of fish oil on VLDL triglyceride kinetics in humans. J Lipid Res. 1990;31(9):1549-58.
57. Durrington PN, Bhatnagar D, Mackness MI, Morgan J, Julier K, Khan MA, et al. An omega-3 polyunsaturated fatty acid concentrate administered for one year decreased triglycerides in simvastatin treated patients with coronary heart disease and persisting hypertriglyceridaemia. Heart. 2001;85(5):544-8.
58. Harris WS, Ginsberg HN, Arunakul N, Shachter NS, Windsor SL, Adams M, et al. Safety and efficacy of Omacor in severe hypertriglyceridemia. J Cardiovasc Risk. 1997;4(5–6):385-91.
59. Bays HE, Ballantyne CM, Kastelein JJ, Isaacsohn JL, Braeckman RA, Soni PN. Eicosapentaenoic acid ethyl ester (AMR101) therapy in patients with very high triglyceride levels (from the Multi-center, plAcebo-controlled, Randomized, double-blINd, 12-week study with an open-label Extension [MARINE] trial). Am J Cardiol. 2011;108(5):682-90.
60. AstraZeneca Pharmaceuticals. (2014) Epanova prescribing information. [online] Available from https://www.accessdata.fda.gov/drugsatfda_docs/label/2014/205060s000lbl.pdf. [Last accessed July 2020].
61. GlaxoSmithKline. (2015) Lovaza prescribing information. [online] Available from https://gskpro.com/en-ca/. [Last accessed July 2020].
62. Amarin Corporation. (2017) Vascepa prescribing information. [online] Available from https://investor.amarincorp.com/news-releases/news-release-details/amarin-receives-fda-approval-vascepar-icosapent-ethyl-reduce. [Last accessed July 2020].
63. Ballantyne CM, Bays HE, Kastelein JJ, Stein E, Isaacsohn JL, Braeckman RA, et al. Efficacy and safety of eicosapentaenoic acid ethyl ester (AMR101) therapy in statin-treated patients with persistent high triglycerides (from the ANCHOR study). Am J Cardiol. 2012;110(7):984-92.
64. Yokoyama M, Origasa H, Matsuzaki M, Matsuzawa Y, Saito Y, Ishikawa Y, et al. Effects of eicosapentaenoic acid on major coronary events in hypercholesterolaemic patients (JELIS): a randomised open-label, blinded endpoint analysis. Lancet. 2007;369(9567):1090-8.
65. Investigators OT, Bosch J, Gerstein HC, Dagenais GR, Diaz R, Dyal L, et al. n-3 fatty acids and cardiovascular outcomes in patients with dysglycemia. N Engl J Med. 2012;367(4):309-18.
66. Roncaglioni MC, Tombesi M, Avanzini F, Barlera S, Caimi V, Longoni P, et al. n-3 fatty acids in patients with multiple cardiovascular risk factors. N Engl J Med. 2013;368(19):1800-8.
67. The ASCEND Study Collaborative Group. The ASCEND Study Collaborative Group. N Engl J Med. 2018;379:1540-50.
68. Bhatt DL, Steg PG, Miller M, Brinton EA, Jacobson TA, Ketchum SB, et al. Cardiovascular risk reduction with icosapent ethyl for hypertriglyceridemia. N Engl J Med. 2018;380:11-22.
69. Lavie CJ, Milani RV, Mehra MR, Ventura HO. Omega-3 polyunsaturated fatty acids and cardiovascular diseases. J Am Coll Cardiol. 2009;54(7):585-94.

70. Fraser DA, Wang X, Alonso C, Lek S, Skjaeret T, Schuppan D. Icosabutate, a novel structurally engineered fatty-acid, exhibits potent anti-inflammatory and anti-fibrotic effects in a dietary mouse model resembling progressive human NASH. 2018.
71. Chapman MJ, Redfern JS, McGovern ME, Giral P. Niacin and fibrates in atherogenic dyslipidemia: pharmacotherapy to reduce cardiovascular risk. Pharmacol Ther. 2010;126(3):314–45.
72. Stein EA, Raal F. Future directions to establish lipoprotein(a) as a treatment for atherosclerotic cardiovascular disease. Cardiovasc Drugs Ther. 2016;30(1):101-8.
73. Gray DR, Morgan T, Chretien SD, Kashyap ML. Efficacy and safety of controlled-release niacin in dyslipoproteinemic veterans. Ann Intern Med. 1994;121(4):252-8.
74. Garg A, Grundy SM. Nicotinic acid as therapy for dyslipidemia in non-insulin-dependent diabetes mellitus. JAMA. 1990;264(6):723-6.
75. Group HTC, Landray MJ, Haynes R, Hopewell JC, Parish S, Aung T, et al. Effects of extended-release niacin with laropiprant in high-risk patients. N Engl J Med. 2014;371(3):203-12.
76. Superko HR, Zhao XQ, Hodis HN, Guyton JR. Niacin and heart disease prevention: engraving its tombstone is a mistake. J Clin Lipidol. 2017;11(6):1309-17.
77. Investigators A-H, Boden WE, Probstfield JL, Anderson T, Chaitman BR, Desvignes-Nickens P, et al. Niacin in patients with low HDL cholesterol levels receiving intensive statin therapy. N Engl J Med. 2011;365(24):2255-67.
78. Group HTC, Landray MJ, Haynes R, Hopewell JC, Parish S, Aung T, et al. Effects of extended-release niacin with laropiprant in high risk patients. N Engl J Med. 2014;371(3):203-12.
79. Ooi EM, Barrett PH, Chan DC, Watts GF. Apolipoprotein C-III: understanding an emerging cardiovascular risk factor. Clin Sci (Lond). 2008;114(10):611-24.
80. Digenio A, Dunbar RL, Alexander VJ, Hompesch M, Morrow L, Lee RG, et al. Antisense-mediated lowering of plasma apolipoprotein C-III by volanesorsen improves dyslipidemia and insulin sensitivity in type 2 diabetes. Diabetes Care. 2016;39(8):1408-15.
81. Gaudet D, Alexander V, Arca M, Jones A, Stroes E, Bergeron J, et al. The APPROACH study: a randomized, double-blind, placebo-controlled, phase 3 study of volanesorsen administered subcutaneously to patients with familial chylomicronemia syndrome (FCS). Atherosclerosis. 2017;283:e10.
82. Dewey FE, Gusarova V, Dunbar RL, O'Dushlaine C, Schurmann C, Gottesman O, et al. Genetic and pharmacologic inactivation of ANGPTL3 and cardiovascular disease. N Engl J Med. 2017;377(3):211-21.
83. Graham MJ, Lee RG, Brandt TA, Tai LJ, Fu W, Peralta R, et al. Cardiovascular and metabolic effects of ANGPTL3 antisense oligonucleotides. N Engl J Med. 2017;377(3):222-32.
84. Jellinger PS, Handelsman Y, Rosenblit PD, Bloomgarden ZT, Fonseca VA, Garber AJ, et al. American Association of Clinical Endocrinologists and American College of Endocrinology guidelines for management of dyslipidemia and prevention of cardiovascular disease. Endocr Pract. 2017;23(Suppl 2):1-87.
85. Jacobson TA, Ito MK, Maki KC, Orringer CE, Bays HE, Jones PH, et al. National lipid association recommendations for patient-centered management of dyslipidemia: part 1–full report. J Clin Lipidol. 2015;9(2):129-69.
86. Grundy SM, Stone NJ, Bailey AL, Beam C, Birtcher KK, Blumenthal RS, et al. 2018 AHA/ACC/AACVPR/AAPA/ABC/ACPM/ADA/AGS/APhA/ASPC/NLA/PCNA guideline on the Management of Blood Cholesterol: executive summary: a report of the American College of Cardiology/American Heart Association Task Force on clinical practice guidelines. J Am Coll Cardiol. 2018.
87. National Cholesterol Education Program Expert Panel on Detection E, Treatment of High Blood Cholesterol in A. Third Report of the National Cholesterol Education Program (NCEP) Expert Panel on detection, evaluation, and treatment of high blood cholesterol in adults (Adult Treatment Panel III) final report. Circulation. 2002;106(25):3143-421.
88. Berglund L, Brunzell JD, Goldberg AC, Goldberg IJ, Sacks F, Murad MH, et al. Evaluation and treatment of hypertriglyceridemia: an Endocrine Society clinical practice guideline. J Clin Endocrinol Metab. 2012;97(9):2969-89.
89. American Diabetes Association. Cardiovascular Disease and Risk Management: : Standards of Medical Care in Diabetesd2020. Diabetes Care. 2020;43(Suppl. 1):S111–S134.
90. Mach F, Baigent C, Catapano AL, Koskinas KC, Casula M, Badimon L, et al. 2019 ESC/EAS Guidelines for the management of dyslipidaemias: lipid modification to reduce cardiovascular risk: The Task Force for the management of dyslipidaemias of the European Society of Cardiology (ESC) and European Atherosclerosis Society (EAS). Eur Heart J 2020;41(1):111-88.
91. Chawla R, Madhu SV, Makkar BM, Ghosh S, Saboo B, Kalra S. RSSDI-ESI clinical practice recommendations for the management of type 2 diabetes mellitus 2020. Indian J Endocr Metab. 2020;24:1-122.

CHAPTER

11

Non-HDL is a Better Marker than LDL for Prediction of ASCVD Risk

Shishir Soni, Dibbendhu Khanra

ABSTRACT

Non-high-density lipoprotein cholesterol (non-HDL-C) which is total cholesterol (TC) minus high density lipoprotein cholesterol (HDL-C) emerged as a useful marker for the prediction of atherosclerotic cardiovascular disease (ASCVD) risk. Clinical situations with normal low-density lipoprotein cholesterol (LDL-C) and unfavorable cardiovascular outcomes on follow-up and inaccuracy in indirect estimation of LDL-C have raised many questions on the reliability of LDL-C as a marker for prediction of ASCVD risk. Non-HDL-C being a more comprehensive marker as it also includes an account of other atherogenic triglyceride (TG)-rich lipoproteins carries an advantage over LDL-C for prediction of ASCVD risk. ASCVD risk prediction and further management require an understanding of the difference between these two markers and here in this chapter we will discuss how non-HDL-C emerged as a better marker than LDL-C for prediction of ASCVD risk.

INTRODUCTION

The predictor of a disease may not always be the primary target of therapy and vice-versa. The cholesterol management for the prevention of atherosclerotic cardiovascular disease (ASCVD) has evolved in the past decade with an approach toward redefining targets of therapy, stratification of high-risk population by risk estimation, and parallel advancements in the measurement of lipid components. As a part of this evolution process, various studies have tried to put some useful shreds of evidence in favor of emerging predictors of ASCVD and in that race, non-high-density-lipoprotein cholesterol (non-HDL-C) appears to be well ahead of low-density lipoprotein cholesterol (LDL-C) in predicting ASCVD risk.[1,2] Although, the primary target of lipid-lowering therapy is LDL-C, and the secondary target is non-HDL-C as per the latest recommendations, one should not underestimate the usefulness of non-HDL-C in predicting cardiovascular outcomes.[2] Moreover, there are certain conditions in which atherogenic risk is high despite achieving target LDL-C levels, where non-HDL-C can define the residual risk.[3]

ROLE OF NON-HDL CHOLESTEROL COMPONENTS IN ATHEROSCLEROSIS

Before considering lipoprotein as a bad entity one must understand that lipoproteins are a transporter of lipids to various tissues for important functions that include energy utilization, hormone production (steroids), bile acid synthesis, and lipid deposition.[4] Major lipoproteins in the blood include lipoprotein-a [LP(a)], high-density lipoprotein (HDL), and four apolipoprotein B (apoB), containing lipoproteins namely chylomicrons (larger sized), low-density lipoprotein (LDL), very low-density lipoprotein (VLDL), and intermediate density lipoprotein (IDL).[5] Out of all major lipoproteins, triglycerides (TGs) (%) are maximum in chylomicrons (90–95%) followed by VLDL (50–65%) and IDL (24–40%).[2,5] Thus, ApoB, TG content, and size of the lipoprotein are three important factors that may help in understanding the role of individual lipoprotein in atherosclerosis. In the presence of endothelial dysfunction, smaller sized ApoB containing lipoprotein and their remnants can easily cross the endothelial barrier and can reside in the arterial wall leading to initiation of atheroma formation via a complex process involving lipid deposition.[5,6] Thus, prolonged exposure of higher concentration of ApoB containing lipoprotein in blood will promote the progression of atherosclerotic plaque depending on the total duration of exposure. Thus, it is not only the LDL that appears to be the culprit albeit all ApoB containing lipoprotein and their remnants are involved in the process of atherosclerosis. This also explains the role of TGs in atherosclerosis as chylomicrons, VLDL, and IDL are having a relatively higher percentage of TGs, thus, in the presence of hypertriglyceridemia, the endothelium is exposed to a higher burden of atherogenic lipoproteins.[7]

After understanding the process of atherosclerosis, non-HDL-C [total cholesterol (TC) minus HDL-C] seems to be a more comprehensive indicator of atherogenic burden in the blood as compared to LDL-C alone.[2,7]

LIPID MEASUREMENTS AND MATHEMATICAL EQUATIONS

In current practice, lipid profile measures concentration of TC, HDL-C, and TGs. However, LDL-C is commonly measured using the Friedewald formula:

LDL-C = TC − HDL-C − (TG/5) (in mg/dL) ......equation 1, and

Non-HDL-C is calculated as:

Non-HDL-C = TC − HDL-C.......equation 2

Using these equations, the actual difference between non-HDL-C and LDL-C is the TG component. Thus, predicting ASCVD risk using LDL-C will underestimate the risk if TG levels are too high (**Fig. 1**). TGs are an important constituent of ApoB containing lipoprotein and their remnants which are potentially atherogenic. Non-HDL-C measurement may not underestimate the atherogenic risk as it also includes the TG which at times become more significant if TG levels are too high.[2,8]

PREDICTING THE ASCVD RISK BY MEASURING LIPIDS

Predicting the increased risk of atherosclerotic disease progression and unfavorable cardiovascular outcomes by measuring the LDL-C concentration in the blood has been the mainstay of the management of dyslipidemia.[1,2,9] However, the measured LDL-C

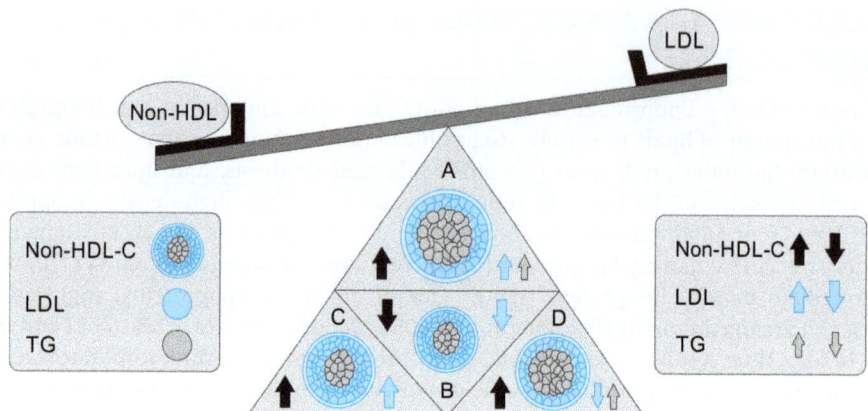

(HDL-C: high-density lipoprotein cholesterol; LDL-C: low-density lipoprotein cholesterol; TG: triglyceride

FIG. 1: Panel A–D showing condition with different levels of non-high-density lipoprotein cholesterol (HDL-C), low-density lipoprotein cholesterol (LDL-C), and triglycerides (TGs). All three (non-HDL-C, LDL-C, and TG) are elevated as shown by colored arrows (Panel A), all three (non-HDL-C, LDL-C, and TG) are below normal cut-off (Panel B), non-HDL-C and LDL-C both are elevated (Panel C), non-HDL-C and TG are elevated but LDL-C is below the cut-off limit (Panel D). Increased risk of ASCVD can be associated with three conditions (Panel A, C, D) and non-HDL-C can predict risk in all these conditions, however, it may remain unrevealed if assessed by LDL-C measurement in the condition shown in panel D.

concentration in the blood reflects the amount of cholesterol-containing LDL particles and not about their number.[10] Moreover, LDL-C concentration measurement does not provide any estimate of other lipoproteins especially IDL, VLDL, and chylomicron remnants which are also involved in atherosclerosis cascade. Therefore, ApoB which is the main content of all four potentially atherogenic lipoproteins (LDL, VLDL, IDL, and chylomicron remnant) appears to be a better indicator for predicting increased risk of atherosclerosis.[11] However, ApoB measurement is not commonly done and therefore, non-HDL-C which also provides an estimate of ApoB containing atherogenic particles including LDL seems more promising as ASCVD risk predictor as compared to LDL-C estimation.[12,13] Apart from the above-mentioned fact, another disadvantage which is being carried out if predicting the ASCVD risk using LDL-C measurement is the error in its correct estimation as mostly it is calculated by the Friedewald formula rather than a direct measurement.[14] Moreover, this error in measurement becomes significant (17–25%) in the presence of elevated TG level in the range of 151–300 mg/dL and above.[15]

EVIDENCE REFLECTING NON-HDL CHOLESTEROL AS A BETTER PREDICTOR OF ASCVD RISK

A retrospective study of 868 patients post-myocardial infarction (MI) by Wongcharoen et al. comparing the non-HDL-C with LDL-C in predicting cardiovascular outcomes found that those with a higher non-HDL-C level (non-HDL-C >130 mg/dL vs. <100 mg/dL) had worst long-term cardiovascular outcomes, whereas such correlation was not found with LDL-C (LDL-C >100 mg/dL vs. <70 mg/dL).[16] Another useful evidence by Liu et al. who analyzed the data from the FHS (Framingham Heart Study) and found a strong and graded association between non-HDL-C and risk of coronary artery disease (CAD) within LDL-C levels while no such association was found between LDL-C and the risk for CAD within non-HDL-C level.[17] Another large retrospective study, analyzing data of 2,406 men and 2,056 women, to determine the usefulness of HDL-C in predicting the cardiovascular outcomes for an average of 19 years found non-HDL-C as a better predictor compared to LDL-C.[18] Similar observations were made by the BARI (Bypass Angioplasty Revascularization Investigation) group investigators evaluating 1,514 patients with multivessel CAD with follow-up for 5 years and they found non-HDL-C as a strong predictor of cardiovascular outcomes (only non-fatal MI and angina, not mortality) at 5 years.[19] A large prospective study in women by Ridker et al. in 2005, with a study population of 15,632 healthy women with follow-up data of 10 years found non-HDL-C as good as apolipoprotein fractions in predicting the first cardiovascular event.[20] Similar observations in patients with diabetes were made by Liu et al. in his study in the year 2005, in which he concluded that non-HDL-C is a better predictor of cardiovascular mortality among those with diabetes as compared to LDL-C.[21] A study in the male population was done by Pischon et al. in 2005, comparing the ApoB, non-HDL-C, and LDL-C as a predictor of CAD in men found both non-HDL-C and ApoB as strong predictors of CAD as compared to LDL-C.[22] Thus, various studies in different study populations found similar results suggesting non-HDL-C as a strong predictor of ASCVD risk.

Apart from all these studies, there were few studies in which inverse relation was found between LDL-C and cardiovascular outcomes, i.e., higher risk in those with lower levels of LDL-C and lower risk in those with higher levels of LDL-C.[10,16,23] Thus, various speculations have been made and the majority of them suggested that probably more atherogenic subclass of LDL might be present in those with lower or normal levels of LDL-C with increased cardiovascular risk and later on various studies have emphasized that small dense LDL particle might be the atherogenic culprit in these subsets of patients.[24] Although LDL particle size is not the new concept and in the year 1992, a study by Campos et al. found that individuals with large LDL particle size are associated with low TG and high HDL.[25] Moreover, corollary of findings made by Campos et al. is now commonly known as atherogenic dyslipidemia (high TG, high small dense LDL, and low HDL).[26] Another study highlighting this fact came in the year 2014 by Hoogeveen et al. in 11,419 subjects to evaluate the relationship between small dense LDL and cardiovascular outcomes events assessed over 11 years and it was concluded that small dense LDL predicts outcome in those with normal LDL-C levels.[27] Thus, various studies have tried to put some emerging risk factors that can predict the risk of atherosclerosis, however, the need of an easily available assay which can be feasible and applicable for a large population and can replace LDL-C as the sole predictor of ASCVD risk has been partially replaced by non-HDL-C and is now well established secondary target of therapy for cholesterol management.

GUIDELINES AND RECOMMENDATIONS

In the recent guidelines by the ACC/AHA (American College of Cardiology and American Heart Association), non-HDL-C is mentioned as a strong indicator of atherogenicity as compared to LDL-C alone. In patients with clinical ASCVD and at very high-risk with LDL-C level (on therapy) 70 mg/dL or higher or non-HDL cut-off level of 100 mg/dL or higher is considered as the threshold for starting additional lipid-lowering therapy with class IIa recommendation from ACC/AHA.[1] Primary hypercholesterolemia with LDL-C level 160–189 mg/dL and non–HDL-C level 190–219 mg/dL is one of the risk enhancing factors that should be considered for risk assessment of the patient. For children and adolescents also non-HDL-C is considered as a reasonable screening test (class IIb recommendation from ACC/AHA) with the cut-off for abnormal values is 145 mg/dL or higher. In the recent ESC (European Society of Cardiology) guidelines, ApoB is mentioned as an accurate marker of the total concentration of atherogenic particles, and discordance between measured values of ApoB and LDL-C can be found in around 20% of patients.[2,28] Therefore, measurement of non-HDL-C is recommended (with class I recommendation from ESC) in those with diabetes, hypertriglyceridemia, or very low LDL-C levels for risk assessment. Moreover, secondary goals have been defined with the cut-off for non-HDL-C should be 30 mg/dL higher than corresponding LDL-C goal and therefore, secondary targets are non-HDL-C <85 mg/dL, <100 mg/dL and <130 mg/dL in people at very high, high, and moderate cardiovascular risk, respectively. It has also been emphasized that dyslipidemia in diabetes mellitus (DM) may remain unrevealed in those with normal LDL-C levels, where non-HDL-C may reveal underlying atherogenic dyslipidemia. In patients with metabolic syndrome, non-HDL-C is a marker of TG-rich lipoprotein and their remnants and is the secondary objective of therapy. For monitoring lipid-lowering therapy, non-HDL-C is labeled as secondary targets of therapy.[1,2]

ADVANTAGES OF USING NON-HDL CHOLESTEROL MEASUREMENT IN PREDICTING THE ASCVD RISK

The measurement of non-HDL-C is simple and cost-effective as compared to ApoB which is another strong predictor of ASCVD risk.[2,29] Using non-HDL-C as a secondary target and its application in day-to-day clinical practice is not cumbersome; moreover, its cut-off value is usually 30 mg/dL higher than corresponding LDL-C targets. Therefore, does not warrant any complicated calculation or scoring. In clinical scenarios, especially DM, hypertriglyceridemia, obesity, and very low LDL-C levels, the use of non-HDL-C can reveal underlying dyslipidemia.[2,30]

CONCLUSION

Non-HDL-C and ApoB both are the strong predictor of the ASCVD risk as compared to LDL-C and have been evidenced in various studies and included in recent guidelines also. Non-HDL-C is recommended as a secondary target of lipid-lowering therapy and is found to be a more useful assay to reveal dyslipidemia in patients with DM, hypertriglyceridemia, obesity, and very low levels of LDL-C. Pathophysiology of atherosclerosis involving TG-rich lipoproteins and pieces of evidence from various studies indicate that LDL-C alone may not always reveal an increased risk of atherosclerosis and at times patients with normal LDL-C are at increased ASCVD risk which may be revealed by non-HDL-C measurement. Thus, non-HDL-C measurement may also prove to be more useful than other available tools for monitoring additional nonstatin lipid-lowering therapy in research trials and screening dyslipidemia in children and adolescents.

KEY POINTS

- Non-HDL-C is a stronger indicator of atherogenicity than LDL-C alone.
- Increased risk of ASCVD in patients with normal LDL-C level and high non-HDL-C can be explained by the presence of higher levels of TG-rich lipoproteins and their remnants.
- Non-HDL-C measurement is recommended for risk assessment in patients with high TG levels, DM, obesity, or very low LDL-C levels.
- At present, non-HDL-C is the secondary treatment target for monitoring lipid-lowering therapy.

REFERENCES

1. Grundy SM, Stone NJ, Bailey AL, Beam C, Birtcher KK, Blumenthal RS, et al. 2018 ACC/AHA Guideline on the Management of Blood Cholesterol: Executive Summary: A Report of the American College of Cardiology/American Heart Association Task Force on Clinical Practice Guidelines. J Am Coll Cardiol. 2019;73(24):3168-209.
2. Mach F, Baigent C, Catapano AL, Koskinas KC, Casula M, Badimon L, et al. 2019 ESC/EAS Guidelines for the management of dyslipidaemias: lipid modification to reduce cardiovascular risk. Eur Heart J. 2020;41(1):111-88.
3. Robinson JG, Wang S, Smith BJ, Jacobson TA. Meta-analysis of the relationship between non-high-density lipoprotein cholesterol reduction and coronary heart disease risk. J Am Coll Cardiol. 2009;53(4):316-22.
4. Walther TC, Farese RV. Lipid droplets and cellular lipid metabolism. Annu Rev Biochem. 2012;81:687-714.
5. Borén J, Williams KJ. The central role of arterial retention of cholesterol-rich apolipoprotein-B-containing lipoproteins in the pathogenesis of atherosclerosis: a triumph of simplicity. Curr Opin Lipidol. 2016;27(5):473-83.
6. Ference BA, Graham I, Tokgozoglu L, Catapano AL. Impact of Lipids on Cardiovascular Health: JACC Health Promotion Series. J Am Coll Cardiol. 2018;72(10):1141-56.
7. Nordestgaard BG, Varbo A. Triglycerides and cardiovascular disease. Lancet. 2014;384(9943):626-35.
8. Martin SS, Blaha MJ, Elshazly MB, Brinton EA, Toth PP, McEvoy JW, et al. Friedewald-estimated versus directly measured low-density lipoprotein cholesterol and treatment implications. J Am Coll Cardiol. 2013;62(8):732-9.
9. Jellinger PS, Handelsman Y, Rosenblit PD, Bloomgarden ZT, Fonseca VA, Garber AJ, et al. American Association Of Clinical Endocrinologists And American College Of Endocrinology Guidelines For Management Of Dyslipidemia And Prevention Of Cardiovascular Disease-Executive Summary. Endocr Pract. 2017;23(4):479-97.
10. Otvos JD, Mora S, Shalaurova I, Greenland P, Mackey RH, Goff DC. Clinical Implications of Discordance Between LDL Cholesterol and LDL Particle Number. J Clin Lipidol. 2011;5(2):105-13.
11. Thanassoulis G, Williams K, Ye K, Brook R, Couture P, Lawler PR, et al. Relations of change in plasma levels of LDL-C, non-HDL-C and apoB with risk reduction from statin therapy: a meta-analysis of randomized trials. J Am Heart Assoc. 2014;3(2):e000759.
12. Zhang Y, Wu N-Q, Li S, Zhu C-G, Guo Y-L, Qing P, et al. Non-HDL-C is a better predictor for the severity of coronary atherosclerosis compared with LDL-C. Heart Lung Circ. 2016;25(10):975-81.
13. Puri R, Nissen SE, Shao M, Elshazly MB, Kataoka Y, Kapadia SR, et al. Non-HDL Cholesterol and Triglycerides: Implications for Coronary Atheroma Progression and Clinical Events. Arterioscler Thromb Vasc Biol. 2016;36(11):2220-8.
14. Lee S-Y, Hahm S-K, Park J-A, Choi S-K, Yoon J-Y, Choi S-H, et al. Measuring low density lipoprotein cholesterol: comparison of direct measurement by HiSens Reagents and Friedewald Estimation. Korean J Fam Med. 2015;36(4):168-73.
15. Bergmann K. Non-HDL cholesterol and evaluation of cardiovascular disease risk. EJIFCC. 2010;21(3):64-7.
16. Wongcharoen W, Sutthiwutthichai S, Gunaparn S, Phrommintikul A. Is non-HDL-cholesterol a better predictor of long-term outcome in patients after acute myocardial infarction compared to LDL-cholesterol? : A retrospective study. BMC Cardiovasc Disord. 2017;17(1):10.
17. Liu J, Sempos CT, Donahue RP, Dorn J, Trevisan M, Grundy SM. Non-high-density lipoprotein and very-low-density lipoprotein cholesterol and their risk predictive values in coronary heart disease. Am J Cardiol. 2006;98(10):1363-8.

18. Cui Y, Blumenthal RS, Flaws JA, Whiteman MK, Langenberg P, Bachorik PS, et al. Non-high-density lipoprotein cholesterol level as a predictor of cardiovascular disease mortality. Arch Intern Med. 2001;161(11):1413-9.
19. Bittner V, Hardison R, Kelsey SF, Weiner BH, Jacobs AK, Sopko G, et al. Non-high-density lipoprotein cholesterol levels predict five-year outcome in the Bypass Angioplasty Revascularization Investigation (BARI). Circulation. 2002;106(20):2537-42.
20. Ridker PM, Rifai N, Cook NR, Bradwin G, Buring JE. Non-HDL cholesterol, apolipoproteins A-I and B100, standard lipid measures, lipid ratios, and CRP as risk factors for cardiovascular disease in women. JAMA. 2005;294(3):326-33.
21. Liu J, Sempos C, Donahue RP, Dorn J, Trevisan M, Grundy SM. Joint distribution of non-HDL and LDL cholesterol and coronary heart disease risk prediction among individuals with and without diabetes. Diabetes Care. 2005;28(8):1916-21.
22. Pischon T, Girman CJ, Sacks FM, Rifai N, Stampfer MJ, Rimm EB. Non-high-density lipoprotein cholesterol and apolipoprotein B in the prediction of coronary heart disease in men. Circulation. 2005;112(22):3375-83.
23. Kastelein JJP, van der Steeg WA, Holme I, Gaffney M, Cater NB, Barter P, et al. Lipids, apolipoproteins, and their ratios in relation to cardiovascular events with statin treatment. Circulation. 2008;117(23):3002-9.
24. Ivanova EA, Myasoedova VA, Melnichenko AA, Grechko AV, Orekhov AN. Small Dense Low-Density Lipoprotein as Biomarker for Atherosclerotic Diseases. Oxid Med Cell Longev. 2017;2017:1273042.
25. Campos H, Blijlevens E, McNamara JR, Ordovas JM, Posner BM, Wilson PW, et al. LDL particle size distribution. Results from the Framingham Offspring Study. Arterioscler Thromb. 1992;12(12):1410-9.
26. Manjunath CN, Rawal JR, Irani PM, Madhu K. Atherogenic dyslipidemia. Indian J Endocrinol Metab. 2013;17(6):969-76.
27. Hoogeveen RC, Gaubatz JW, Sun W, Dodge RC, Crosby JR, Jiang J, et al. Small dense low-density lipoprotein-cholesterol concentrations predict risk for coronary heart disease: the Atherosclerosis Risk In Communities (ARIC) study. Arterioscler Thromb Vasc Biol. 2014;34(5):1069-77.
28. Sniderman AD, Islam S, Yusuf S, McQueen MJ. Discordance analysis of apolipoprotein B and non-high density lipoprotein cholesterol as markers of cardiovascular risk in the INTERHEART study. Atherosclerosis. 2012;225(2):444-9.
29. Lavie CJ, Milani RV, O'Keefe JH. To B or Not to B: Is Non–High-Density Lipoprotein Cholesterol an Adequate Surrogate for Apolipoprotein B? Mayo Clin Proc. 2010;85(5):446-50.
30. Sarat Chandra K, Bansal M, Nair T, Iyengar SS, Gupta R, Manchanda SC, et al. Consensus statement on management of dyslipidemia in Indian subjects. Indian Heart J. 2014;66(Suppl 3):S1-51.

CHAPTER
12

Approach to the Patient with Lipid Disorders

Partha Pratim Chakraborty, Avivar Awasthi

ABSTRACT
Dyslipidemia or disorders of lipoprotein metabolism is broadly classified into two groups: primary (genetic/inherited) and secondary (acquired). There are certain clues which usually point toward a genetic etiology such as grossly deranged lipid profile, onset of lipid abnormality at a young age, strong family history of premature coronary heart disease (CHD) or pancreatitis, or similar lipid abnormalities and/or clinical manifestations in one or more first degree relatives. Diet, medications, and a number of chronic systemic diseases may alter lipid parameters, giving rise to secondary dyslipidemia. To add to this complexity, secondary causes may coexist with or may exacerbate underlying primary disorder of lipoprotein metabolism. Depending on abnormal lipid parameters, dyslipidemia increases the risk for atherosclerotic cardiovascular disease or acute pancreatitis. In addition, they may have characteristic systemic manifestations. Monogenic defects of lipoprotein metabolism are very rare and effort should be made to rule out appropriate secondary causes in every patient with dyslipidemia. A stepwise approach that starts with characterization of abnormal lipid parameters, followed by meticulous history and vigilant systemic examination is often sufficient to reach a diagnosis of dyslipidemia in the majority. Special or advanced investigations are required rarely, and are usually considered in suspected genetic etiologies of dyslipidemia.

INTRODUCTION
Lipids are hydrophobic compounds having diverse roles in normal physiology and are essential for life. Lipoproteins are large macromolecular complexes of lipids and proteins (apolipoproteins) that are necessary for transport of the poorly soluble lipids and fat soluble vitamins in the circulation. The lipoproteins facilitate absorption of dietary lipids in the small intestine and regulate transport of lipids from intestine and liver to peripheral tissues. A robust transport system, facilitated by lipoproteins, is also present for delivery of lipids from peripheral tissues to the liver and intestine, which is called the "reverse cholesterol transport". Disorders of lipoprotein metabolism are collectively termed as "dyslipidemias". In modern day clinical practice, "lipid profile" is a commonly ordered laboratory test. It is most frequently used in determining the atherosclerotic cardiovascular

disease (ASCVD) event risk in appropriate population. The test is also advised in patients with established ASCVD for evaluating the efficacy of treatment with lipid-lowering therapies and to monitor adherence to the same. The test is also instrumental in identifying the underlying etiology of certain clinical conditions such as acute pancreatitis and different forms of xanthomas. Individuals with a positive family history of grossly abnormal lipid parameters in one or more first degree relatives often are tested to rule out monogenic defects in lipoprotein metabolism. These individuals, even if asymptomatic, are at a very high risk of developing similar lipid abnormality and consequences thereof; hence, require screening. There are various systemic disorders which may also have an impact on the lipid profile. Dyslipidemia in the majority of patients, however, is interplay between genetic predisposition, often polygenic in nature, and environmental contribution. In this chapter, we shall focus on evaluation of primary (genetic causes) and secondary (acquired form) dyslipidemias. The management of dyslipidemia has been discussed in details in the section entitled "Therapeutic Lipidology".

EVALUATION

Dyslipidemia is a biochemical diagnosis. Evaluation for dyslipidemia starts in a retrograde fashion once an abnormal lipid profile is encountered. Most of the laboratories measure lipid profile from serum. Anticoagulants exert osmotic effects, in which, water leaves the cells and enters the plasma, thus diluting the plasma and lowering the concentrations of the non-diffusible components. The magnitude of this effect depends on the anticoagulant used (oxalate or citrate exert greater osmotic effect) and its concentration. Thus, the serum concentrations of lipids and lipoproteins probably reflect more accurately the subjects' physiological state at the time of venipuncture. A standard lipid profile therefore includes measurements of serum concentrations of total cholesterol (TC), triglyceride (TG), high-density lipoprotein cholesterol (HDL-C), and low-density lipoprotein cholesterol (LDL-C). TC and TG concentrations are about 3–5% higher in serum than in EDTA-plasma, although no significant serum-plasma difference is observed for HDL-C. TC, HDL-C and TG are measured directly, while LDL-C can either be measured directly or calculated indirectly by the Friedewald equation, only if TG concentration is less than 400 mg/dL. Traditionally, the Friedewald equation has been applied to a fasting lipid profile using the following formula with values in mg/dL:

$$LDL\text{-}C = [TC-(HDL\text{-}C + TG/5)]$$

These standard parameters of a lipid profile report may be supplemented with non-HDL-C, calculated as (TC-HDL-C) and remnant cholesterol/TG-rich lipoprotein cholesterol/very low-density lipoprotein (VLDL) cholesterol (VLDL-C), calculated as (TG in mg/dL/5). This calculation is based on the principle that VLDL particles contain ≈1 mg of cholesterol for every 5 mg of TG. Both non-HDL-C and VLDL-C are obtained from the existing lipid profile report. Measurements of apolipoprotein B (ApoB) and apolipoprotein AI (ApoA-I) are used as alternatives to non-HDL-C and HDL-C concentrations respectively. ApoB also includes Lp(a), a lipoprotein produced by the liver, consisting of an LDL particle with which apolipoprotein A has been covalently linked to its ApoB-100 moiety by a disulfide linkage. Lp(a) is used as an additional marker for ASCVD risk prediction and is suggested to be used in patients with premature CHD, a strong family history of premature CHD, resistance to LDL-C lowering with statins, familial hypercholesterolemia (FH), aortic valvular stenosis of uncertain cause and in patients with an unknown cause of ischemic stroke or in patients with intermediate risk profiles.[1]

Two most important consequences of dyslipidemia are ASCVD, associated with hypercholesterolemia and acute pancreatitis, related to hypertriglyceridemia. Very severe hypertriglyceridemia (TG >2,000 mg/dL) increases the risk of pancreatitis by manifold, where as severe hypertriglyceridemia (TG: 1,000–1,999 mg/dL), although usually not causative for pancreatitis, indicates risk for development of very severe form, and hence, pancreatitis. On the other hand, TG concentration between 150–1,000 mg/dL is associated with ASCVD risk.[2] In clinical practice, lipid profile has conventionally been measured after fasting for at least 8 hours. Dietary lipids are transported by chylomicrons, a TG rich lipoprotein, which is subsequently converted to chylomicron remnants and are cleared from circulation by the liver. Few, if any, chylomicrons or chylomicron remnants are normally present in circulation after a 12-hour fast, except in patients with abnormal lipid metabolism. However, post-prandial states normally predominate over a 24-hour period and therefore, fasting lipid parameters may not reflect the daily average plasma lipid and lipoprotein concentrations and its associated ASCVD risks. Interestingly, evidence regarding superiority of fasting lipid profile over nonfasting sample for ASCVD risk assessment is lacking, where as simplicity, convenience, and compliance are obvious advantages of sampling in nonfasting state. A number of studies have compared nonfasting with fasting lipid profiles, and noticed clinically insignificant minor increases in plasma TG, minor decreases in TC and LDL-C and no changes in HDL-C concentrations in the former.[3] Hence, some societies have recommended nonfasting lipid measurement in routine practice of late. UK NICE (National Institute for Health and Care Excellence) guidelines have endorsed nonfasting lipid testing, that include measurement of TC, HDL-C, non-HDL-C, and TG concentrations for the primary prevention of ASCVD since 2014.[4] The 2013 American College of Cardiology/American Heart Association (ACC/AHA) guidelines also do not require fasting for ASCVD risk estimation.[5] However, the Endocrine Society strictly recommends fasting lipid profile for evaluation of hypertriglyceridemia.[2]

To summarize, a non-fasting sample is good enough for ASCVD risk calculation if the TG concentration is <450 mg/dL. However, if a primary abnormality of lipid metabolism is suspected, or an individual is being evaluated specifically for hypertriglyceridemia like those with acute pancreatitis or with insulin resistance or if the TG concentration in a non-fasting sample is ≥450 mg/dL, a fasting lipid profile should be obtained. It also needs to be remembered that the diagnosis of hypertriglyceridemia is based on fasting levels where the length of fast is recommended to be 12 hours and during this period, intake of liquids without caloric content is acceptable. Patients with known hypertriglyceridemia, who are being followed up, should also be tested in fasting state.

Once in a while, lipid parameters may be altered artifactually due to assay interference. For an example, grossly abnormal lipid profile has been encountered with severe hyperbilirubinemia (>20 mg/dL), hemolysis, and paraproteinemia. Such preanalytical errors need to be excluded before embarking on thorough evaluation.[6]

The first step in evaluating a patient with dyslipidemia is to identify the underlying primary defect, i.e., whether the predominant abnormality is hypercholesterolemia or hypertriglyceridemia. The second step is to rule out relevant secondary causes of such lipid abnormalities. The third step is to determine whether the patient needs further evaluation or not for that abnormality and the last step is to decide on treatment. Evaluation of dyslipidemia starts with a detailed history and thorough clinical examination. A genetic cause needs to be considered, if lipid profile is grossly deranged, the onset of the lipid abnormality is at a young age, or if there is positive family history of premature CHD, or pancreatitis or clinical manifestations similar to the index case. A previous normal lipid profile, if available, effectively rules out genetic defects of lipoprotein metabolism and secondary cause(s) for dyslipidemia should be actively searched for. Patients of primary dyslipidemia may also experience worsening of their lipid profile due to coexistent secondary causes such as diet, drugs, infections, endocrine diseases, chronic systemic illnesses, rheumatological disorders, and malignancies.

The supernatant serum is milky or creamy and whole blood appears pinkish in patients with severe hypertriglyceridemia and the laboratory, testing the sample, usually mentions this in the report. If the fasting TG is >1,000 mg/dL, the patient is likely to have chylomicronemia. Peak TG (in mg/dL): Cholesterol (in mg/dL) ratio of more than 5 indicates, while a value of >10 strongly suggests familial chylomicronemia syndrome.[7] Very high LDL-C (>200 mg/dL), in absence of a secondary etiology, points toward genetic causes of hypercholesterolemia. In absence of lipid lowering therapy, severe malnutrition or serious chronic illness, LDL-C of <60 mg/dL in an otherwise healthy individual also suggests an inherited condition.[8] Similarly HDL-C <20 mg/dL in absence of hypertriglyceridemia is a consequence of severe alterations in HDL metabolism of genetic origin.[9]

History

A meticulous history for possible offending drug use/misuse/abuse not only helps to reach an early diagnosis, it also avoids unnecessary investigations. Addressing the underlying cause also greatly improves the lipid abnormality without lipid lowering agents. Androgenic anabolic steroids and testosterone are often associated with low HDL-C, profound at times, and increase in LDL-C. This history should be actively sought for in sportsmen, weight lifters and body builders in particular, and those suffering from catabolic diseases such as malignancy, human immunodeficiency virus (HIV) infection and chronic kidney disease (CKD). Antiretroviral protease inhibitors are associated with hypertriglyceridemia. Elevated TG is also encountered with excess alcohol intake, anti-hypertensive medications, such as thiazide diuretics and β-blockers, steroids such as estrogen and glucocorticoid, other immunosuppressants, bile acid-binding resins and isotretinoin and antipsychotics. Elevated TG is almost always associated with low HDL-C, and the unusual combination of elevated TG and elevated HDL-C suggests chronic alcoholism or estrogen administration. High carbohydrate diet, especially those rich in simple carbohydrates, drives hepatic VLDL secretion, and thus, elevates TG. Diet, high in saturated and trans fats, on the other hand, reduces LDL receptor activity and raises LDL-C. As discussed earlier, family history of premature CHD, pancreatitis, lipid abnormalities, particularly in younger age group, should be actively sought for.

Clinical Examination

Insulin resistance is characterized by elevated fasting TG [secondary to excess hepatic VLDL production and decreased lipoprotein lipase (LPL) activity], low HDL-C, variable elevations of LDL-C, elevated small dense LDL, and elevated non-HDL-C or ApoB. Similarly, patients of Cushing's syndrome, either exogenous or endogenous, frequently have high TG, low HDL-C, and variable elevation of LDL-C. Hypothyroidism, cholestatic liver disease, and nephrotic syndrome are associated with predominant elevation of TC, LDL-C, at times severe, with mild hypertriglyceridemia (**Fig. 1**). Hepatitis of any etiology gives rise to mild to moderate hypertriglyceridemia and liver failure is associated with grossly reduced cholesterol and TG. Altered lipid parameters, thus, may be utilized for a more focused clinical examination. Body mass index (BMI) needs to be measured in every case. A thorough general and systemic examination for signs of insulin resistance (acanthosis nigricans, skin tags), hirsutism, lipodystrophy, hypothyroidism, Cushing's syndrome, chronic liver disease, CKD, nephrotic syndrome, and autoimmune disorders should be undertaken. Partial forms of lipodystrophy are often missed; hence, patients should be evaluated after proper exposure (**Fig. 2**). The type of xanthomas, at times, gives an idea about the nature of underlying lipid disorders.[10,11] Severe elevation of TG is associated with eruptive xanthomas, which are small yellowish-white papules arising from an erythematous base over the trunk, back, buttocks, thighs, and extensor surfaces of arms and legs in clusters (**Fig. 3**). These lesions usually are asymptomatic, but may become pruritic. Tendon, planar (wide-based yellowish macules or plaques) or tuberous (painless firm nodules over extensor surfaces of joints) xanthomas, particularly in children, suggest elevated cholesterol and LDL-C as encountered in FH (**Fig. 4**). Dorsum of the hands and the achilles tendons are the common sites for tendon xanthomas (**Fig. 5**). Xanthoma striata palmaris appearing as yellow-orange macular discoloration of the palmar and finger creases and tuberoeruptive xanthomas are pathognomonic of familial dysbetalipoproteinemia (FDBL). Severe cholestasis, however, may also have xanthomas identical to those seen in FDBL.

A detailed ophthalmological examination is also of help. About 60% of patients with planar xanthomas over the upper eyelids (xanthelasma palpebrarum) (**Fig. 6**) have an

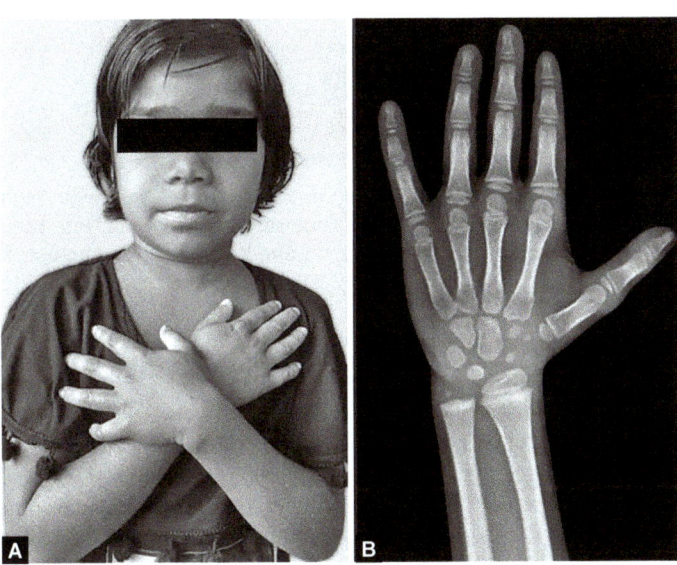

FIG. 1: Grossly elevated total cholesterol (>500 mg/dL) and low-density lipoprotein cholesterol (>300 mg/dL) and raised triglyceride (630 mg/dL) in a 9-year-old girl with primary hypothyroidism due to chronic Hashimoto's thyroiditis having a bone age of around 5 years. Lipid parameters normalized completely, once euthyroidism was achieved with levothyroxine.

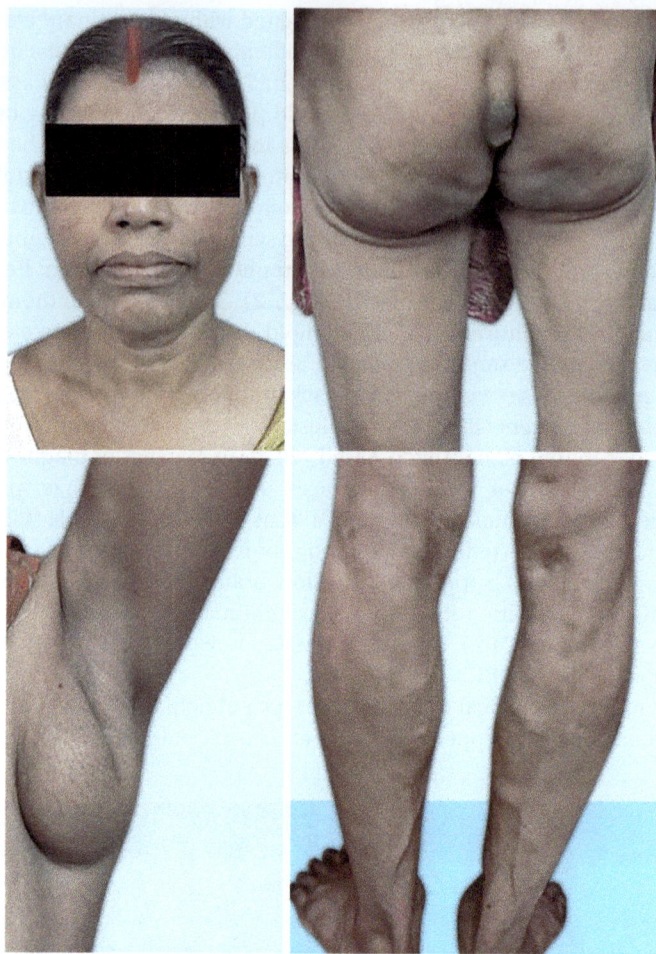

FIG. 2: Partial lipodystrophy was diagnosed in a lady presenting with high triglyceride (532 mg/dL) and low high-density lipoprotein cholesterol (23 mg/dL). Note the normal facial appearance, loss of subcutaneous fat from buttocks and lower limbs and ectopic fat deposition in axilla.

association with hypercholesterolemia. Corneal arcus in a young also suggest underlying primary hypercholesterolemia. Corneal opacification is the hallmark physical finding of lecithin–cholesterol acyltransferase (LCAT) deficiency, both complete deficiency and partial deficiency (fish eye disease). These patients develop slowly progressive corneal clouding accompanied by a white or gray ring in the corneal margin, similar to arcus senilis. Despite the dramatic appearance of the cornea, vision remains intact in most cases. Lipemia retinalis (opalescent or creamy retinal blood vessels) on direct fundoscopy is encountered in some patients with severe hypertriglyceridemia. Patients with abetalipoproteinemia develop progressive pigmented retinopathy with decreased night and color vision, followed by reduction in daytime visual acuity and ultimately progressing to near-blindness.

Approach to the Patient with Lipid Disorders

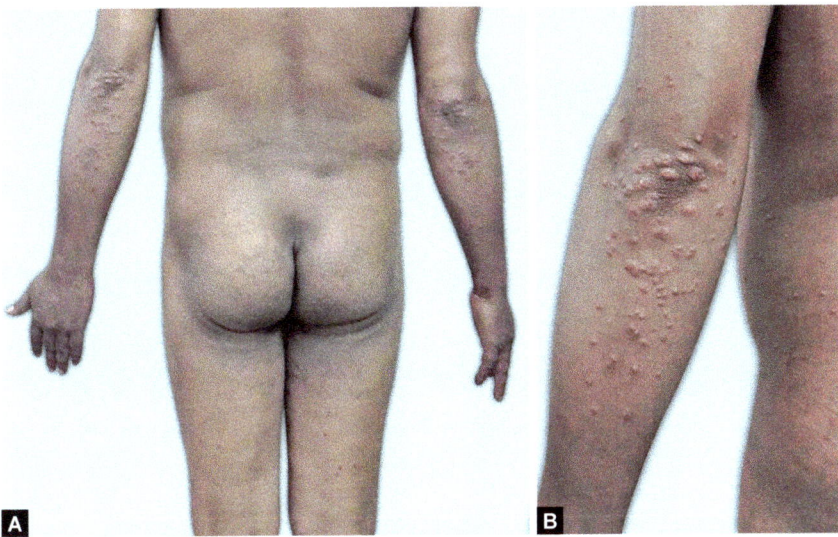

FIG. 3: Eruptive xanthomas in a young with newly diagnosed type 2 diabetes mellitus (venous plasma glucose: 653 mg/dL). His fasting triglyceride was 1,278 mg/dL.

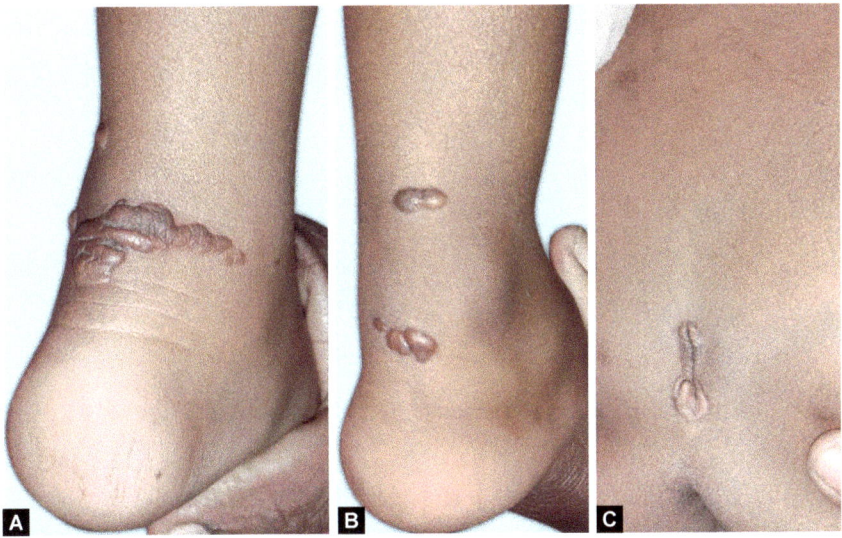

FIG. 4: Planar and tuberous xanthomas in a 3-year-old girl with familial hypercholesterolemia (FH) (homozygous/compound heterozygous) with total cholesterol: 1,076 mg/dL, triglyceride: 158 mg/dL, low-density lipoprotein cholesterol: 737 mg/dL and high-density lipoprotein cholesterol: 134 mg/dL. Both parents and both of her siblings (aged 10 years and 14 years) had deranged lipid parameters, though to a lesser extent.

Patients with ApoA-I deficiency, may present with cutaneous planar xanthomas and mild corneal opacification due to accumulation of cholesterol from impaired peripheral cellular efflux. The hallmark physical findings in Tangier disease [homozygous adenosine

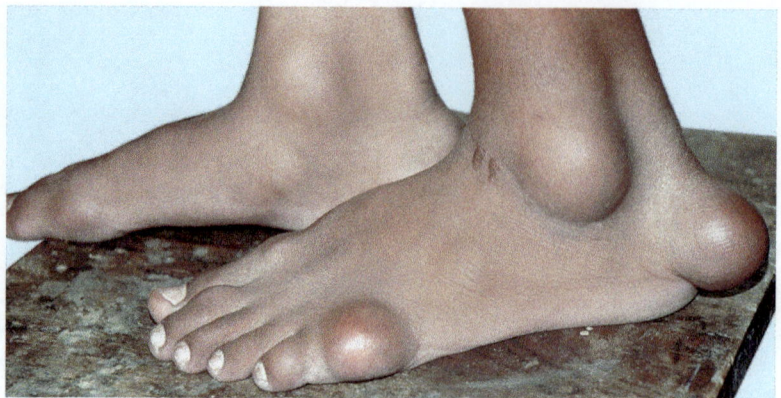

FIG. 5: Tendon xanthomas in a girl with familial hypercholesterolemia (FH).

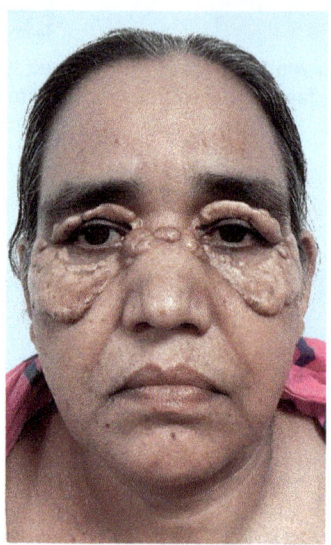

FIG. 6: Xanthelasma palpebrarum in a 45-year-old lady with suspected familial combined hyperlipidemia (FCHL) with total cholesterol: 252 mg/dL, triglyceride: 245 mg/dL, low-density lipoprotein cholesterol: 146 mg/dL and high-density lipoprotein cholesterol: 57 mg/dL.

triphosphate (ATP)-binding cassette transporter protein 1 (*ABCA1*) gene mutation] are enlarged yellow-orange tonsils. Similar discoloration of rectal mucosa is also observed due to intracellular accumulation of lipophilic retinyl esters and carotenoids. Mucocutaneous hyperpigmentation of primary adrenal insufficiency, if coexists with hypercholesterolemia suggests Wolman's disease, the severe form of cholesteryl ester storage disease, due to deficiency of lysosomal acid lipase.

Some forms of primary lipid disorders may be associated with neurological manifestations. Cerebrotendinous xanthomatosis (CTX) develops secondary to an abnormal *CYP27A1* gene and resultant deficiency of the mitochondrial enzyme sterol 27-hydroxylase, an enzyme required for bile acid synthesis. These patients often have cerebellar ataxia, dementia, and spinal cord involvement with tendon xanthomas. Abetalipoproteinemia is initially manifested by loss of deep tendon reflexes, followed by decreased lower extremity large fiber senses, dysmetria, ataxia, and development of a spastic gait due to deficiency of vitamin E and other fat soluble vitamins. Abetalipoproteinemia, thus, mimics spinocerebellar ataxia and Friedreich's ataxia. Peripheral neuropathy commonly manifests as a remitting-relapsing multifocal neuropathy (mononeuritis multiplex) with sensory abnormalities in almost half of patients with Tangier disease. Rarely, Tangier disease is associated with a syringomyelia like presentation.

Laboratory Parameters

Lipid profile, if abnormal, should always be supplemented with complete blood count and examination of peripheral blood smear, serum creatinine, liver function tests, urine protein estimation, measurement of thyroid stimulating hormone (TSH), and fasting plasma glucose. Sitosterolemia is associated with hemolytic anemia, anisocytosis/poikilocytosis, and megathrombocytes due to accumulation of plant sterols into cell membranes. Complete LCAT deficiency may also be associated with hemolytic anemia, secondary to increased erythrocyte fragility due to abnormal membrane lipid composition. Acanthocytes in blood smears are suggestive of abetalipoproteinemia. In complete LCAT deficiency, proteinuria manifests in early childhood, which gradually progresses to nephrotic syndrome and end-stage CKD.

Patients of familial combined hyperlipidemia (FCHL), FH, familial defective ApoB-100 (FDB), autosomal dominant hypercholesterolemia type 3, autosomal recessive hypercholesterolemia, sitosterolemia, FDBL, ApoA-I deficiency, and Tangier disease are at very high risk of premature CHD; hence, should be evaluated with echocardiography, carotid artery doppler study and coronary angiography, either invasive or noninvasive. Atherosclerosis often starts with involving the aortic root; hence, echocardiography should focus on that area to look for valvular or supravalvular stenosis.

Special Tests

These tests are not done routinely and prescribed only in special circumstances. Though genetic analysis is confirmatory for the monogenic form of primary dyslipidemia, mutational analysis is seldom performed. The list of the special tests includes the following:

Advanced Lipoprotein Tests to Estimate ASCVD Risk: Lipoprotein(a), ApoB concentration (denotes atherogenic lipoprotein burden), ApoA-I concentration, LDL size, and LDL particle number (useful in insulin resistance and other conditions associated with elevated small dense LDL; LDL-C concentrations in these situations are not reliable marker of LDL burden).

Differentiating Lipoprotein Lipase Deficiency from ApoC-II Deficiency in Patients with Chylomicronemia: Triglyceride lipolytic activity is assayed after intravenous administration of heparin (60 U/Kg), which releases the endothelial bound LPL. LPL activity is profoundly reduced in both LPL and ApoC-II deficiency. However, after addition of normal plasma, that provides a source of ApoC-II, activity normalizes in the later, but, not in the former. The diagnosis is confirmed with molecular sequencing of genes.

Diagnosis of Sitosterolemia in Patients with Grossly Elevated LDL-C: Diagnosis is suspected when LDL-C elevation is associated with episodes of hemolysis and splenomegaly. A dramatic response to dietary modification and/or ezetimibe, but not with statins is characteristic and diagnosis is confirmed by elevated plasma plant sterols such as sitosterol and campesterol.

Diagnosis of FDBL in Patients with Elevated Total Cholesterol and TG: Non-HDL-C/ApoB ratio is a simple screening test for FDBL.[12] Remnant lipoproteins may be identified by different methods such as β-quantification by ultracentrifugation, lipoprotein electrophoresis or nuclear magnetic resonance lipoprotein profiling. High levels of remnant lipoproteins are suggested by higher ratio (>0.3) of directly measured VLDL-C to total plasma TG or a broad β band on electrophoresis. The diagnosis is confirmed by homozygosity for ApoE-2.

Primary monogenic defect of lipoprotein metabolism is rarely encountered in routine practice. However, the patterns of lipid abnormality are often useful to narrow down differential diagnoses of primary dyslipidemia and have been summarized in **Table 1**.

TABLE 1: Possible lipid alterations in monogenic cause of abnormal lipid profile.						
Very high TG, low HDL-C, normal/low LDL-C	High LDL-C, normal TG	High TG, high cholesterol	Isolated low HDL-C	Low LDL-C	Grossly elevated HDL-C
LPL deficiency	Familial hypercholesterolemia	Familial dys-betalipoproteinemia	ApoA-I deficiency	Abetalipoproteinemia	Loss of function of CETP
ApoC-II deficiency	Familial defective ApoB-100	Familial combined hyperlipidemia	ApoA-I structural mutations	Familial hypobetalipoproteinemia	
ApoC-III overproduction	Autosomal dominant Hypercholesterolemia type 3 (gain-of-function mutation of PCSK9)	Hepatic lipase deficiency	ABCA1 deficiency (Tangier disease)	PCSK9 deficiency	
ApoA-V deficiency	Autosomal recessive hypercholesterolemia		LCAT deficiency		
GPIHBP1 deficiency	Sitosterolemia		Elevated CETP activity		

Continued

Approach to the Patient with Lipid Disorders

Continued

TABLE 1: Possible lipid alterations in monogenic cause of abnormal lipid profile.					
Very high TG, low HDL-C, normal/ low LDL-C	High LDL-C, normal TG	High TG, high cholesterol	Isolated low HDL-C	Low LDL-C	Grossly elevated HDL-C
LMF-1 deficiency					
Familial hypertri-glyceridemia					

[ABCA1: adenosine triphosphate (ATP)-binding cassette transporter protein 1; APOA-V: apolipoprotein A-V (facilitates association of LPL with its substrates; ApoC-II: apolipoprotein C-II (Cofactor for LPL); CETP: cholesteryl ester transfer protein (It transfers HDL-cholesteryl to apoB-containing lipoproteins in exchange for TG; GPIHBP1: glycosylphosphatidylinositol anchored high-density lipoprotein binding protein 1 (A protein, that anchors LPL to vascular endothelium); LMF-1: Lipase maturation factor 1 (An intracellular protein required for correct intracellular folding and activation of LPL); HDL-C: high-density lipoprotein (HDL) cholesterol; LCAT: lecithin-cholesterol acyltransferase (A plasma enzyme associated with HDL, that esterifies free cholesterol); LDL-C: low-density lipoprotein cholesterol; LPL: lipoprotein lipase; PCSK9: proprotein convertase subtilisin/kexin type 9; TG: triglyceride].

KEY POINTS

- Dyslipidemia is widely prevalent in routine practice and develops due to primary (genetic) or secondary (acquired) defects in lipoprotein metabolism. An acquired insult in a genetically predisposed individual is the most common underlying cause of abnormal lipid parameters.
- Dyslipidemia is a biochemical diagnosis and evaluation starts with detailed history and thorough clinical examination, once abnormal lipid profile is encountered. Secondary etiologies such as drug, endocrine disorders, and other chronic systemic illnesses should be actively searched for in each and every case of dyslipidemia to avoid unnecessary investigations and therapeutic misadventures.
- A primary (genetic) cause of dyslipidemia is rare and should be suspected in all young patients with grossly deranged lipid parameters, in individuals with strong family history of premature ASCVD, family history of pancreatitis, family history of dyslipidemia in a young relative or similar clinical manifestations in one or more relatives.
- Clinical examination should focus on BMI, features of insulin resistance, lipodystrophy, and presence of xanthomas after adequate exposure of the patient. Signs suggestive of hypothyroidism, chronic liver disease, and chronic kidney disease need to be searched for in every case.
- A fasting lipid profile comprising of TC, TG, LDL-C and HDL-C is sufficient to reach an etiological diagnosis in most cases of dyslipidemia. Advanced tests are required rarely, to reach a definite diagnosis, only when a monogenic defect of lipoprotein metabolism is strongly suspected. All patients with dyslipidemia should be advised for complete blood count, venous plasma glucose, renal function tests, LFT, and thyroid profile.

REFERENCES

1. Cegla J, Neely RDG, France M, Ferns G, Byrne CD, Halcox J, et al. HEART UK consensus statement on Lipoprotein(a): A call to action. Atherosclerosis. 2019;291:62-70.
2. Berglund L, Brunzell JD, Goldberg AC, Goldberg IJ, Sacks F, Murad MH, et al. Evaluation and treatment of hypertriglyceridemia: an Endocrine Society clinical practice guideline. J Clin Endocrinol Metab. 2012;97(9):2969-89.
3. Nordestgaard BG, Langsted A, Mora S, Kolovou G, Baum H, Bruckert E, et al. Fasting is not routinely required for determination of a lipid profile: clinical and laboratory implications including flagging at desirable concentration cut-points—a joint consensus statement from the European Atherosclerosis Society and European Federation of Clinical Chemistry and Laboratory Medicine. Eur Heart J. 2016;37(25):1944-58.
4. NICE clinical guideline. Cardiovascular disease: risk assessment and reduction, including lipid modification. [online] Available from www.nice.org.uk/guidance/cg181. [Last accessed June, 2020].
5. Stone NJ, Robinson JG, Lichtenstein AH, Bairey Merz CN, Blum CB, Eckel RH, et al. 2013 ACC/AHA guideline on the treatment of blood cholesterol to reduce atherosclerotic cardiovascular risk in adults: a report of the American College of Cardiology/American Heart Association Task Force on Practice Guidelines. J Am Coll Cardiol. 2014;63:2889-934.
6. Tsai LY, Tsai SM, Lee SC, Liu SF. Falsely low LDL-cholesterol concentrations and artifactual undetectable HDL-cholesterol measured by direct methods in a patient with monoclonal paraprotein. Clin Chim Acta. 2005; 358(1-2):192-5.
7. Stroes E, Moulin P, Parhofer KG, Rebours V, Löhr JM, Averna M. Diagnostic algorithm for familial chylomicronemia syndrome. Atheroscler Suppl. 2017;23:1-7.
8. Rader DJ, Hobbs HH. Disorders of lipoprotein metabolism. In. Jameson JL (Ed). Harrison's Endocrinology, 4th edition. New York: McGraw Hill Education (India); 2017. pp. 339-60.
9. Rader DJ, deGoma EM. Approach to the patient with extremely low HDL-cholesterol. J Clin Endocrinol Metab. 2012;97(10):3399-407.
10. White LE. Xanthomatoses and lipoprotein disorders. In: Wolff K, Goldsmith LA, Katz SI, Gilchrest BA, Paller AS, Leffell DJ (Eds). Fitzpatrick's Dermatology in General Medicine, 7th edition. New York: McGraw-Hill; 2008. pp. 1272-81.
11. Flynn PD. Xanthomas and abnormalities of lipid metabolism and storage. In: Burns T, Breathnach S, Cox N, Griffiths C (Eds). Rook's Textbook of Dermatology, 8th edition. Oxford: Blackwell Science; 2010. pp. 59.81-59.103.
12. Boot CS, Middling E, Allen J, Neely RDG. Evaluation of the Non-HDL cholesterol to apolipoprotein B ratio as a screening test for dysbetalipoproteinemia. Clin Chem. 2019;65:313-20.

CHAPTER
13

Laboratory Assessment of Lipoproteins: Principles and Pitfalls

Rana Bhattacharjee

ABSTRACT

Lipoproteins are measured in the routine clinical practice for the diagnosis of a number of conditions including the assessment of risk for cardiovascular diseases (CVDs). They are also useful for monitoring the effectiveness of therapeutic interventions. Electrophoresis, ultracentrifugation, precipitation, and other methods are used to measure lipoproteins in lipid research laboratories and clinical laboratories. Recently, point of care assessment of lipoproteins has also become available. However, one must be aware of various pitfalls associated with these assays.

INTRODUCTION

As cholesterol and triglycerides are insoluble in water, they are carried in plasma by binding with lipoproteins which also contains phospholipids and proteins called apolipoprotein. The various types of lipoproteins are high-density lipoprotein (HDL), intermediate-density lipoprotein (IDL), low-density lipoprotein (LDL), very low-density lipoprotein (VLDL), chylomicron and its remnant, and lipoprotein(a) [Lp(a)]. Some of them are further classified into various subclasses (discussed later). In this chapter, we will mainly discuss the issues related to the measurement of the cholesterol component of lipoproteins. We will also briefly discuss the methods of assessment of size, and number of the circulating lipoproteins and the concentration of apolipoproteins.

TOTAL CHOLESTEROL

Cholesterol is commonly measured by automated analyzers using enzymatic methods. In National Health and Nutrition Examination Survey (NHANES) 2003–2004, cholesterol was measured enzymatically by a series of reactions catalyzed by cholesteryl ester hydrolase and cholesterol oxidase. One of the byproducts of the reactions, H_2O_2, was measured by the intensity of color production by peroxidase catalyzed reaction. The intensity of the color was proportional to the concentration of cholesterol.[1] A similar principle is followed in the analyzers used in the recent NHANES.[2,3]

- Cholesteryl ester + H_2O ------------> cholesterol + fatty acid (by cholesteryl ester hydrolase)
- Cholesterol + O_2 ------------> cholest-4-en-3-one + H_2O_2 (by cholesterol oxidase)
- $2H_2O_2$ + 4-aminophenazone + phenol ------------> 4-(p-benzoquinonemonoimino)-phenazone + $4 H_2O$ (by Peroxidase)

TRIGLYCERIDES

Triglycerides are also commonly measured enzymatically in serum or plasma using a series of reactions in which glycerol is produced from hydrolyzation of triglycerides. H_2O_2 is produced by oxidation of glycerol, which is measured by the intensity of color production by peroxidase reaction.[1]

- Triglycerides + $3H_2O$ ------------> glycerol + fatty acids (by lipase)
- Glycerol + ATP ------------> glycerol-3-phosphate + ADP (by glycerokinase)
- Glycerol-3-phosphate + O_2 ------------> dihydroxyacetone phosphate + H_2O_2 (by glycerophosphate oxidase)
- H_2O_2 + 4-aminophenazone + 4-chlorophenol ------------> 4-(p-benzoquinone-monoimino)-phenazone + $2H_2O$ + HCl (by peroxidase)

HIGH-DENSITY LIPOPROTEIN CHOLESTEROL

Electrophoresis

Back in the 1940s and 1950s, paper electrophoresis was used to separate the lipoproteins. The HDL migrates to the alpha position.[4] Fredrickson classification of dyslipidemia (types I, IIa, IIb, III, IV, and V) was based on semiquantitative estimation of serum lipoproteins by the paper electrophoresis pattern along with the ultracentrifugation technique (discussed later).[5,6] Two-dimensional gel electrophoresis uses agarose and polyacrylamide gels to separate lipoproteins by the charge and the size along the horizontal and vertical axis respectively. Twelve ApoA1 containing subpopulation of HDL has been found using this method. This method can also detect pre-β migrating HDL particles, which has been found to be associated with the increased risk of CVD.[7,8] Separation of HDL subfractions was also done with HPLC using gel separation column. Subsequently, they were identified and quantified with cholesterol or choline-containing phospholipids estimation.[9]

Ultracentrifugation

This method separates the lipoproteins according to their density. HDL has a density of 1.063–1.21 g/mL. It is subdivided into HDL2 (density 1.063–1.125 g/mL) and HDL3 (density 1.125–1.21 g/mL).[10] (**Fig. 1**) However, separation of lipoproteins by *sequential* floatation ultracentrifugation is technically cumbersome and time-consuming (several hours to days). Therefore, it is rarely performed outside the research laboratory setting. *Single spin density gradient ultracentrifugation* is a relatively rapid method of isolation of major lipoprotein classes. In this method, a step gradient is constructed from NaCl/KBr solutions of variable density. Single ultracentrifugation using swinging bucket rotor leads to separation of bands which are identified and quantified by densitometric scanning. Sub-fractionation of HDL is possible in this method.[11] *Vertical auto profile (VAP)* is a modification of single spin density gradient technique where a vertical rotor is used. As the centrifuge tube remains vertical to the ground, the lipoproteins are separated along

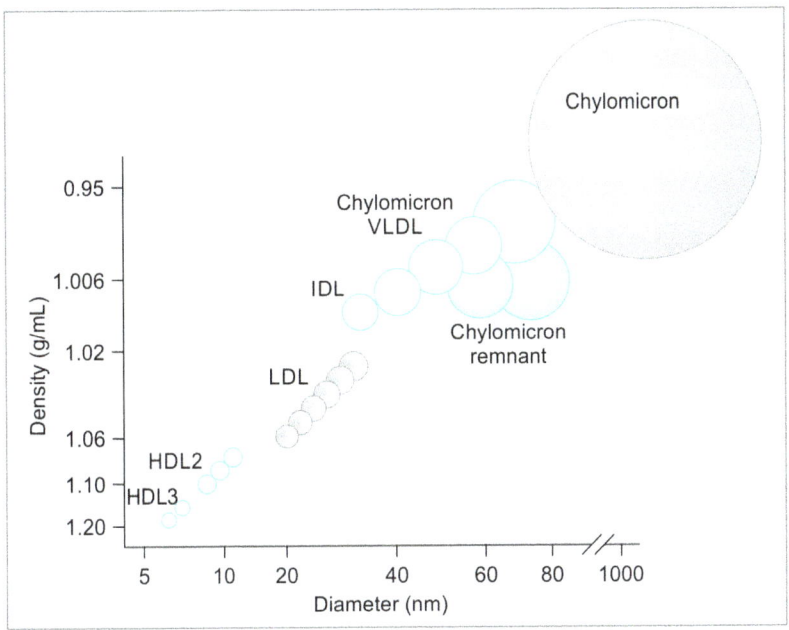

(HDL: high density lipoprotein; IDL: intermediate-density lipoprotein; LDL: low density lipoprotein; VLDL: very low-density lipoprotein)

FIG. 1: Classification of lipoproteins according to the density and size.
Source: Zipes DP, Libby P, Bonow R, Braunwald E. Braunwald's Heart Disease: A Textbook of Cardiovascular Medicine, 7th edition. Philadelphia, PA: Saunders; 2004.

the shorter horizontal axis of the centrifuge tube. It leads to the rapid separation of lipoproteins and their subclasses (typically <1 h). These separated layers are measured for their cholesterol content by the enzymatic reaction.[12]

Precipitation

In this method, lipoproteins containing positively charged ApoB (chylomicrons, chylomicron remnants, VLDL, LDL, and Lp(a) are precipitated by bridge formation with sulfated polysaccharides which is facilitated by divalent cations. These precipitated lipoproteins are subsequently sedimented by centrifugation and HDL is measured in the supernatant. The cholesterol content of the HDL is then measured by the enzymatic method. Heparin manganese and dextran sulfate magnesium have commonly been used for the precipitation.[13,14] Combined heparin and dextran precipitation method has been used to measure total HDL and HDL3 subfraction. This helps in the calculation of HDL2 subfraction value, which is known to have a protective role against CVD.[15]

More recently, a variant of precipitation method, *homogeneous assays* have been developed to directly estimate HDL-C level. These methods do not need any pretreatment of samples making the automation easier. The initial homogeneous method was a four-step process where reagents were added sequentially. As a first step, polyethylene glycol (PEG) was used to precipitate ApoB containing lipoproteins. Antibodies against ApoB and C were used as the second step to block them. Cholesterol reaction enzymes were added next, which acted on free HDL-C (containing ApoA), and subsequently color was generated with 4-aminoantipyrine. As the last step, guanidine salt was used to stop the enzymatic reaction and to solubilize the complexes of ApoB-containing lipoproteins, which would otherwise interfere with the reading of absorbance.[16] Subsequent homogeneous assays used PEG-modified enzyme with cyclodextrin, synthetic polymer with detergents, or catalase with surfactants to reduce the number of steps to make automation even easier.[17]

Some of these rapid methods have a very good correlation with older standard assays. These homogeneous methods are the most commonly used assay for measuring HDL-C in clinical laboratories worldwide. However, lipid research laboratories still prefer standard precipitation methods.

LOW-DENSITY LIPOPROTEIN CHOLESTEROL

Indirect Methods

Indirect methods of measuring cholesterol assume that total cholesterol composed mainly of cholesterol circulating in HDL, LDL, and VLDL. This includes the Friedewald equation and the beta-quantification method.

Friedewald Equation

This equation states: LDL-C = Total cholesterol – HDL-C - VLDL cholesterol. The blood is drawn in a fasting state. The total and HDL cholesterol are measured directly. The approximated value of VLDL cholesterol is calculated by dividing the triglyceride level by 5. This approximation is based on the fact that the fasting plasma contain very little chylomicrons and the usual triglyceride: Cholesterol ratio of VLDL[18] is 5:1. This method should only be applied to a fasting blood sample. In nonfasting patient, the contribution of postprandial chylomicrons to the total lipoprotein pool makes the formula less reliable. Moreover, this formula is valid only when the triglyceride level is <400 mg/dL because such samples can also contain chylomicrons, chylomicron remnants, or VLDL remnants. It has also been found that this formula is most reliable when the triglyceride level is <200.[19] Despite these limitations, Friedewald equation is commonly used in clinical practice for the measurement of LDL-C.

Beta-Quantification

This indirect method of LDL estimation is used when the use of the Friedewald equation is inappropriate. In this method, plasma is first ultracentrifuged at 10°C temperature at 105,000 × g for 18 hours. Under these conditions, VLDL (and, if present, chylomicrons plus β-VLDL) accumulate in a floating layer. The infranatant contains LDL and HDL [plus IDL and Lp(a)]. The infranatant solution is remixed and its cholesterol content measured. HDL is measured in a different aliquot of plasma. LDL value is derived by subtracting HDL-C from cholesterol measured in the infranatant. Although this is a gold standard method of measuring LDL-C, due to its complexity, it is almost exclusively performed in lipid research laboratories.[17]

Direct Methods

Ultracentrifugation and *gel electrophoresis*, as described in the HDL measurement section, can be used for direct measurement of LDL-C too. These are usually not used in routine clinical practice because of the technical complexity of the former and incomplete resolution of VLDL, LDL, and Lp(a) in the latter method. In *selective precipitation method*, polyvinyl sulfate or heparin are used to precipitate LDL at low pH. The precipitated LDL can be directly measured. Alternatively, LDL can be derived from subtracting cholesterol in the supernatant from the total plasma cholesterol level. In the antibody-based selective precipitation method, a polyclonal antibody against ApoE and ApoA-1 linked to resin is used to bind VLDL, IDL, and HDL. The cholesterol in LDL is measured in the filtrate by enzymatic assay. *Homogeneous assays* of LDL-C have been developed with principles similar to HDL-C. In one method, α-cyclodextrinsulfate with $MgCl_2$ has been used to mask cholesterol in chylomicrons and VLDL. Then a second reagent containing polyoxyethylene-polyoxypropylenepolyether (POE-POP) has been used to block cholesterol in HDL. The unblocked LDL-C can be measured by enzymatic reactions. In another method, two types of detergents are used in two steps, one to solubilize non-LDL-C and the other to release cholesterol from LDL to facilitate its measurement. Other methods use catalase (along with other agents) to remove non-LDL particles to directly measure the cholesterol in LDL.[17] Automated homogenous assays are the most commonly used methods of direct LDL measurement in clinical laboratories.

Nuclear magnetic resonance is used to measure LDL particle number, which may be a superior marker for the prediction of the adverse cardiovascular events in comparison to LDL-C. Particle concentrations of lipoproteins of different sizes were calculated from the measured amplitudes of their spectroscopically distinct lipid methyl group NMR signals. This method can also measure the particle number and size of HDL and VLDL.[20] However, it cannot reliably distinguish the Lp(a) from the LDL particle of the same size. However, this method is rarely performed outside the lipid research laboratories because the procedure is technically challenging, and the instruments required for this assay is very expensive.[17]

Small Dense Low-density Lipoprotein

This atherogenic lipoprotein has been measured with a precipitation method where a combination of 150 USP unit/mL of heparin and 90 mmol/L of Mg^{2+} has been used to selectively precipitate lipoproteins with density <1.044 g/mL, i.e., VLDL, IDL, and large, buoyant LDL fractions. After precipitation, the supernatant contains small dense LDL (sdLDL) and HDL (which have a density of >1.044 g/mL). LDL-C in the supernatant (sdLDL) can be directly measured by the homogenous methods.[21]

Oxidized Low-density Lipoprotein

Oxidized LDL (OxLDL) can be measured by enzyme-linked immunosorbent assay (ELISA) method. However, the clinical relevance of this measurement is uncertain.[17]

LIPOPROTEIN A

Initially, Lp(a) cholesterol had been measured by *lectin affinity* and *electrophoresis* followed by immunoblotting. However, because of a lack of sensitivity, these methods are not recommended. Subsequently, *immunoassays* for Lp(a) have been developed. They have been reasonably well standardized for use in clinical and research purposes.[22]

INTERMEDIATE-DENSITY LIPOPROTEIN CHOLESTEROL

Intermediate-density lipoprotein level can be measured either by sequential ultracentrifugation or antibody-based methods.[17]

POINT OF CARE LIPID ANALYSIS

These assays are done in small devices that are capable of measuring cholesterol, triglycerides, and HDL-C, and calculated LDL-C rapidly in nonlaboratory settings from a few microliters of whole blood, serum, or plasma. In one of these methods, HDL lipoprotein is separated in the presence of phosphotungstic acid and magnesium chloride. Total cholesterol, HDL, and triglycerides are measured using dry enzymatic methods.[23]

APOLIPOPROTEINS

Measurement of apolipoprotein has certain advantages over the measurement of the cholesterol content of the lipoproteins. A high level of ApoB-100 indicates a high risk of cerebrovascular (CBD) in an individual with a high triglyceride level or with borderline LDL level. Similarly, in a patient with borderline HDL cholesterol level, low ApoA-1 may identify a patient with higher cardiovascular risk. Apolipoproteins are measured with different types of immunoassays such as radioimmunoassay (RIA), ELISA, radial immunodiffusion (RID), immunoturbidimetric assay, and immunonephelometric assay. Unlike most of the circulating proteins, the antigenic epitopes of apolipoproteins are masked by lipids. Nonionic detergents are used to remove those lipids for the facilitation of antibody binding to these epitopes. Polyclonal, monoclonal, and a mixture of monoclonal (panmonoclonal) antibodies are used for immunoassays.[17]

NATIONAL CHOLESTEROL EDUCATION PROGRAM RECOMMENDATIONS

National Cholesterol Education Program (NCEP) formulated a set of guidelines for the measurement of lipoproteins.[24,25]

- Either serum or plasma can be used for lipoprotein measurements.
- Equivalent serum value = Plasma value × 1.03.
- Triglycerides should preferably be made in samples collected after a 12-hour fast. However, for the sake of the convenience, sampling may be done after a 9-hour fast without introducing unduly large errors into the measurements.
- The blood samples should be drawn in a seated position whenever possible.
- Serum or plasma should be separated from cells within 3 hours of collection. This can be stored at 4°C for up to 3 days or at -20°C for several weeks.
- To nullify the physiologic and analytic variability of measured lipids and lipoproteins, at least two samples should be drawn one week apart, and their values averaged.

PITFALLS

The major pitfall of lipoprotein assays is an intra-individual variation of values over time. This can be due to several analytical or physiological factors.

Analytical Factors

Analytical variations are inherent in the measurements themselves. They may arise from the variations in sample collection procedures, volume measurements, instrument function, reagent related issues, etc. Few examples of variation due to analytic factors include alteration of lipoprotein molecules during ultracentrifugation procedure, falsely elevated HDL value in ultracentrifugation due to density overlap with Lp(a), falsely elevated HDL value in precipitation methods due to either residual manganese interfering with cholesterol enzymatic assays or in a sample with high triglycerides, retention of some ApoB containing lipoproteins in the supernatant.[17,22] Problems with the calculated LDL level in presence of triglyceride >400 have already been discussed.

Physiological Factors

These factors can lead to the variability of lipoprotein measurements independent of analytics factors. For example, standing posture for 30 minutes can elevate LDL, HDL, VLDL, and triglyceride value to 9–12% in comparison to supine posture due to hemoconcentration. The concentration of triglyceride prominently increases while the levels of total cholesterol and HDL-C decrease during infectious and inflammatory illnesses. After myocardial infarction and stroke, the total cholesterol level remains unchanged for 24–48 hours. After that, the value declines progressively and remains low for 6–8 weeks. Total cholesterol levels can also show seasonal variation with the highest levels observed in winter but lower values in summer. Although total cholesterol, HDL-C (and non-HDL-C), and directly measured LDL-C can be measured in a non-fasting state, fasting for 10–12 hours is traditionally recommended for measurement of triglyceride and calculated LDL-C. However, prolonged fasting for 6 days resulted in an 18% increase in plasma total cholesterol and triglyceride levels and a 22% decrease in serum HDL-C level.[26,27]

CONCLUSION

Accurate and precise laboratory measurement of lipoproteins are required for prediction and treatment of cardiovascular disorders in clinical and research settings. Lipoprotein concentration and their cholesterol content can be measured in laboratories by various methods. Multiple physiological and analytical factors can influence the results.

> **KEY POINTS**
> - Lipoproteins are classified according to their densities and size.
> - Total, HDL (and non-HDL), and directly measured LDL-C can be measured in a non-fasting state. However, the measurement of triglyceride and calculated LDL should ideally be done in the fasting state.
> - Earlier, electrophoresis and ultracentrifugation were used to separate the lipoproteins followed by the measurement of their cholesterol content by enzymatic assays.
> - Precipitation technique is most commonly use nowadays for the measurement of lipoproteins. A variation of this technique, homogeneous method, is most commonly used in the clinical laboratories.

- Calculated value of LDL-C by Friedewald equation is valid only when the blood is drawn in fasting state and the triglyceride value is <400 mg/dL. Otherwise, LDL should be measured directly.
- Apolipoprotein number and density can also be measured which can be useful in fine-tuning the CVD risk stratification.
- Pitfalls arising out of physiological and analytical factors should be kept in mind while interpreting lipid profile data.

REFERENCES

1. Centers for Disease Control and Prevention. (2004) Centers for Disease Control and Prevention Laboratory procedure manual: total cholesterol, HDL-cholesterol, triglycerides, and LDL-cholesterol. [online] Available from http://www.cdc.gov/nchs/data/nhanes/nhanes_03_04/l13_c_met_lipids.pdf. [Last accessed June, 2020]
2. Cholesterol T, Studies C. Laboratory procedure manual, 1972. Abstr Hosp Manag Stud. 1976;13(2).
3. National Health and Nutrition Examination Survey. [online] Available from https://wwwn.cdc.gov/Nchs/Nhanes/2015-2016/TCHOL_I.htm. [Last accessed June, 2020]
4. Barr DP, Russ EM, Eder HA. Protein-lipid relationships in human plasma. II. In atherosclerosis and related conditions. Am J Med. 1951;11(4):480-93.
5. Fredrickson DS, Levy RI LR. Fat transport in lipoproteins—an integrated approach to mechanisms and disorders. N Engl J Med. 1967;276:34-42.
6. Beaumont JL, Carlson LA, Cooper GR, Fejfar Z, Fredrickson DS, Strasser T. World Health Organization Memorandum Classification of hyperlipidemias and hyperlipoproteinemias. Circulation. 1972;45(1):501-8.
7. Asztalos BF, Sloop CH, Wong L, Roheim PS. Two-dimensional electrophoresis of plasma lipoproteins: Recognition of new apo A-I-containing subpopulations. Biochim Biophys Acta (BBA)/Lipids Lipid Metab. 1993;1169(3):291-300.
8. Asztalos BF, Cupples LA, Demissie S, Horvath K V, Cox CE, Batista MC, et al. High-density lipoprotein subpopulation profile and coronary heart disease prevalence in male participants of the Framingham Offspring Study. Arterioscler Thromb Vasc Biol. 2004;24(11):2181-7.
9. Okazaki M, Hagiwara N, Hara I. Heterogeneity of human serum high density lipoproteins on high performance liquid chromatography. J Biochem. 1982;92(2):517-24.
10. Havel RJ, Eder HA, Bragdon JH. The distribution and chemical composition of ultracentrifugally separated lipoproteins in human serum. J Clin Invest. 1955;34(9):1345-53.
11. Chapman MJ, Goldstein S, Lagrange D, Laplaud PM. A density gradient ultracentrifugal procedure for the isolation of the major lipoprotein classes from human serum. J Lipid Res. 1981;22(2):339-58.
12. Kulkarni KR. Cholesterol profile measurement by vertical auto profile method. Clin Lab Med. 2006;26(4):787-802.
13. Warnick GR, Albers JJ. A comprehensive evaluation of the heparin-manganese precipitation procedure for estimating high density lipoprotein cholesterol. J Lipid Res. 1978;19(1):65-76.
14. Warnick GR, Benderson J, Albers JJ. Dextran sulfate-Mg2+ precipitation procedure for quantitation of high-density-lipoprotein cholesterol. Clin Chem. 1982;28(6):1379-88.
15. Gidez LI, Miller GJ, Burstein M, Slagle S, Eder HA. Separation and quantitation of subclasses of human plasma high density lipoproteins by a simple precipitation procedure. J Lipid Res. 1982;23(8):1206-23.
16. Nauck M, März W, Haas B, Wieland H. Homogeneous assay for direct determination of high-density lipoprotein cholesterol evaluated. Clin Chem. 1996;42(3):424-9.
17. Rifai N WGR. Lipids, lipoproteins, apoliporoteins, and other cardiovascular risk factors. In: Burtis CA, Ashwood ER, Bruns DE (Ed). Tietz Textbook of Clinical Chemistry and Molecular Diagnostics, 4th edition. Philadelphia: Elsevier Saunders; 2005. pp. 903-81.
18. Friedewald WT, Levy RI, Fredrickson DS. Estimation of the concentration of low-density lipoprotein cholesterol in plasma, without use of the preparative ultracentrifuge. Clin Chem. 1972;18(6):499-502.

19. Warnick GR, Knopp RH, Fitzpatrick V, Branson L. Estimating low-density lipoprotein cholesterol by the Friedewald equation is adequate for classifying patients on the basis of nationally recommended cutpoints. Clin Chem. 1990;36(1):15-9.
20. Mora S, Otvos JD, Rifai N, Rosenson RS, Buring JE, Ridker PM. Lipoprotein particle profiles by nuclear magnetic resonance compared with standard lipids and apolipoproteins in predicting incident cardiovascular disease in women. Circulation. 2009;119(7):931-9.
21. Hirano T, Ito Y, Saegusa H, Yoshino G. A novel and simple method for quantification of small, dense LDL. J Lipid Res. 2003;44(11):2193-201.
22. Schaefer EJ, Tsunoda F, Diffenderfer M, Polisecki E, Thai N, Asztalos B. The Measurement of Lipids, Lipoproteins, Apolipoproteins, Fatty Acids, and Sterols, and Next Generation Sequencing for the Diagnosis and Treatment of Lipid Disorders. Endotext. 2000.
23. Xavier HT, Ruiz RM, Kencis Jr. L, Melone G, Costa W, Fraga RF, et al. Clinical correlation between the Point-of-care testing method and the traditional clinical laboratory diagnosis in the measure of the lipid profile in patients seen in medical offices. J Bras Patol Med Lab. 2016:387-90.
24. Third Report of the National Cholesterol Education Program (NCEP) Expert Panel on Detection, Evaluation, and Treatment of High Blood Cholesterol in Adults (Adult Treatment Panel III) final report. Circulation. 2002;106(25):3143-421.
25. Bachorik PS, Ross JW. National Cholesterol Education Program recommendations for measurement of low-density lipoprotein cholesterol: executive summary. The National Cholesterol Education Program Working Group on Lipoprotein Measurement. Clin Chem. 1995;41(10):1414-20.
26. Cooper GR, Myers GL, Smith SJ, Schlant RC. Blood lipid measurements: Variations and practical utility. JAMA. 1992;267(12):1652-60.
27. Ockene IS, Chiriboga DE, Stanek III EJ, Harmatz MG, Nicolosi R, Saperia G, et al. Seasonal variation in serum cholesterol levels: Treatment implications and possible mechanisms. Arch Intern Med. 2004;164(8):863-70.

CHAPTER

14

Statins Diabetogenicity: Are All Same?

Lohit Kumbar, Chitra Selvan

ABSTRACT

Atherosclerotic cardiovascular disease (ASCVD) is one of the supreme causes of morbidity and mortality. Statins are the mainstay of treatment with significant cardiovascular benefit. Statins are associated with increased risk of diabetes with an annual risk of 0.1%. The mechanisms causing diabetes may be related to beta cell dysfunction and insulin resistance. High-intensity statins are associated with increased risk of diabetes. Lipophilic statins such as atorvastatin are associated with higher risk of diabetes than hydrophilic statins. Newer statins such as pitavastatin and pravastatin may be associated with less deleterious effect on glycemia. Frequent reiteration of lifestyle changes and surveillance for diabetes are necessary. The risk-benefit ratio of statin use in reducing ASCVD-related morbidity and mortality should guide their usage in high-risk individuals. Further understanding of the mechanisms causing diabetes could potentially pave the way for the development of newer statins lacking this side effect.

INTRODUCTION

One of the leading causes of mortality worldwide is atherosclerotic cardiovascular disease (ASCVD) and statin therapy plays a major role in primary prevention and development of consequences in secondary prevention.[1] Despite which, statins remain one of the most vilified class of medications with data from surveys, registries, and insurance claims suggesting widespread discontinuation or irregular intake at their recommended doses for fear of adverse outcomes.[2]

Diabetes or hyperglycemia is one of the feared outcomes that deter the usage of statins. The risk of new onset diabetes while on statins is more likely with intensive treatment regimens and in individuals with features of the metabolic syndrome or prediabetes.[3]

PREVALENCE

The risk of development of diabetes in patients on statin therapy is about 0.1% annually.[4] Reduction of major coronary events in patients on statin therapy has been seen to be approximately 0.42% annually.[1] A network meta-analysis (NMA) of 29 trials involving 163,039 participants, revealed that there was a 12% increased risk of diabetes, with statins (pooled OR 1.12; 95% CI 1.05–1.21; I^{2}36%; p = 0.002; 18 RCTs). In the analysis, development of diabetes was highest with atorvastatin 80 mg (OR of 1.34 (95% CI 1.14–1.57) which was followed by rosuvastatin (OR 1.17; 95% CI 1.02–1.35). The analysis also showed that simvastatin 80 mg, simvastatin, atorvastatin, pravastatin, lovastatin, and pitavastatin were associated with higher odds of developing diabetes with OR of 1.21 (0.99–1.49), 1.13 (0.99–1.29), 1.13 (0.94–1.34), 1.04 (0.93–1.16), 0.98 (0.69–1.38) and 0.74 (0.31–1.77), respectively. The development of diabetes increased with higher dose of atorvastatin than with low dose of atorvastatin and pravastatin, and simvastatin.[5]

PROPOSED MECHANISMS FOR NEW-ONSET DIABETES INDUCED BY STATIN (TABLE 1 AND FLOWCHART 1)[6,7]

Mechanisms have been postulated for increased risk of diabetes with statins ranging from a decline in insulin secretion by pancreatic cells or by an increase in insulin resistance.

The following have been suggested as possible mechanisms:
- Inhibition of the 3-hydroxyl-3-methylglutaryl-CoA reductase receptor (HMGCR)
  Studies have shown *HMGCR* gene mutations as proxies of HMGCR inhibition by statins.[8] An extra-allele was associated with lower LDL-C by about a 2.3 mg/dL (0.06 mmol/L), increased body weight by 0.3 kg, mild increase in waist circumference, blood glucose levels and insulin concentration with increased risk of diabetes by 2–6% diabetes.[9]
- Ca^{2+} channels in ß-cells
  Improper organization of membrane lipid-raft bound proteins or any change in Ca^{2+} channel subunits confirmation may lead to a defect in insulin secretion.[10]
- Glucose transporter 4 (GLUT4)
  Insulin resistance has been shown to be induced by atorvastatin and lovastatin probably via reduction in GLUT4 translocation to the cell membrane.[9] The downregulation of GLUT4 membrane expression, caused by atorvastatin, is in part via induced Rab4 dysfunction, a G protein.[11]
- Caveolin
  Atorvastatin suppresses GLUT4 and caveolin-1, which leads to a decrease in glucose uptake and an increase in insulin resistance in muscle, liver, and adipose tissue.[12]
- Insulin receptor substrate-1 (IRS-1) and AKT
  Atorvastatin therapy reduces IRS-1 and AKT phosphorylation in a dose-dependent manner. Atorvastatin therapy also leads to a downregulation of IRS-1 and IR-ß levels in the differentiation process in adipocytes with use of atorvastatin.[12]
- Inhibition of isoprenoid, coenzyme Q10 (CoQ10), and dolichol biosynthesis
  Suppression of isoprenoid synthesis plays a key role in lovastatin-induced insulin resistance. Lovastatin significantly decreases dolichol and CoQ10. Lower CoQ10, and hence a decrease of mitochondrial ATP production leads to closure of ATP-dependent potassium channels on ß-cells and inhibits insulin secretion due to Ca^{2+} influx inhibition.[13]
- Adiponectin and leptin
  Studies have shown that use of some statin (atorvastatin and simvastatin) were associated with a reduction in adiponectin levels and the subsequent attenuation

TABLE 1: Studies comparing risk of incident diabetes with various statin groups and doses.[6]

Study, year	Type of study	Number of trials, sample size	Study design—comparison	Risk of incident diabetes
Preiss, 2011	Meta-analysis	5,32,752	Intensive vs. moderate dose statin	↑ incidence of diabetes in intensive dose statin (OR 1.12, 1.04–1.22)
Waters, 2014	Meta-analysis	2,15,056	Atorvastatin 80 mg vs. low dose statin	↑ risk of diabetes in atorvastatin group (HR 1.16, 1.03–1.30)
Zaharen, 2013	Retrospective cohort	197,138	Risk of diabetes among various statin	↑ risk of diabetes (HR 1.18, 1.15–1.22) with rosuvastatin, atorvastatin and simvastatin
Carter, 2013	Retrospective cohort	471,250	Pravastatin vs. other statins	Atorvastatin (HR 1.22, 1.15–1.29); Rosuvastatin (HR 1.18, 1.10–1.26); Simvastatin (HR 1.10, 1.04–1.17); Lovastatin and fluvastatin—NS
Navarese, 2013	Meta-analysis (17 RCT)	17,113,394	Different statin doses vs. Placebo	Rosuvastatin vs. placebo (OR 1.25, 0.82–1.90) Pravastatin vs. placebo (OR 1.07, 0.86–1.30)
LIVES extension study, 2011	Prospective post-marketing	~20,000	Efficacy and safety of pitavastatin	↓ HbA1c by 0.4, $p < 0.001$.
Sukhija, 2009	Observational	345,417	Statin vs. no statin	Δ fasting plasma glucose in nondiabetic statin users = 7 mg/dL ($p < 0.0001$), and in diabetic statin users = 39 mg/dL ($p < 0.0001$) respectively
Simsek, 2012	Randomized, parallel-group	263	High-dose statin therapy vs. baseline glycemic levels	↑ HbA1c levels in both atorvastatin (p 0.003) and rosuvastatin groups ($p < 0.001$)

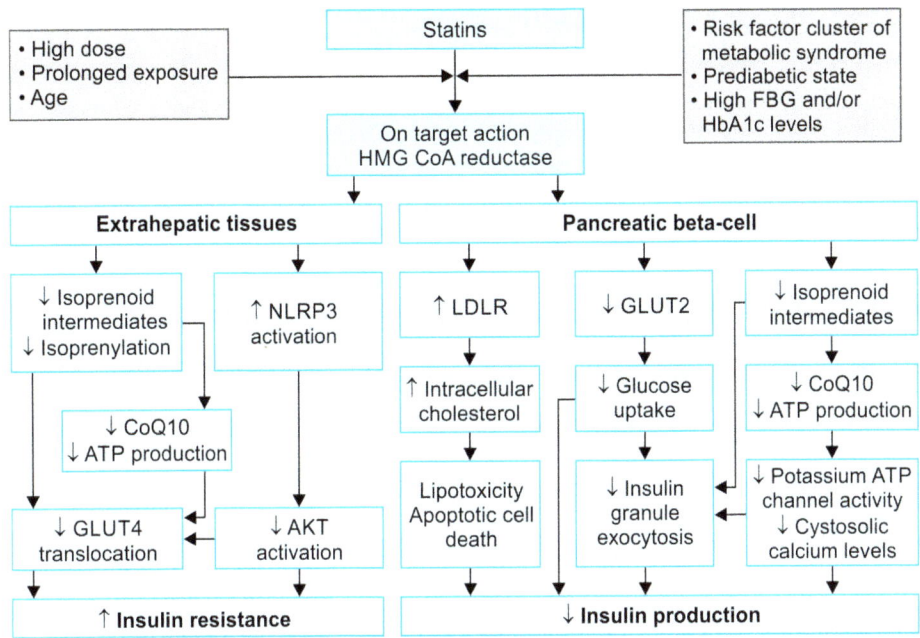

FLOWCHART 1: Mechanism of statins-induced diabetes.[7]
Source: Mach F, Ray KK, Wiklund O, Corsini A, Catapano AL, Bruckert E, et al. Adverse effects of statin therapy: perception vs. the evidence focus on glucose homeostasis, cognitive, renal and hepatic function, haemorrhagic stroke and cataract. Eur Heart J. 2018;39:2526-39.

in insulin sensitivity[14] while other study has suggested that the simvastatin-reduced leptin is a result of the inhibition of PI3K which leads to elevation of intracellular cAMP, a protein kinase A (PKA) activator.[15]

- Uncoupling protein 3 (UCP3)
  In simvastatin treated individuals, glucose intolerance and decrease in UCP3 occur concurrently and the statin-induced insulin resistance is a consequence of statin-induced UCP3 reduction.[16]
- Effects on lipoprotein particle size
  Statins increase the size of very low-density lipoprotein cholesterol (VLDL-C) particles and decreases the size of LDL-C, HDL-C, and the lipoprotein insulin resistance (LPIR) score, which contributes to the development of new onset diabetes.[17]
- Genetic polymorphisms
  A study with patients with following SNPs: TCF7L2, SLC30A8, HHEX, CDKN2B, CDKALI, MTNR1B, KCNJ11, and IGF2BP2 had impaired release of early-phase insulin or impaired conversion from proinsulin to insulin.[18]
- miRNA
  A group of miRNAs, especially miR-375, miR-7, miR-195, miR-126, miR-9, miR-96, and miR-34a, have been shown to play a role in development of pancreas and insulin secretion, and other miRNAs such as miR-7, miR-139, miR-145, and miR-1 have been shown to play a role in the insulin growth factor-1 receptor expression and hence differences in susceptibility to diabetes secondary to statin use.[19]

Factors Affecting Risk of Statin-induced Diabetes—Type, Dose, and Age

Some studies have shown little difference in risk of diabetes between patients on medium or high-statin doses, while other studies have shown a greater risk of diabetes in patients on rosuvastatin, atorvastatin, and simvastatin, and no excess risk with other statins such as pravastatin, fluvastatin, lovastatin, and pitavastatin.[14-16] In this context, it is important to note that the possible increased risk of development of diabetes may be related to the potency of each statin. This has been shown by a meta-analysis where >30% reduction in LDL-C was associated with risk of new-onset diabetes.[20]

Statins may be lipophilic or hydrophilic. The lipophilic statins include atorvastatin, simvastatin, lovastatin, fluvastatin and pitavastatin, and hydrophilic statins include pravastatin and rosuvastatin.[21] It has shown that lipophilic statins can readily penetrate extrahepatic cell membranes such as beta cells, adipocytes, and skeletal muscles and are more diabetogenic while hydrophilic statins such as pravastatin are more hepatocyte specific and are at lesser risk of development of diabetes and they are less likely to enter beta cells or adipocytes. A meta-analysis has supported this hypothesis, where pravastatin was found to improve insulin sensitivity while simvastatin worsened insulin sensitivity.[22] However there are conflicting results with various trials, where WOSCOP trial supported this hypothesis, but the risk of NOD with rosuvastatin as noted in JUPITER trial do not sit well with this hypothesis. Another, meta-analysis by Sattar et al. failed to show any difference between lipophilic and hydrophilic statins.[4]

Pitavastatin, which is a newer statin, has shown to be as effective as atorvastatin in reducing LDL-C levels, and also has the benefit of improving high-density lipoprotein cholesterol levels. It lacks the adverse effect on glycemic control. Pitavastatin, has been shown to improve glycosylated hemoglobin (HbA1c) levels in patients with pre-existing diabetes mellitus in some studies.[23]

Patients aged >65 years may be particularly susceptible to this effect of statin therapy. The mechanism mostly is related to induction of β-cell dysfunction and apoptosis rather than reduction of insulin sensitivity in this population. Yet, the favorable ratio of statin therapy between cardiovascular benefits and risk of developing diabetes (9:1) supports the current approach of widespread use of statin in high-risk populations. For low-risk patients, particularly if age is >65 years, the risk of diabetes should be considered high when considering to start a statin.[24] Individuals who have other risk factors for developing diabetes should be prescribed a lesser diabetogenic statin, such as pravastatin.[13]

RECOMMENDATIONS

There is a four-fold decrease in cardiovascular risk compared to the risk of new onset diabetes and hence there is strong suggestion of initiating or continuing statin therapy in patients who have demonstrated benefit from statin use.[4]

Clinicians can consider statins which has more neutral effects on insulin sensitivity in patients who are at higher risk for developing incident diabetes, although further studies are required to fully support this recommendation. In addition, there should be a heightened clinical surveillance of patients who are at risk for development of new onset diabetes. Education of all the patients who are at high risk of development of diabetes should be done about frequent laboratory evaluation of glycemic status, dietary modifications, and increased physical activity.

FUTURE CONSIDERATIONS

Cholesteryl ester transfer protein (CETP) inhibitors, such as dalcetrapib, anacetrapib, torcetrapib, and evacetrapib, cause an increase in plasma concentration of HDLs and ApoA-I, both of which have been found to increase beta cell function and glucose uptake by the skeletal muscle. Improved glycemic control and the decreased risk of development of new onset diabetes have been observed in people treated with CETP inhibitors. If the beneficial effects of CETP inhibitors are confirmed in a future investigation, there will be an added benefit of improving cardiovascular outcome as well as decreased risk of new onset diabetes in such population.[25]

CONCLUSION

Although a number of mechanisms have been postulated, there is a need for further investigations to explore the understandings of statin therapy and new onset of diabetes. Risk-benefit ratio of statin therapy should be weighed, so as to direct the type and dose of statins the patient receives. Novel preventative or therapeutic approaches to be instituted and guide the production of a new generation of statins lacking this side effect.[26]

KEY POINTS

- Statins are the first-line therapy for the prevention of cardiovascular diseases by its lipid lowering effect.
- Mechanism of statin-induced diabetes include specific to beta cell function and insulin resistance.
- Newer statins such as pravastatin appears to reduce risk of development of diabetes while increased risk has been seen with atorvastatin, rosuvastatin, and simvastatin.
- Intensive statin therapy has been associated with higher risk of diabetes; hence statins should be initiated with a lower dose. High-intensity statins should be avoided in elderly as much as possible.
- Better understanding of the mechanisms associated with increased risk of diabetes with statin therapy is needed, as it may enable for better novel preventative or therapeutic approaches and production of newer generation of statins which lacks this deleterious effect.

REFERENCES

1. Cholesterol Treatment Trialists C, Mihaylova B, Emberson J, et al. The effects of lowering LDL cholesterol with statin therapy in people at low risk of vascular disease: meta-analysis of individual data from 27 randomised trials. Lancet. 2012;380:581-90.
2. Collins R, Reith C, Emberson J, Armitage J, Baigent C, Blackwell L et al. Interpretation of the evidence for the efficacy and safety of statin therapy. Lancet. 2016;388:2532-61.
3. Carmena R, Betteridge DJ. Diabetogenic Action of Statins: Mechanisms. Curr Atheroscler Rep. 2019;21:23.
4. Sattar N, Preiss D, Murray HM, Welsh P, Buckley BM, de Craen AJM, et al. Statins and risk of incident diabetes: a collaborative meta-analysis of randomised statin trials. Lancet. 2010;375:735-42.
5. Thakker D, Nair S, Pagada A, Jamdade V, Malik A. Statin use and the risk of developing diabetes: a network meta-analysis. Pharmacoepidemiol Drug Saf. 2016;25(10):1131-49.
6. Rochlani Y, Kattoor AJ, Pothineni NV, Palagiri RDR, Romeo F, Mehta JL. Balancing Primary Prevention and Statin-Induced Diabetes Mellitus Prevention Am J Cardiol. 2017;120:1122-8.

7. Mach F, Ray KK, Wiklund O, A, Catapano AL, Bruckert E, et al. Adverse effects of statin therapy: perception vs. the evidence focus on glucose homeostasis, cognitive, renal and hepatic function, haemorrhagic stroke and cataract. Eur Heart J. 2018;39:2526-39.
8. Swerdlow DI, Preiss D, Kuchenbaecker KB, Holmes MV, Engmann JEL, Shah T, Sattar N. HMG-coenzyme A reductase inhibition, type 2 diabetes, and bodyweight: Evidence from genetic analysis and randomised trials. Lancet. 2015;385(9965):351-61.
9. Robinson JG. Statins and diabetes risk: How real is it and what are the mechanisms? Curr Opin Lipidol. 2015;26(3):228-35.
10. Xia F, Xie L, Mihic A, Gao X, Chen Y, Gaisano HY, et al. Inhibition of cholesterol biosynthesis impairs insulin secretion and voltage-gated calcium channel function in pancreatic β-cells. Endocrinology. 2008;149(10):5136-45.
11. Takaguri A, Satoh K, Itagaki M, Tokumitsu Y, Ichihara K. Effects of atorvastatin and pravastatin on signal transduction related to glucose uptake in 3T3L1 adipocytes. J Pharmacol Sci. 2008;107(1):80-9.
12. Nakata M, Nagasaka S, Kusaka I, Matsuoka H, Ishibashi S, Yada T. Effects of statins on the adipocyte maturation and expression of glucose transporter 4 (SLC2A4): Implications in glycaemic control. Diabetologia. 2006;49(8):1881-92.
13. Paseban M, Butler AE, Sahebkar A. Mechanisms of statin induced new onset diabetes. Review article. J Cell Physiol. 2019;1-11.
14. Koh KK, Quon MJ, Sakuma I, Lee Y, Lim S, Han SH, et al. Effects of simvastatin therapy on circulating adipocytokines in patients with hypercholesterolemia. Int J Cardiol. 2011;146(3):434-7.
15. Maeda T, Horiuchi N. Simvastatin suppresses leptin expression in 3T3-L1 adipocytes via activation of the cyclic AMP–PKA pathway induced by inhibition of protein prenylation. J Biochem. 2009;145(6):771-81.
16. Larsen S, Stride N, Hey-Mogensen M, Hansen CN, Bang LE, Bundgaard H, Dela F. Simvastatin effects on skeletal muscle: Relation to decreased mitochondrial function and glucose intolerance. J AmColl Cardiol. 2013; 61(1):44-53.
17. Mackey RH, Mora S, Bertoni, AG, Wassel CL, Carnethon MR, Sibley C,T, et al. Lipoprotein particles and incident type 2 diabetes in the multi-ethnic study of atherosclerosis. Diabetes Care. 2015; 38(4):628-36.
18. Stancakova A, Kuulasmaa T, Paananen J, Jackson AU, Bonnycastle LL, Collins FS, et al. Association of 18 confirmed susceptibility loci for type 2 diabetes with indices of insulin release, proinsulin conversion, and insulin sensitivity in 5,327 nondiabetic Finnish men. Diabetes. 2009;58(9):2129-36.
19. Guay C, Roggli E, Nesca V, Jacovetti C, Regazzi R. Diabetes mellitus, a microRNA-related disease? Translational Research. 2011;157(4):253-64.
20. Ciosek CP Jr, Magnin DR., Harrity TW, Logan JV, Dickson JK Jr, Gordon E, et aL. Lipophilic 1, 1-bisphosphonates are potent squalene synthase inhibitors and orally active cholesterol lowering agents in vivo. J Biol Chem. 1993;268(33):24832-7.
21. Schachter M. Chemical, pharmacokinetic and pharmacodynamic properties of statins: An update. Fundam Clin Pharmacol. 2005;19:117-25.
22. Baker WL, Talati R, White CM, Coleman CI. Differing effect of statins on insulin sensitivity in non-diabetics: A systematic review and meta-analysis. Diabetes Res Clin Pract. 2010;87:98-107.
23. Teramoto T. Pitavastatin: clinical effects from the LIVES study. Atheroscler Suppl 2011;12:285-8.
24. Sampson UK, Linton MF, Fazio S. Are statins diabetogenic? Curr Opin Cardiol. 2011;26:342-7.
25. Barter PJ, Cochran BJ, Rye K-A. CETP inhibition, statins and diabetes. Atherosclerosis. 2018;278:143-6.
26. Agarwala A, Kulkarni S, Maddox T. The Association of Statin Therapy with Incident Diabetes: Evidence, Mechanisms, and Recommendations. Current Cardiology Reports. 2018;20:50.

Section 3

Therapeutic Lipidology

CHAPTER 15

The Effects of Diet and Exercise in Hyperlipidemia

Partha Sarathi Choudhury

ABSTRACT

Hyperlipidemia is a dominant risk factor for atherosclerotic cardiovascular disease (ASCVD). Dietary modification and physical activity are the first-line approach for primary prevention in dyslipidemia. The cumulative effect of replacing the intake of saturated and trans fats by polyunsaturated and monounsaturated fats in diet, fortifying foods with plant stanols or sterols, and adopting a Mediterranean, Portfolio, or low-carbohydrate diet can result in the most beneficial changes during treatment of hyperlipidemia. Because these dietary approaches tend to reduce total cholesterol, low-density lipoprotein (LDL) cholesterol, and triglycerides (TGs) and exercise tends to increase high-density lipoprotein (HDL) cholesterol and decrease TGs, it seems quite logical to combine these approaches to get a synergistic or at least additive effect in the treatment of hyperlipidemia.

INTRODUCTION

Atherosclerotic cardiovascular disease (ASCVD) has become a significant cause of mortality in developed countries. Due to a gradually expanding population, economic wellbeing, urbanization with a greater intake of junk food, and an increasingly sedentary lifestyle, the burden of ASCVD is also rising in India. Of all the risk factors, hyperlipidemia is the most important risk factor for ASCVD. A review of various Indian population studies demonstrates increasing mean total cholesterol levels. The prevalence of hypercholesterolemia in these studies varies from 10–15% in rural to 25–30% in the urban Indian population.[1]

LIPIDS AND LIPOPROTEINS

Lipids are a group of substances that share the common property of being soluble in fat solvents. Biologically important lipids are phospholipids, cholesterol, and triglycerides (TGs). Cholesterol, however, is a sterol compound without any fatty acid but it is derived from fat. Cholesterol is used to help to build plasma membranes, synthesis of steroid hormones (cortisol, aldosterone, estrogen, progesterone, and testosterone), production of vitamin D, and formation of bile acids that help in the process of digestion.

Lipoproteins are complexes of lipids and proteins [apolipoproteins (Apo)] that act as transport vehicles for poorly soluble lipids such as cholesterol, TGs, and fat-soluble vitamins through body fluids (plasma, interstitial fluid, and lymph) to and from tissues. The plasma lipoproteins are classified into five major classes according to their relative density:
1. Chylomicrons
2. Very low-density lipoprotein (VLDL)
3. Intermediate density lipoprotein (IDL)
4. Low-density lipoprotein (LDL)
5. High-density lipoprotein (HDL)

Chylomicrons

They are the largest lipoproteins and are the most lipid-rich lipoproteins. These comprise approximately 85% TGs and carry ApoB-48 as the major apolipoprotein. TGs are the predominant forms of lipid in adipose tissue and diet. With the removal of TGs from the chylomicron at receptor sites in the body, the chylomicron particles gradually shrink in size and are converted into chylomicron remnants. The chylomicron remnant is then returned to the liver for further metabolism.[2]

Very Low-density Lipoprotein

Very low-density lipoprotein is also a triglyceride-rich lipoprotein but it contains ApoB-100 rather than ApoB-48 as the major apolipoprotein and is smaller and less buoyant than chylomicron.

Intermediate Density Lipoprotein

After the TGs of VLDL are hydrolyzed by lipoprotein lipase (LPL) in peripheral tissues, the TG-depleted VLDL remnants are called IDLs.

Low-density Lipoprotein

Low-density lipoprotein is the predominant transport carrier of cholesterol in the circulation. Approximately 50–60% of cholesterol is delivered to the peripheral tissues by LDL. Evidence suggests that LDL may directly contribute to the formation of atheroma in the inner wall of arteries leading to the development of ASCVD. Small dense LDL is found in approximately 50% of men with ASCVD and is known as LDL pattern B. This pattern is often found in individuals with raised TG and reduced HDL, a combination known as the dyslipidemic triad, noticed in individuals with type 2 diabetes mellitus (T2DM) and other insulin resistance syndromes.

High-density Lipoprotein

Studies demonstrate that HDL has an inverse relationship with coronary artery disease (CAD), as it offers some protection against the development of CAD. Cholesterol in the peripheral tissues is transported from the plasma membranes of peripheral cells to the liver and intestine for excretion by a process termed "reverse cholesterol transport" that is facilitated by HDL. HDL is predominantly composed of phospholipids and is subclassified into HDL2 and HDL3, based on size and particle density. Evidence suggests that HDL2-cholesterol (HDL2-C) provides greater protection against CAD and that higher concentration of HDL2-C in women than men helps to protect women from developing CAD.[2]

There are multiple health conditions associated with hyperlipidemia apart from CAD including obesity, hypertension, and T2DM. The management of hyperlipidemia needs a comprehensive plan to control lipid levels and address associated metabolic abnormalities and other risk factors. The first-line approach to primary prevention in individuals with hyperlipidemia should be the implementation of lifestyle modification in the form of medical nutrition therapy and physical activity.

DIETARY APPROACHES

Therapeutic Lifestyle Changes Diet

The Adult Treatment Panel III (ATP-III) of the National Cholesterol Education Program (NCEP) has recommended nonpharmacological treatment as the initial therapy in most of the patients with hyperlipidemia for the last decade.[3] The therapeutic lifestyle changes (TLCs) approach was developed on the panel's review of the gathered evidence in 1999 that inferred dietary modification and exercise can have a favorable influence on serum levels of total cholesterol, LDL, HDL, and TGs. ATP-III emphasizes TLC as a fundamental therapy for individuals at risk for CAD. The cumulative effect of the TLC diet components in decreasing LDL by 25–30% compared to a typical US diet is similar to the effect of hypolipidemic drug therapy. Moreover, a TLC-like diet has also been found to positively influence blood pressure and serum TG levels with minimal or no effect on HDL levels.[4] This effect of the TLC diet is further improved by an increase in physical activity in overweight individuals, particularly significant for those with T2DM and/or metabolic syndrome.[5]

Components of TLC

Therapeutic lifestyle change recommends a reduction in the intake of saturated fat, trans fats, and cholesterol, while it emphasizes on increasing consumption of monounsaturated and polyunsaturated fatty acids (**Table 1**).

Saturated Fat

Saturated fat is known to be the chief dietary constituent to raise LDL. It has been found that for every 1% rise in total calories from saturated fats, there is an associated 2% increase in LDL.[6] Compared with a Western diet, reducing saturated fat intake to 7% of total calories and cholesterol intake to 200 mg/day decreases LDL cholesterol level by 9–12%.[7]

Trans Fats

Trans fatty acids are a type of unsaturated fat produced due to industrial hydrogenation of natural polyunsaturated fats.[8] Trans fats raise LDL cholesterol and lower HDL cholesterol. This dual effect on cholesterol levels eventually raises the total cholesterol

TABLE 1: Dietary recommendations of therapeutic lifestyle change (TLC) diet.	
Components	Recommended amount
Total fat	25–35% of total calories
Saturated fat	<7% of total calories
Polyunsaturated fatty acid	Up to 10% of total calories
Monounsaturated fatty acid	Up to 20% of total calories
Trans fats	<1% of total calories
Cholesterol	<200 mg/day
Carbohydrate	50–60% of total calories
Protein	15–25% of total calories
Sodium	<2,300 mg/day
Soluble fiber	5–10 g/day
Plant sterols/stanols	Up to 2 g/day

to HDL ratio leading to an increase in the risk of CAD.[9] Most of the dietary trans fats come from prepared foods such as cakes, cookies, and commercially synthesized fried foods. Substitution of dietary trans fats by cis polyunsaturated fatty acids reduces total cholesterol, LDL cholesterol, TGs, ApoB, and increases HDL cholesterol.[10]

Monounsaturated Fatty Acids and Polyunsaturated Fatty Acids

A recent meta-analysis of 60 randomized controlled trials (RCTs) demonstrated that substituting trans fats with polyunsaturated fatty acids is the most effective technique for improving blood lipid profile.[11] Evidence from a 20-year follow-up of the Nurses' Health Study revealed that a higher polyunsaturated fat intake (approximately 7% vs. 5% of total calories) in women was associated with a diminished risk of CAD.[12] In an earlier examination of the Nurses' Health Study group, monounsaturated fatty acids were found to be inversely associated with the risk for CAD.[13] Based on this cardioprotective effect of monounsaturated and polyunsaturated fatty acids, TLC recommends the use of monounsaturated and polyunsaturated fats with intakes ranging up to 20% and 10% of total calories, respectively.

Higher intake of carbohydrate (>60% of calories) in overweight/obese individuals can significantly increase blood glucose and TG levels and decrease HDL cholesterol. Diets that substitute a fraction of the carbohydrates with monounsaturated fats and/or polyunsaturated fats can reduce serum TG levels with minimal or no decrease in HDL levels.[14]

Dietary Cholesterol

On average, a 100-mg increase in dietary cholesterol intake raises total serum cholesterol levels by 2.2 mg/dL. A recent meta-analysis showed that dietary cholesterol intake elevates the total cholesterol to HDL ratio. Based on this evidence, the TLC diet recommends limiting dietary cholesterol intake to <200 mg/day.

Plant Stanols/Sterols

Plants contain several sterols, stanols, and esters that can decrease serum cholesterol. Plant sterols and stanols decrease cholesterol by inhibiting intestinal cholesterol absorption mostly through disruption of intraluminal solubilization.[15] Moreover, plant sterols and stanols may also reduce TGs.

Naturally occurring sterols and stanols are found in nuts, legumes, whole grains, fruits, vegetables, and plant oils. Also, margarine enriched with sterols and stanols is commercially available (containing Benecol and Take Control spreads). Benecol spread contains sitostanol and campestanol; and Take Control spread contains sitosterol and campesterol.[16] Foods fortified with plant sterols and/or stanols may be utilized as an adjunct to dietary modification and statins in individuals with elevated LDL cholesterol, although long-term safety information and data on reduction in cardiovascular outcome are still lacking.

Dietary Fibers

Intake of certain soluble fibers, including psyllium, pectin, certain beans (navy, pinto, and black beans), lentils, nuts, citrus fruits, strawberries, apple pulp, and oat products, can reduce both total cholesterol and LDL cholesterol.

Soluble fibers act by slowing gastric emptying, enhancing satiety, inhibiting hepatic cholesterol synthesis, and/or increasing fecal excretion of cholesterol and bile salts. Improvement in lipid profile and glucose homeostasis is attributed to the gel-forming property of soluble fiber. In a meta-analysis of randomized trials of patients with both normal and elevated cholesterol levels, the addition of 10.2 g/day of psyllium reduced LDL cholesterol by a mean of 12.8 mg/dL.[17]

Omega-3 Fatty Acids

Intake of omega-3 fatty acids can reduce TGs but may also elevate total cholesterol and LDL cholesterol, particularly in those with hypertriglyceridemia. Rich dietary sources of omega-3 fatty acids include fatty fish especially salmon, herring, mackerel, and trout [containing eicosapentaenoic acid (EPA) and docosahexaenoic acid (DHA)]. Plant products such as flaxseeds, chia seeds, canola oil, soybean oil, and some nuts are also rich sources of omega-3 fatty acids.

In a meta-analysis of 55 trials, each 1 g/day increase in EPA + DHA decreased TGs by 5.9 mg/dL and this effect was even more potent when baseline TG levels were higher, i.e., above the median TG level of 83 mg/dL, leading to the reduction of TGs by 8.4 mg/dL.[18] As many omega-3 preparations reduce plasma TGs (which is a determinant of small dense LDL particles), it is expected that these can lower the concentration of small dense LDL particles and this view is supported by the observed reduction of cholesteryl ester transfer (CET) activity following fish oil therapy.[19]

Based upon these demonstrated benefits of omega-3 fatty acids, the American Heart Association (AHA) supports the intake of two servings of fatty fish per week for the general population.

Mediterranean Diet

Mediterranean diet is characterized by high consumption of monounsaturated fats (primarily from olive oil), fruits, vegetables, whole grains, beans associated with typically low-to-moderate quantity of fish, poultry, dairy products, and very little red meat. A meta-analysis of six randomized trials comparing a Mediterranean diet with a low-fat diet showed that a Mediterranean diet led to a greater decrease in total cholesterol and a non-significant decrease in LDL cholesterol.[20] Adherence to a Mediterranean diet can also improve TG levels in T2DM individuals.

Portfolio Diet

The portfolio diet is a form of TLC diet that comprises of four LDL cholesterol-reducing constituents—(1) Soluble fiber, (2) Soy and other vegetable proteins, (3) Plant sterols, and (4) Almonds. Based upon the evidence, it has been found to decrease LDL cholesterol levels by 29-35%, which is proportionate to a combination of a diet with reduced saturated fats and cholesterol plus 20 mg of lovastatin daily.[21]

Low Carbohydrate Diet

This type of diet varies in the quantity and types of carbohydrates they contain but usually, such a diet is limited to <130 g/day of carbohydrates (with <60 g/day considered a very low carbohydrate diet). In a randomized trial comparing low fat, the Mediterranean, and low carbohydrate diets in over 300 overweight adults, maximal weight loss and HDL cholesterol increase occurred among those on low carbohydrate diet (−5.5 kg and 8.4 mg/dL, respectively). A low carbohydrate diet is of particular benefit in individuals with hypertriglyceridemia as it significantly reduces TGs.

Food Additives and Supplements

Soy

Soy is a good source of protein and contains isoflavones (phytoestrogens). Many soy foods and food products (e.g., tofu, soy butter, edamame) are found to have beneficial effects on lipid levels, as they contain low saturated fats, high-unsaturated fats, and dietary fibers.

Chinese Red Yeast Rice

Chinese red yeast rice extracts have multiple active constituents including naturally occurring lovastatin (in the form of monacolin K, 5-10 mg per standard daily dose). A meta-analysis showed that red yeast rice reduced total cholesterol, LDL cholesterol, and TGs. However, some preparations may have harmful byproducts of fermentation and there is no standardization of the available products.

EXERCISE AND LIPIDS

The influence of exercise on serum lipid profile has been extensively studied. Published data subjected to meta-analysis demonstrated that regular aerobic exercise raises HDL cholesterol levels by a mean of 1.9-2.5 mg/dL and lowers total cholesterol, LDL cholesterol, and TG levels by a mean of 3.9, 3.9, and 7.1 mg/dL, respectively.[22]

Potential Mechanisms of Benefit

The mechanisms by which aerobic exercise has a beneficial effect on lipid profiles are not clearly understood. Exercise increases LPL activity and decreases hepatic lipase (HL) activity.[23] Enhanced LPL activity indicates that increased lipolysis of triglyceride-rich lipoproteins may be an initial step in higher HDL cholesterol levels. Moreover, as HL catalyzes the conversion of larger HDL-2 particles to smaller HDL-3 particles, diminished HL activity with exercise results in slower catabolism of cardioprotective HDL-2 particles.

Exercise is also associated with a decrease in the serum CET protein concentration[24] and an increase in the serum lecithin cholesterol acyltransferase (LCAT) concentration,[25] eventually leading to increased HDL concentration, particularly HDL-2 particles.

Muscle contraction during exercise can recruit glucose transporter type 4 (GLUT4) to the cell surface by activating a signaling system that is independent of the insulin signaling system. Insulin resistance is exhibited by T2DM patients and a high proportion of obese individuals, where inadequate quantities of GLUT4 are recruited to the cell surface by insulin due to defect in the insulin-signaling cascade. Therefore, regular aerobic exercise helps in ameliorating insulin resistance in T2DM and obese individuals.

Exercise Regimens

American Association of Clinical Endocrinologists (AACE) recommends that exercise regimens should involve at least 30 minutes of moderate-intensity physical activity (equivalent to a consumption of 4–7 kcal/min) 4–6 times every week. Activities may involve brisk walking, stationary bike ridings, scrubbing, lawn-mowing, and sporting activities such as skiing, basketball, or volleyball with light effort. Total 60 minutes or more of daily exercise is recommended for weight loss programs; however, the minimum recommendation is 30 minutes daily, as a more stringent exercise regimen may lead to poor adherence for some persons.

Evidence-based studies have also demonstrated that weight and resistance training may help some persons with insulin resistance syndrome by improving their lipid profiles. Resistance training has resulted in a reduction of total cholesterol and LDL-C in both men and women, with women also demonstrating a remarkable reduction in TGs.[2] Therefore, muscle-strengthening activity is suggested for at least 2 days a week over and above the usual aerobic exercise.

Effects of Aerobic Exercise on HDL

Many studies have shown that HDL cholesterol levels are more sensitive to aerobic exercise than LDL cholesterol and TGs. A meta-analysis including 19 RCTs by Kelley et al.[26] found that HDL2-C increased by about 11% in subjects undergoing aerobic exercise.

Effects of Aerobic Exercise on LDL

Unlike HDL, the effect of aerobic exercise on LDL-C in humans is inconsistent. In some individuals with mild-to-moderate dyslipidemia undergoing a few months of aerobic exercise, LDL-C level did not change significantly but atherogenic small LDL particles decreased and the average size of LDL particles increased. Kamani et al. in his study found a significant decrease in average proprotein convertase subtilisin kexin 9 (PCSK-9) levels and average LDL-C levels in volunteers undergoing 3 months exercise and concluded that aerobic exercise is independently associated with a decrease in PCSK-9 levels over time.[27]

Effects of Aerobic Exercise on Triglyceride

Studies reported that exercise lowers plasma TG concentrations. But, researchers have found that, after exercise, those participants having lower baseline levels of TG showed only a slight decline in TG level, while those having higher baseline TG levels showed a significant TG reduction.

CONCLUSION

The efficacy of dietary modification and exercise in hyperlipidemia has been elaborately discussed. However, many counseled patients will fail to make persistent changes and responses to lifestyle modification also differ among individual patients. So, it will be logical to motivate patients about the lifestyle changes, which are most expected to improve their lipid-related cardiovascular risk and then assess the results that occur as they make those changes.

KEY POINTS

- Reduced intake of saturated and trans fats, increased intake of polyunsaturated and monounsaturated fatty acids, and supplementation with plant sterols or stanols and soluble fibers are likely to produce the most favorable changes in lipid levels.
- Diets enriched in fruits and vegetables, legumes, high-fiber cereals, low-fat dairy products, fish, lean meats, and skinless poultry are the recommended diets for the reduction of cardiovascular risk through lipid management.
- Exercise training programs in the form of aerobic exercise bring about favorable changes in the lipoprotein profile in CAD patients that help to slow the progression of CAD.

REFERENCES

1. Farzadfar F, Finucane MM, Danaei G, Pelizzari PM, Cowan MJ, Pacioreket CJ, et al. National, regional and global trends in serum total cholesterol since 1980: systematic analysis of health examination surveys and epidemiological studies with 321 country-years and 3.0 million participants. Lancet. 2011;377:578-86.
2. Kravitz L, Heyward V. The Exercise & Cholesterol Controversy. IDEA Today. 1994;12:38-42.
3. Expert Panel on Detection, Evaluation, and Treatment of High Blood Cholesterol in Adults. Executive Summary of the Third Report of the National Cholesterol Education Program (NCEP) Expert Panel on Detection, Evaluation, and Treatment of High Blood Cholesterol in Adults (Adult Treatment Panel III). JAMA. 2001;285(19):2486-97.
4. Appel LJ, Sacks FM, Carey VJ, Obarzanek E, Swain JF, Miller ER III, et al. Effects of protein, monounsaturated fat, and carbohydrate intake on blood pressure and serum lipids: Results of the Omniheart Randomized Trial. JAMA. 2005;294(19):2455-64.
5. Villareal DT, Miller BV III, Banks M, Fontana L, Sinacore DR, Klein S. Effect of lifestyle intervention on metabolic coronary heart disease risk factors in obese older adults. Am J Clin Nutr. 2006;84(6):1317-23.
6. Third Report of the National Cholesterol Education Program (NCEP) Expert Panel on Detection, Evaluation, and Treatment of High Blood Cholesterol in Adults (Adult Treatment Panel III). Final report. Circulation. 2002;106(25):3143-421.

7. Van Horn L, McCoin M, Kris-Etherton PM, Burke F, Carson JAS, Champagne CM, et al. The evidence for dietary prevention and treatment of cardiovascular disease. J Am Diet Assoc. 2008;108(2):287-331.
8. Lichtenstein AH, Appel LJ, Brands M, Carnethon M, Daniels S, Franch HA, et al. American Heart Association Nutrition Committee. Diet and lifestyle recommendations revision 2006: a scientific statement from the American Heart Association Nutrition Committee. Circulation. 2006;114:82.
9. Mozaffarian D, Katan MB, Ascherio A, Stampfer MJ, Willett WC. Trans fatty acids and cardiovascular disease. N Engl J Med. 2006;354(15):1601-13.
10. Lichtenstein AH, Ausman LM, Jalbert SM, Schaefer EJ. Effects of different forms of dietary hydrogenated fats on serum lipoprotein cholesterol levels. N Engl J Med. 1999;340:1933.
11. Mensink RP, Zock PL, Kester AD, Katan MB. Effects of dietary fatty acids and carbohydrates on the ratio of serum total to HDL cholesterol and on serum lipids and apolipoproteins: a meta-analysis of 60 controlled trials. Am J Clin Nutr. 2003;77(5):1146-55.
12. Oh K, Hu FB, Manson JE, Stampfer MJ, Willett WC. Dietary fat intake and risk of coronary heart disease in women: 20 years of follow-up of the Nurses' Health Study. Am J Epidemiol. 2005;161(7):672-9.
13. Hu FB, Stampfer MJ, Manson JE, Rimm E, Colditz GA, Rosner BA, et al. Dietary fat intake and the risk of coronary heart disease in women. N Engl J Med. 1997;337(21):1491-9.
14. Wister AP, Loewen NMD, Kennedy-Symonds HM, McGowan BMD, McCoy BMA, Singer JP. One-year follow-up of a therapeutic lifestyle intervention targeting cardiovascular disease risk. CMAJ. 2007;177(8):859-65.
15. Jones PJ, MacDougall DE, Ntanios F, Vanstone CA. Dietary phytosterols as cholesterol-lowering agents in humans. Can J Physiol Pharmacol. 1997;75:217.
16. Katan MB, Grundy SM, Jones P, Law M, Miettinen T, Paoletti R, et al. Efficacy and safety of plant stanols and sterols in the management of blood cholesterol levels. Mayo Clin Proc. 2003;78:965.
17. Jovanovski E, Yashpal S, Komishon A, Zurbau A, Mejia SB, Ho HVT, et al. Effect of psyllium (Plantago ovata) fiber on LDL cholesterol and alternative lipid targets, non-HDL cholesterol, and apolipoprotein B: a systematic review and meta-analysis of randomized controlled trials. Am J Clin Nutr. 2018;108:922.
18. Mozaffarian D, Wu JH. Omega-3 fatty acids and cardiovascular disease: effects on risk factors, molecular pathways, and clinical events. J Am Coll Cardiol. 2011;58:2047.
19. Abbey M, Clifton P, Kestin M, Belling B, Nestel P. Effect of fish oil on lipoproteins, lecithin: cholesterol acyltransferase, and lipid transfer protein activity in humans. Arteriosclerosis. 1990;10:85.
20. Nordmann AJ, Suter-Zimmermann K, Bucher HC, Shai I, Tuttle KR, Estruch R, et al. Meta-analysis comparing Mediterranean to low-fat diets for modification of cardiovascular risk factors. Am J Med. 2011;124:841.
21. Katcher HI, Hill AM, Lanford JL, Yoo JS, Kris-Etherton PM. Lifestyle approaches and dietary strategies to lower LDL-cholesterol and triglycerides and raise HDL-cholesterol. Endocrinol Metab Clin North Am. 2009;38(1):45-78.
22. Halbert JA, Silagy CA, Finucane P, Withers RT, Hamdorf PA. Exercise training and blood lipids in hyperlipidemic and normolipidemic adults: a meta-analysis of randomized, controlled trials. Eur J Clin Nutr. 1999;53(7):514-22.
23. Thompson PD, Cullinane EM, Sady SP, Flynn MM, Bernier DN, Kantor MA, et al. Modest changes in high-density lipoprotein concentration and metabolism with prolonged exercise training. Circulation. 1988;78:25-34.
24. Seip RL, Moulin P, Cocke T, Tall A, Kohrt WM, Mankowitz K, et al. Exercise training decreases plasma cholesteryl ester transfer protein. Arterioscler Thromb. 1993;13:1359-67.
25. Marniemi J, Dahlström S, Kvist M, Seppanen A, Hietanen E. Dependence of serum lipid and lecithin: cholesterol acyltransferase levels on physical training in young men. Eur J Appl Physiol Occup Physiol. 1982;49:25-35.
26. Kelley GA, Kelley KS. Aerobic exercise and HDL2-C: a meta-analysis of randomized controlled trials. Atherosclerosis. 2006;184:207-15.
27. Kamani CH, Gencer B, Montecucco F, Courvoisier D, Vuilleumier N, Meyer P, et al. Stairs instead of elevators at the workplace decreases PCSK9 levels in a healthy population. Eur J Clin Investig. 2015;45(10):1017-24.

CHAPTER 16

Nonalcoholic Fatty Liver Disease: The Lipid Disease of the Liver and the Effect of Lipid-lowering Drugs

Anshita Aggarwal, Deep Dutta

ABSTRACT

Nonalcoholic fatty liver disease (NAFLD) is growing at pandemic proportions worldwide, following closely on the heels of the rising prevalence of obesity and type 2 diabetes mellitus (T2DM). It has an intricate association with metabolic risk factors and thus control of these metabolic risk factors remains the mainstay of the treatment. Dyslipidemia is commonly associated with NAFLD, warranting the use of drugs such as statins. However, the role of anti-lipid agents for treating NAFLD/nonalcoholic steatohepatitis (NASH) per se is still controversial. Studies with statins, dual peroxisome proliferator-activated receptor (PPAR) α and γ agonists, ezetimibe, and omega-3-fatty acids have shown mixed results, wherein few studies showed promising data while others did not. PPARα agonists have not shown much benefit. Stearoyl-CoA desaturase-1 inhibitors are novel anti-lipid agents that are currently being evaluated for their role in NAFLD.

INTRODUCTION

Nonalcoholic fatty liver disease (NAFLD) is not an isolated disease per se, but rather a spectrum of disease. This spectrum ranges from simple steatosis to Nonalcoholic steatohepatitis (NASH), through to advanced fibrosis and cirrhosis. NAFLD is defined as the development of steatosis in >5% of hepatocytes identified either histologically or radiologically, after ruling out secondary causes such as viral hepatitis, alcohol, or hereditary liver diseases.[1] Over the years, NAFLD has emerged as the most common form of chronic liver disease in both the developed as well as the developing countries, mirroring the exponential rise in the prevalence of obesity and type 2 diabetes mellitus (T2DM).[2] Moreover, NASH is not considered innocuous anymore. An increased risk of hepatocellular carcinoma (HCC) has been documented in patients of NASH, that too with a worse prognosis.

EPIDEMIOLOGY AND RISK FACTORS

According to a very recent study, obesity prevalence increased in every country between 1975 and 2016.[3] The exact prevalence varies from country to country, ranging from 3.7% in Japan to 38.2% in the United States.[4] India has the dubious distinction of being one of the countries that have shown an accelerated increase in body mass index (BMI). The prevalence of NAFLD is also highly variable, ranging from 25% in the general population to >90% in very obese individuals.[5] Thus obesity remains as one of the most important risk factors for NAFLD. However, the true prevalence may be even higher as many patients of NAFLD go undiagnosed. Other than obesity, T2DM is another important risk factor for NAFLD, and vice-versa. The prevalence of NAFLD and NASH in patients with T2DM is over 60%.[5] Older age, female sex, and Hispanic ethnicity have also been established as risk factors for NAFLD.[5,6] NAFLD has also been associated with other cardiovascular disease risk factors such as dyslipidemia, hypertension, and smoking.[7]

PATHOGENESIS

The pathogenesis of NAFLD could perhaps be described aptly as the "dyslipidemia of the liver". There is an increase in intrahepatic lipid accumulation, which can be attributed to an interplay between various factors such as dyslipidemia, insulin resistance, and obesity. A specific pattern of "atherogenic" dyslipidemia has been observed in patients of NAFLD, which is characterized by elevated triglycerides (TGs) and small-dense low-density lipoprotein cholesterol (sdLDL-C) levels and by decreased high-density lipoprotein cholesterol (HDL-C) concentrations.[8] Insulin resistance in adipose tissue (due to increased visceral adiposity in obese individuals) impairs the ability of insulin to suppress lipolysis; as a result of which free fatty acids (FFAs) released from adipose tissue are transported into the liver. These FFAs contribute to ~60% of the hepatic lipid accumulation. Not only this, but the insulin resistance also leads to increased hepatic very-low-density lipoprotein (VLDL) production and enhanced hepatic gluconeogenesis. VLDL consists of almost all of the measurable TGs. Visceral adiposity also increases the drainage of FFAs into the liver via increased levels of portal blood FFAs. All of these mechanisms combine to culminate into unrestricted hepatic steatosis.[9] Hepatic steatosis leads to oxidative stress, mitochondrial dysfunction, and generation of reactive oxygen species. All of this culminates in hepatic ballooning, inflammation, and cell death. It also triggers the activation of Kupffer cells and hepatic stellate cells, which are the main drivers of fibrosis and then cirrhosis and proliferate in response to TNF-alpha, IL-1 beta, and other proinflammatory cytokines.

DIAGNOSIS

The diagnosis of NAFLD requires the exclusion of both secondary causes and a daily alcohol consumption ≥30 g for men and ≥20 g for women. All patients with metabolic risk factors should be evaluated for NAFLD. While the diagnosis of NAFLD is mostly made radiologically, the diagnosis of NASH requires a liver biopsy. Though elevated liver enzymes can be seen with NAFLD, they may be normal in some patients even in the presence of steatosis. Conversely, liver enzymes may be deranged even in the absence of steatosis. Surrogate markers of fibrosis (NFS, FIB-4, ELF, or FibroTest) are also available to rule out significant fibrosis.

MANAGEMENT

Since obesity and insulin resistance are the main factors behind NAFLD, lifestyle modifications remain the mainstay of the treatment of NAFLD. A weight loss of 7–10% for overweight and obese individuals is optimal, which can be achieved via dietary restriction and aerobic exercise along with resistance training. Diet should be devoid of processed foods and fructose-containing food items.

Pharmacotherapy is primarily recommended for progressive NASH, though it can be tried in early-stage NASH too. While there is no robust data, but existing guidelines say that pioglitazone, vitamin E, or their combination can be tried for NASH. The role of the various anti-lipid agents in NAFLD/NASH has been elaborated upon below.

Statins

Many of the NAFLD patients are on statin therapy, not because of the disease per se, but because of the associated comorbidities such as dyslipidemia, T2DM, and coronary artery disease (CAD). At the same time, many physicians are wary of initiating NAFLD patients on statins, fearing the worsening of transaminitis. Management of atherogenic dyslipidemia in NAFLD is warranted and should be done as per the standard guidelines for the general population. Statins are safe even in patients with baseline elevated liver enzymes and have low hepatotoxicity, in fact, they can lead to an improvement in liver function tests.[10] As per the available literature, the incidence of acute liver failure in patients on statin therapy is not any different from that seen in the general population (1:130,000 vs. 1:114,000).[11]

All of the existing guidelines agree on the safety of statins in NAFLD, even in cirrhosis. Decompensated cirrhosis and acute liver failure are the only contraindications to its use.

Patients may experience an elevation in alanine aminotransferase (ALT) levels with statin therapy, but this increase is not synonymous with liver injury and remains asymptomatic in most cases. Meta-analyses[12] have shown that treatment with low-moderate intensity statins is safe but high-intensity statin treatment may lead to hepatotoxicity, particularly hydrophilic statins (pravastatin and atorvastatin). It is usually recommended that if ALT <3xULN, statin therapy can be continued with annual monitoring of liver enzymes. If values rise ≥3xULN, therapy should be withheld or the dose should be reduced and liver enzymes should be rechecked after 4–6 weeks. Though all statins are equally efficacious in lowering cholesterol levels in NAFLD patients, the experience is richer with atorvastatin, and it is the only statin that has shown a reduction in cardiovascular events in these patients.

Though there is a plethora of data on the safety of statins in NAFLD, the same cannot be said for their beneficial effects in NAFLD/NASH. Owing to the pleiotropic effects of statins, which include anti-inflammatory and antioxidant effects, they have been proposed for the treatment of NAFLD/NASH. In a recent meta-analysis that included two trials, an improvement in liver enzymes and ultrasound findings was seen in NAFLD patients with statin therapy, but there was no data on histology. These trials included a small number of patients and had a high risk of bias.[13]

In a study by Hyogo et al.[14] the role of pitavastatin in NASH was evaluated. About 20 patients with biopsy-proven NASH received pitavastatin (2 mg/day) for 12 months. Liver enzymes improved at the end of 12 months but there was no significant improvement in the NAFLD activity score or fibrosis stage. Similar results were seen in another study by Nakahara et al.[15] wherein 19 patients of biopsy-proven NASH received rosuvastatin (2.5 mg/day) for 24 months.

However, few studies have shown improvement in liver histology with statins. In a study from Italy and Finland with biopsy-proven NASH (n = 107), statin therapy was associated with protection from steatosis, NASH, and advancement of fibrosis stage.[16] Similar results were reported in a study from Italy and France which analyzed 346 diabetics with biopsy-proven NAFLD.[17] The retrospective nature of both the studies, however, makes it difficult to draw a causal association between statin therapy and liver histology. Also, in the former study, the proportion of patients with fibrosis was small, which is again a limitation. The post-hoc analysis of two large studies [GREACE (Greek Atorvastatin and Coronary Heart Disease Evaluation) study[18] and IDEAL (Identification of the Determinants of the Efficacy of Arterial Blood Pressure Lowering Drugs) study[19]] investigated the efficacy and safety of statins in CAD patients with NASH/NAFLD (total number of participants combined >10,000). Apart from a significant reduction in secondary cardiovascular (CV) events with atorvastatin, a significant improvement in liver enzymes was seen with atorvastatin. Interestingly, the statin-related relative risk reduction in CV events was greater in patients with deranged liver enzymes than in those with normal liver tests. However, neither of the studies had data on histopathology, which was a major drawback.

In a retrospective analysis of 18,080 NAFLD patients, the use of statins was seen as an independent factor associated with a decreased risk of HCC development. However, the study had several limitations such as lack of prospective data and lack of histopathological data.[20]

Based on all the available literature thus, we can conclude that the safety of statins in NAFLD, even with deranged liver enzymes is unquestionable. The data on efficacy in amelioration of NAFLD on the other hand is controversial, with different studies showing varied results. Not surprisingly, none of the guidelines (EASL/NICE/Asia-Pacific/AISF/AASLD) have said that statins are beneficial for NASH per se. Large randomized-controlled trials (RCTs) with data on biopsy are needed to further study the same.

Peroxisome Proliferator-activated Receptor Agonists

Fibrates are agonists of the peroxisome proliferator-activated receptor alpha (PPARα) isoform that is mainly expressed in the hepatocytes, and are used commonly for the treatment of hypertriglyceridemia. A protective role of PPARα against liver steatosis and inflammation in NASH has been suggested based on the enhanced vulnerability of PPARα-knock out mice to NASH. Their clinical role in NAFLD/NASH has only been studied in small studies and has not shown promising results.

In an RCT of NAFLD patients (n = 186), with three arms (atorvastatin/fenofibrate/both), atorvastatin fared significantly better than fenofibrate in improving liver enzymes and ultrasonography features.[21] In yet another RCT on NASH patients, fibrates demonstrated no significant benefit on the biochemical, radiological, or histological outcome.[22] Saroglitazar is a dual PPARα and PPARγ agonist. PPARγ agonists are potent insulin sensitizers. In mouse studies, saroglitazar has shown biopsy-proven improvement in NASH.[23]

Saroglitazar is a predominant PPARα agonist with some PPARγ activity also. It has been found to be useful and is approved for use for managing hypertriglyceridemia in T2DM in India, not controlled by statins. Initial data has suggested some benefit of saroglitazar on NAFLD. EVIDENCES II is an on-going biopsy-driven trial of saroglitazar

4 mg versus placebo in patients with NASH. The primary endpoint will be evaluated by the histological improvement of NASH using liver biopsy at the end of 52 weeks. The phase III of the trial has shown promising results and after this, the Drug Controller General of India (DCGI) has approved the drug for NASH. Two other drugs, elafibranor (dual PPARα and PPARγ agonist) and lanifibranor (pan PPAR agonist) are currently under phase 3 and phase 2 of clinical trials, respectively to study their efficacy in NASH.

Ezetimibe

Ezetimibe, which is a potent inhibitor of cholesterol absorption, has been used as a treatment for NAFLD/NASH, but the results have been highly variable. In a study of 10 NASH patients, ezetimibe was given at a dose of 10 mg daily for 6 months.[24] At the end of 6 months, a significant improvement was seen in the liver enzymes. Also, the histological features of steatohepatitis were reduced, though there was no significant improvement in the fibrosis stage. However, the small sample size and the lack of a control group in the aforementioned study somewhat abrogated the impact of the findings.

Similar findings were observed in yet another study wherein 45 patients of NASH were treated with ezetimibe (10 mg OD) for 24 months, though this study too, lacked a control arm.[25] Various possible mechanisms of action of ezetimibe have been postulated. These include inhibition of absorption of biliary cholesterol in the liver [via inhibition of the transporter Niemann-Pick C1-Like 1 (NPC1L1)], improvement in insulin resistance, and increased sterol regulatory element-binding protein-1c (SREBP-1c) mRNA expression in the liver. The recent MOZART trial was a randomized, double-blind, placebo-controlled trial wherein 50 patients with biopsy-proven NASH were randomized to either ezetimibe 10 mg OD or placebo for 24 weeks.[26] The primary outcome was a change in liver fat as measured by magnetic-resonance-imaging derived proton-density-fat-fraction (MRI-PDFF). Ezetimibe did not fare better than the placebo in achieving the primary outcome.

As per a meta-analysis[27] which included three other studies apart from the three aforementioned ones, it was concluded that though ezetimibe improved hepatic steatosis, it did not improve hepatic fibrosis and inflammation. The meta-analysis itself had many limitations, including the fact that the duration of therapy was highly variable between the studies. Also, it included only two RCTs. The only positive change observed in these RCTs was an amelioration of hepatocyte ballooning.

To summarize, in light of the conflicting evidence in support of ezetimibe for NAFLD, its clinical use in this context remains controversial and has not found a place in any of the guidelines yet. Large scale RCTs are needed to approve or disprove its benefits.

Omega-3 Polyunsaturated Fatty Acids

It has been postulated that omega-3 polyunsaturated fatty acids (ɷ-3 PUFAs) may ameliorate hepatic lipogenesis and inflammation and thus may have a role in the treatment of NASH. The results of the clinical trials examining the therapeutic benefit of ɷ-3 PUFAs in NAFLD have been inconsistent. In a recent meta-analysis,[28] which included 18 RCTs (n = 1424), an improvement in liver enzymes and reduction in liver fat was seen with ɷ-3 PUFAs. These RCTs included studies on pediatric NAFLD as well. The postulated mechanism of action is via the downregulation of SREBP-1c and upregulation of PPAR-α. Out of these 18 RCTs, only 5 had data on liver biopsy. Out of these five, three studies did not show any significant change in the histological parameters while two did. The rest of the studies evaluated liver fat radiologically.

Because of the varied results with ω-3 PUFAs, their role in NAFLD is not established yet and is not endorsed by any of the guidelines. More large-scale RCTs are needed to study hard endpoint such as progression to cirrhosis, and resolution of hepatic inflammation.

Stearoyl-CoA Desaturase-1 Inhibitors (Aramchol)

It is a novel inhibitor that inhibits the catalysis of the rate-limiting step in the synthesis of monounsaturated fatty acids, thereby leading to decreased liver injury in response to a high-fat diet and it also decreases insulin resistance. These results suggest that stearoyl-CoA desaturase (SCD)-1 inhibitor may be a promising agent for the treatment of NASH. It has shown attenuation of liver injury and inflammation in mouse models of NASH. Currently, there is a multicentre Phase 2b trial[29] ongoing to study its efficacy in NASH. The initial phase 3-month study showed a reduction in hepatic steatosis on MRI.

CONCLUSION

While NAFLD continues to grow as a pandemic, the need for good therapeutic agents in now greater than ever before. The various old, and a few novel anti-lipid agents may have a promising role to play, but there is still a lot of scope for further research in this area which can help in expanding the therapeutic armamentarium for NAFLD/NASH.

KEY POINTS

- The prevalence of NAFLD is accelerating the worldover, mirroring the exponential increase in the prevalence of obesity and type 2 diabetes mellitus.
- Though NAFLD and dyslipidemia are intricately connected, the role of lipid-lowering drugs in the treatment of NAFLD per se is not yet well established.
- Certain lipid-lowering agents like Statins, PPAR-alpha agonists, Cholesterol absorption inhibitors, and Omega-3 fatty acids have shown benefits in NAFLD in a few studies, but more robust data is needed to establish their usefulness.
- There are newer drugs such as Stearoyl-CoA denaturase-1 inhibitors in the pipeline that may prove to be efficacious in NAFLD in the coming future.

REFERENCES

1. Huang TD, Behary J, Zekry A. Non-alcoholic fatty liver disease (NAFLD): a review of epidemiology, risk factors, diagnosis and management. Intern Med J. 2019.
2. Mahady SE, Adams LA. Burden of non-alcoholic fatty liver disease in Australia. J Gastroenterol. Hepatol. 2018;33:1-11.
3. NCD Risk Factor Collaboration (NCD-RisC). Worldwide trends in body-mass index, underweight, overweight, and obesity from 1975 to 2016: a pooled analysis of 2416 population-based measurement studies in 128.9 million children, adolescents, and adults. Lancet. 2017;390:2627-42.
4. Organisation for Economic Co-operation and Development. (2017). Obesity update 2017. [online] Available from https:// www.oecd.org/els/health-systems/Obesity-Update2017.pdf. [Last accessed June, 2020]
5. Younossi Z, Tacke F, Arrese M, Sharma BC, Mostafa I, Bugianesi E, et al. Global perspectives on non-alcoholic fatty liver disease and non-alcoholic steatohepatitis. Hepatology. 2018.

6. Younossi Z, Anstee QM, Marietti M, Hardy T, Henry L, Eslam M, et al. Global burden of NAFLD and NASH: trends, predictions, risk factors and prevention. Nat Rev Gastroenterol Hepatol. 2018;15:11-20.
7. Katsiki N, Mikhailidis DP, Mantzoros CS. Non-alcoholic fatty liver disease and dyslipidemia: An update. Metab Clin Exp. 2016;65(8):1109-23.
8. Chatrath H, Vuppalanchi R, Chalasani N. Dyslipidemia in patients with nonalcoholic fatty liver disease. Semin Liver Dis. 2012;32(1):22-9.
9. Birkenfeld AL, Shulman GI. Nonalcoholic fatty liver disease, hepatic insulin resistance, and type 2 Diabetes. Hepatology. 2014;59,:713-23.
10. Pastori D, Polimeni L, Baratta F, Pani A, del Ben M, Angelico F. The efficacy and safety of statins for the treatment of non-alcoholic fatty liver disease. Dig Liver Dis. 2015;47:4-11.
11. Onofrei MD, Butler KL, Fuke DC, Miller HB. Safety of statin therapy in patients with preexisting liver disease. Pharmacotherapy. 2008;28:522-9.
12. Dale KM, White CM, Henyan NN, Kluger J, Coleman CI. Impact of statin dosing intensity on transaminase and creatine kinase. Am J Med. 2007;120:706-12.
13. Eslami L, Merat S, Malekzadeh R, Nasseri-Moghaddam S, Aramin H. Statins for nonalcoholic fatty liver disease and nonalcoholic steatohepatitis. Cochrane Database of Syst Rev. 2013;12(12):CD008623.
14. Hyogo H, Ikegami T, Tokushige K, Hashimoto E, Inui K, Matsuzaki Y, et al. Efficacy of pitavastatin for the treatment of non-alcoholic steatohepatitis with dyslipidemia: an open-label, pilot study. Hepatol Res. 2011;41:1057-65.
15. Nakahara T, Hyogo H, Kimura Y, Ishitobi T, Arihiro K, Aikata H, et al. Efficacy of rosuvastatin for the treatment of non-alcoholic steatohepatitis with dyslipidemia: an open-label, pilot study. Hepatol Res. 2012;42:1065-72.
16. Dongiovanni P, Petta S, Mannisto V, Mancin RM, Pipitone R, Karja V, et al. Statin use and non-alcoholic steatohepatitis in at risk individuals. J Hepatol. 2015;63(3):705-12.
17. Nascimbeni F, Aron-Wisnewsky J, Pais R, Tordjman J, Poitou C, Charlotte F, et al. Statins, antidiabetic medications and liver histology in patients with diabetes. BMJ Open Gastroenterol. 20168;3(1):e000075.
18. Athyros VG, Tziomalos K, Gossios TD, Griva T, Anagnostis P, Kargiotis K, et al. Safety and efficacy of long-term statin treatment for cardiovascular events in patients with coronary heart disease and abnormal liver tests in the Greek Atorvastatin and Coronary Heart Disease Evaluation (GREACE) Study: A post-hoc analysis. Lancet. 2010;376(9756):1916-22.
19. Pedersen TR, Faergeman O, Kastelein JJ, Olsson AG, Tikkanen MJ, Holme I, et al. High-dose atorvastatin vs usual-dose simvastatin for secondary prevention after myocardial infarction: the IDEAL study: A randomized controlled trial. JAMA. 2005;294(19):2437-45.
20. Lee T-Y, Wu J-C, Yu S-H, Lin J-T, Wu M-S, Wu C-Y. The occurrence of hepatocellular carcinoma in different risk stratifications of clinically noncirrhotic nonalcoholic fatty liver disease. Int J Cancer. 2017;141(7):1307-14.
21. Athyros VG, Mikhailidis DP, Didangelos TP, Giouleme OI, Liberopoulos EN, Karagiannis A, et al. Effect of multifactorial treatment on nonalcoholic fatty liver disease in metabolic syndrome: a randomised study. Curr Med Res Opin. 2006;22:873-83.
22. Basaranoglu M, Acbay O, Sonsuz A. A controlled trial of gemfibrozil in the treatment of patients with nonalcoholic steatohepatitis. J Hepatol. 1999;31:384.
23. Jain MR, Giri SR, Bhoi B, Trivedi C, Rath A, Rathod R, et al. Dual PPARα/γ agonist saroglitazar improves liver histopathology and biochemistry in experimental NASH models. Liver Int. 2018;38(6):1084-94.
24. Yoneda M, Fujita K, Nozaki Y, Endo H, Takahashi H, Hosono K, et al. Efficacy of ezetimibe for the treatment of non-alcoholic steatohepatitis: an open-label, pilot study. Hepatol Res. 2010;40:566-73.
25. Park H, Shima T, Yamaguchi K, Mitsuyoshi H, Minami M, Yasui K, et al. Efficacy of long-term ezetimibe therapy in patients with nonalcoholic fatty liver disease. J Gastroenterol. 2011;46:101-7.
26. Loomba R, Sirlin CB, Ang B, Bettencourt R, Jain R, Salotti J, et al. San Diego Integrated NAFLD Research Consortium (SINC). Ezetimibe for the treatment of nonalcoholic steatohepatitis: assessment by novel magnetic resonance imaging and magnetic resonance elastography in a randomized trial (MOZART trial). Hepatology. 2015;61:1239-50.
27. Nakade Y, Murotani K, Inoue T, Kobayashi Y, Yamamoto T, Ishii N, et al. Ezetimibe for the treatment of non-alcoholic fatty liver disease: A meta-analysis. Hepatol Res. 2017;47(13):1417-28.
28. Yan J-H, Guan B-J, Gao H-Y, Peng X-E. Omega-3 polyunsaturated fatty acid supplementation and non-alcoholic fatty liver disease: A meta-analysis of randomized controlled trials. Medicine. 2018;97(37):e12271.
29. Clinical Trials.gov. A Clinical Trial to Evaluate the Efficacy and Safety of Two Aramchol Doses Versus Placebo in Patients With NASH. [online] Available from https://clinicaltrials.gov/ct2/show/NCT02279524. [Last accessed June, 2020].

CHAPTER 17

Beyond Statins: Who and When to Prescribe?

Subhodip Pramanik

ABSTRACT

Cardiovascular disease (CVD) is a major challenge for the 21st century, exacerbated by the pandemics of metabolic syndrome, obesity, and type 2 diabetes mellitus. While best standards of care, including high-dose statins, can reduce the risk of vascular complications, patients still remain at high-risk of cardiovascular (CV) events. Newer studies reveal lowering low-density lipoprotein cholesterol (LDL-C) beyond traditional cutoffs is possibly beneficial and lower the LDL-C, better the outcome. Ezetimibe and monoclonal antibody therapy targeting proprotein convertase subtilisin/kexin type 9 (PCSK9) inhibitors reduce LDL-C and lowered the CV events when used on top of statins. Similarly, many other LDL-C-lowering therapies (mipomersen, lomitapide, bempedoic acid, gemcabene, and others) are now under development. Additionally, one has to look beyond statins where we find atherogenic dyslipidemia—a combination of elevated triglyceride (TG) and low high-density lipoprotein (HDL), which is an important modifiable contributor to residual CV risk. Efforts are required to improve the awareness and clinical management of atherogenic dyslipidemia to decrease residual CV risk seen beyond the usage of statins. Fibrates in selected patients (high TG ≥204 mg/dL and low HDL ≤34 mg/dL) and high-dose omega-3 fatty acid (icosapent ethyl) have shown to improve CV outcomes. Several emerging treatments offer promise. These also include the cholesteryl ester transfer protein inhibitors and saroglitazar. However, long-term outcomes and safety data are needed before clinical use.

INTRODUCTION

Atherosclerotic cardiovascular disease (ASCVD), including coronary heart disease (CHD), stroke, and peripheral arterial disease, is the most prominent cause of death worldwide. About 75% of the cases are due to abnormal blood lipids (dyslipidemia), smoking, and high blood pressure. But among the nine independent risk factors for myocardial infarction (MI), dyslipidemia is associated with the highest risk that can be attributed to a population.[1]

Additionally, the evidence from randomized controlled trials (RCTs) supporting the importance of aggressive low-density lipoprotein cholesterol (LDL-C) lowering is the strongest for various Cardiovascular disease (CVD) therapies, particularly for statins.[2] However, despite use of proven therapies, event rates in RCTs are relatively high, emphasizing and supporting the search for novel treatments. In addition to trying and testing of various LDL-C-lowering therapies, there has been an increased emphasis on therapies that can favorably influence high-density lipoprotein cholesterol (HDL-C) and triglyceride (TG) concentrations. Such treatments may have particular advantage in people with diabetes, which is related with both an increased case fatality rate after acute coronary events and a worsened long-term prognosis.

CARDIOVASCULAR RISK REDUCTION WITH STATIN AND ITS LIMITATIONS

A linear relationship is clearly distinguishable between the degree of LDL-C lowering achieved with statins and benefits, with a 10% and 21% reduction seen in all-cause mortality and major vascular events, respectively per 39 mg/dL reduction in LDL-C levels. Despite the immense evidence supporting the use of statins, other LDL-C-lowering therapies are needed. This is because the LDL-C target suggested in various guidelines, <70 mg/dL in highest-risk patients, is often not attained. Patients may be intolerant to statins. Finally, in addition to the high "residual risk" despite reaching the proposed LDL-C target, more aggressive therapy may be required in special conditions.[3]

Extensive evidence points to the fact that both elevated TGs and low HDL-C are predictive for CVD, independent of LDL-C concentration. Various guidelines and expert consensus have also recognized the importance of atherogenic dyslipidemia as a key driver of cardiovascular (CV) risk in insulin-resistant states, even if LDL-C levels are well-controlled.[4] Thus, targeting atherogenic dyslipidemia secondary to LDL-C reduction has the potential for reducing this risk.

LIPID GUIDELINES AND THE RATIONALE FOR THERAPIES BEYOND STATINS

A meta-analysis of 26 large statin trials provided the firm evidence that statins reduce major adverse cardiovascular events (MACEs) by 20–40% over 5 years in both secondary and primary preventions and its dependent on the dose.[5] So, not surprisingly, most lipid guidelines are more concerned about LDL-C lowering either in the form of percentage lowering or lowering below a specific cutoff. The current lipid guidelines by the AHA/ACC (American Heart Association/American College of Cardiology)[6] recommends that statins should be employed as the primary drug in cardiovascular events (CVE) reduction and more intensively for all patients with prior CVE, those with familial hypercholesterolemia (FH), LDL-C ≥190 mg/dL, or those with diabetes, aged 40–75 years, and 10-year CVE risk ≥7.5%. Those with moderate risk, i.e., 10-year CVE risk 5–7.5% should be treated with moderate-intensity statin. Here the intensive- and moderate-dose statin therapy is defined as LDL-C reduction ≥50% and 30–49% from baseline. However, this percentage reduction of LDL-C approach is not followed by many other organizations as the response to intensive statin therapy may be highly variable, partly due to ethnicity, gender, and pharmacogenetic factors. The American Association of Clinical Endocrinologists (AACE)[7]

TABLE 1: The ASCVD risk categories and LDL-C treatment goals (adapted from American College of Clinical Endocrinology recommendations 2017).

Risk category	Risk factors/10-year risk	Treatment goals		
		LDL-C (mg/dL)	Non-HDL-C (mg/dL)	ApoB (mg/dL)
Extreme risk	• Progressive ASCVD including unstable angina in individuals after achieving an LDL-C <70 mg/dL • Established clinical cardiovascular disease in individuals with DM, stage 3 or 4 CKD, or HeFH • History of premature ASCVD (<55 male, <65 female)	<55	<80	<70
Very high-risk	• Established or recent hospitalization for ACS, coronary, carotid, or peripheral vascular disease, and 10-year risk >20% • DM or stage 3 or 4 CKD with one or more risk factor(s) • HeFH	<70	<100	<80
High-risk	• ≥2 risk factors and 10-year risk 10–20% • DM or stage 3 or 4 CKD with no other risk factors	<100	<130	<90
Moderate risk	≤2 risk factors and 10-year risk <10%	<100	<130	<90
Low-risk	Zero risk factors	<130	<160	NR

(ACS: acute coronary syndrome; ApoB: apolipoprotein B; ASCVD: atherosclerotic cardiovascular disease; CKD: chronic kidney disease; DM: diabetes mellitus; HDL-C: high-density lipoprotein cholesterol; HeFH: heterozygous familial hypercholesterolemia; LDL-C: low-density lipoprotein cholesterol; NR: not recommended)

in 2017 has recommended an LDL-C goal of <55 mg/dL and a non-HDL-C goal of <80 mg/dL in those with an "extreme risk", defined by factors such as progressive ASCVD, established ASCVD with diabetes mellitus (DM), chronic kidney disease (CKD) stage 3 or 4, or FH (**Table 1**). This level of LDL-C reduction often not possible with use of only statin in many cases due to limited efficacy and poor tolerability issues, thus requires additional LDL-C-lowering agents. Finally, even after adjustments for other risk factors (diabetes, albuminuria, smoking, and family history), many patients may have residual risk for ASCVD events from other lipid factors such as triglyceride-rich lipoproteins (TRLs).[4]

LOW-DENSITY LIPOPROTEIN CHOLESTEROL-LOWERING THERAPIES

Additional LDL-C-lowering therapies are required when specified LDL-C target for an individual is not reached with use of maximal tolerated dose of statin. The agents are enumerated in **Table 2**.

TABLE 2: Therapeutic agents to lower LDL-C beyond statins.		
Groups	**Agents**	**Mechanism of action**
Cholesterol absorption inhibitor	Ezetimibe	Inhibits the absorption of cholesterol and plant sterols
Nicotinic acid	Niacin	Inhibits a hormone-sensitive lipase in adipose tissue which reduces the breakdown of triglycerides
Bile acid sequestrants	Cholestyramine	Nonspecifically bind cholesterol-rich bile acids in the gut
PCSK9-targeted therapy	Alirocumab and Evolocumab	Prevents LDL receptor degradation in liver
	Inclisiran	siRNA that inhibits PCSK9 synthesis
Apolipoprotein B synthesis inhibitor	Mipomersen	Binds to messenger RNA to prevent its translation and formation of apolipoprotein B
Microsomal transfer protein (MTP) inhibitor	Lomitapide	Inhibits MTP, thus reduces assembly of apolipoprotein B
Newer agents in development	Bempedoic acid	ATP citrate lyase inhibitor
	Gemcabene	Reduced synthesis and increased clearance of hepatic TG-rich lipoproteins
	Angiopoietin-like protein 3 (ANGPLP3) inhibitors	Inhibits lipoprotein lipase
	THR-β agonist	Agonist of thyroid hormone β receptor

(ATP: adenosine triphosphate; LDL-C: low-density lipoprotein cholesterol; PCSK9: proprotein convertase subtilisin/kexin type 9; RNA: ribonucleic acid; TG: triglyceride; THR-β: thyroid hormone receptor-beta)

Ezetimibe

Ezetimibe inhibits the absorption of cholesterol and plant sterols targeting the Niemann-Pick C1-Like 1 (NPC1L1) receptor in the intestinal cells. Ezetimibe lowers LDL-C in the tune of 20% and its combination is rational as statins increase cholesterol absorption. Combination therapy with statin plus ezetimibe is an attractive alternative approach to the management of residual CV risk.

In a meta-analysis of 27 trials (>21,000 patients), there was impressive lowering of LDL-C (by 15%), non-HDL-C (by 13%) and TGs (by 5%), and raising of HDL-C (by 1.6%) with the combination of ezetimibe plus statin as compared with statin alone.[8]

In 2015, ezetimibe became the first nonstatin drug to show CV benefits in a large RCT, the IMPROVE-IT (Improved Reduction of Outcomes: Vytorin Efficacy International Trial).[9] This trial included >18,000 post-acute coronary syndrome (ACS) patients and randomized to simvastatin 40 mg plus ezetimibe or placebo. The median LDL-C was reduced significantly by additional 20% with the addition of ezetimibe versus placebo (69.5 mg/dL vs. 54.7 mg/dL) and the primary MACEs were reduced by a modest but significant 6.6% [hazard ratio (HR) 0.936; 95% confidence interval (CI) 0.89–0.99; p = 0.016] at 7.0 years. The data was even stronger in patients suffering from DM who revealed a 15% reduction in the primary endpoint (HR 0.85; 95% CI 0.78–0.94) as compared with only 2% in the non-DM subgroup overall (HR 0.98; 95% CI 0.91–1.04; p = 0.02). Currently, ezetimibe is recommended as the preferred initial drug in combination with statins or in those with statin intolerance.

Niacin

Recent trials with niacin (nicotinic acid) have not shown encouraging results. The AIM-HIGH (Atherothrombosis Intervention in Metabolic Syndrome With Low HDL/High Triglycerides) trial (n = 3,414) evaluated the effect of extended-release (ER) niacin (1.5-2 g/day) on residual CV risk in patients with CVD who were optimally treated with a statin but with residual atherogenic dyslipidemia (median HDL-C 35 mg/dL and median fasting TGs 161 mg/dL).[10]

The study was suspended 18 months earlier than planned due to futility. However, a subgroup analysis in individuals with TGs >200 mg/dL and HDL-C <32 mg/dL showed a 36% relative reduction in the primary composite endpoint (p = 0.032), consistent with findings from the fibrate meta-analysis.

More recently, the much larger HPS2-THRIVE[11] (Heart Protection Study 2-Treatment of HDL to Reduce the Incidence of Vascular Events) (n = 25,673) where niacin was used with laropiprant (which attenuates the niacin flushing response) failed to show a benefit on clinical outcomes. Moreover, there were safety issues with niacin/laropiprant, especially significant increase in diabetes complications, new-onset diabetes, infections, and musculoskeletal, gastrointestinal, bleeding, and skin adverse events, leading to subsequent worldwide withdrawal of this therapy.

Bile Acid Sequestrants

Cholestyramine (nonabsorbable anion-exchange resin) nonspecifically bind cholesterol-rich bile acids in the gut, thereby increasing the breakdown of cholesterol and fecal excretion of bile salts by interrupting the enterohepatic circulation. They have a moderate LDL-C-lowering capacity by 10-20% and were used particularly when statins were contraindicated or not tolerated. They may increase TG levels. LRC-CPPT (Lipid Research Clinics-Coronary Primary Prevention Trial)[12] demonstrated a significant reduction in CHD events with use of cholestyramine in primary prevention. However, upper and lower gastrointestinal side effects are relatively common and the use of bile acid sequestrants has been largely supplanted by other agents.

Proprotein Convertase Subtilisin/Kexin Type 9-targeted Therapy

Novel therapy targeting proprotein convertase subtilisin/kexin type 9 (PCSK9) has gained interest recently. PCSK9 has a pivotal role in promoting degradation of hepatic LDL receptors and hence in LDL homeostasis. Thus, inhibition of PCSK9 prevents LDL receptor degradation which in turn helps scavenging LDL-C from circulation, a rationale which is now supported by clinical trials.

Indeed, monoclonal antibody therapy targeting PCSK9 is an excellent therapeutic strategy to lower LDL-C. Both of the most advanced therapies, alirocumab and evolocumab (AMG 145), have been shown to improve LDL-C in the tune of 60-70%, even on top of statin therapy. It helped attainment of LDL-C goals in high-risk statin-treated patients and have beneficial effects on other lipids, including lowering atherogenic TRLs and lipoprotein(a) and modestly raising HDL-C.[13]

A 2.2-year follow-up study with evolocumab (the FOURIER outcome study)[14] did not reveal any mortality benefit however, the risk for MI or stroke was significantly reduced. The ODYSSEY OUTCOMES[15] trial followed-up patients after ACS for a median 2.8 years and demonstrated that alirocumab on top of intensive statin treatment reduced all-cause mortality in long-term (HR 0.85; 95% CI 0.73-0.98; p = 0.03).

Inclisiran, a synthetic small interfering ribonucleic acid (siRNA) molecule, has been shown to significantly decrease hepatic production of PCSK9 and cause a marked reduction in LDL-C levels.[16] It has an advantage of less frequent dosing, although CV outcome trials are still ongoing.

Antisense Oligonucleotides

Mipomersen is an antisense oligonucleotide which binds to messenger ribonucleic acid (RNA) and prevents its translation and formation of apolipoprotein B (ApoB). This in turn decreases formation of ApoB-containing lipoproteins, including LDL-C. It demonstrated effective reduction in LDL-C levels of 40–50%. However, it was associated with multiple side effects, including injection-site reactions in almost all, flu-like illness in many, and hepatic steatosis with transaminitis in up to 15% of patients. However, mipomersen has now approved in the US for use in patients with homozygous FH in whom need for apheresis can be minimized with use of this agent.[17]

HIGH-DENSITY LIPOPROTEIN CHOLESTEROL AND TRIGLYCERIDE

As discussed above, a significant residual risk for MACE persists even after achieving presently defined targets for LDL-C with help of lifestyle changes, statin therapy, and administration of ezetimibe and PCSK9 inhibitors. None of the LDL-C-lowering drugs have appreciable effects on TG or HDL-C. High TG and low HDL-C are characteristic components of diabetic dyslipidemia which is often seen in the setting of insulin resistance and metabolic syndrome. Various guidelines and expert consensus have also recognized the importance of atherogenic dyslipidemia as a key driver of CV risk in insulin-resistant states, even if LDL-C levels are well-controlled.[4]

Fibrates

Fibrates [peroxisome proliferator-activated receptor-α (PPARα) agonists] are one of the most potent option to lower TG, especially in the context of insulin-resistant conditions.

In subgroup analyses of the FIELD[18] (Fenofibrate Intervention and Event Lowering in Diabetes) study, type 2 diabetes mellitus (T2DM) patients who had marked atherogenic dyslipidemia, defined as TGs ≥204 mg/dL and low HDL-C, obtained greater benefit from fenofibrate treatment than those without this dyslipidemia. In this subgroup, which comprised nearly one-fifth of the total study population, there was a 27% relative reduction in CV risk (versus 11% in all patients). Subsequently, in the ACCORD[19] (Action to Control Cardiovascular Risk in Diabetes) lipid trial (n = 5,518), a predefined subgroup (with TG ≥204 mg/dL and HDL ≤34 mg/dL) analysis revealed substantial benefit with combining fenofibrate with background simvastatin in T2DM patients. In these patients (comprising 17% of total study population), there was ~30% relative reduction in CV events versus nil benefit in 83% of the study population without this dyslipidemia.

An ancillary study of the ACCORD lipid trial, compared with simvastatin monotherapy, the combination of fenofibrate plus simvastatin led to significant reductions in postprandial rise of TGs and ApoB-48, indicating a decrease in the accumulation of intestinally-derived TRL remnants. This effect of fenofibrate was selective to patients with elevated fasting TGs at baseline. This provides a possible explanation as to why the benefit of fibrates is specific to patients with atherogenic dyslipidemia. Additionally, there is also evidence of benefit with fenofibrate on few diabetes-related complications, including diabetic retinopathy and nephropathy.[20]

Omega-3 Fatty Acids

There were positive findings from JELIS[21] (Japan Eicosapentaenoic Acid Lipid Intervention Study) (n = 18,645) where the combination of omega-3 fatty acids [1.8 g/day of eicosapentaenoic acid (EPA)] plus statin versus statin alone was compared for treatment of dyslipidemia. A post hoc analysis of JELIS evaluated the effect of EPA treatment in individuals with atherogenic dyslipidemia as defined by TGs ≥150 mg/dL and HDL-C <40 mg/dL, representing 5% of the total study population. In this dyslipidemic group, EPA treatment reduced the risk of coronary artery disease by 53% (p = 0.043), compared with individuals without this dyslipidemia.

In REDUCE-IT (Reduction of Cardiovascular Events with Icosapent Ethyl-Intervention Trial),[22] a randomized trial designed to evaluate the effect of icosapent ethyl on CVD outcomes in patients with hypertriglyceridemia, over 8,000 patients with elevated TG levels (fasting levels 135–499 mg/dL) who are already on statin and with either established CVD or diabetes plus other CV risk factors were randomized to receive either icosapent ethyl 4 g daily or mineral oil. Icosapent ethyl reduced the primary outcome (CV death, nonfatal MI, nonfatal stroke, coronary revascularization, or unstable angina) (17.2% vs. 22.0%; HR 0.75; 95% CI 0.68–0.83) after median follow-up of 4.9 years. The reduction of TG was modest (18%) in the treatment group whereas it was increased 2.2% in the control group. A low-dose omega-3 fatty acid primary prevention trial in patients with T2DM [ASCEND (A Study of Cardiovascular Events in Diabetes)][23] was recently published but it did not reveal any CV benefits. This differential effect was possibly due to the dose of icosapent ethyl as the dose was 4 g daily in REDUCE-IT vis-à-vis only 1 g daily in ASCEND.

Cholesteryl Ester Transfer Protein Inhibitors

Cholesteryl ester transfer protein (CETP) inhibition has also gained attention as a therapeutic approach. The initial result with the first two CETP inhibitors, torcetrapib and dalcetrapib, is not encouraging. The former was terminated due to safety issues in ILLUMINATE[24] (Investigation of Lipid Level Management to Understand Its Impact in Atherosclerotic Events), possibly due to off-target effects on blood pressure and/or the artery wall whereas dalcetrapib was terminated due to futility in dal-OUTCOMES in ACS patients. The reason for this is unknown, but may relate to the patient characteristics, given the lack of association between baseline HDL.

Next-generation Peroxisome Proliferator-activated Receptor Agonists

The dual PPAR agonists with selective activity for PPAR-α/-γ or PPAR-α/-δ offer opportunities to concomitantly manage several areas of dysmetabolism. Saroglitazar is currently the only dual PPAR-α and -γ agonist approved for the treatment of diabetic dyslipidemia (India, June 2013). Saroglitazar has shown to provide a comprehensive management of diabetic dyslipidemia by reducing TGs by 46.7% and non-HDL-C and also provides glycemic control by reducing glycosylated hemoglobin levels and fasting blood glucose. It is a safer option for atherogenic dyslipidemia.[25]

CONCLUSION

Stringent LDL-C reduction is required for those with high-risk for CVD. So, newer effective and safe therapies to reduce LDL are the need of the hour. Ezetimibe and PCSK9 inhibitors are two agents which are very effective in lowering LDL-C and shown to have CV benefit in long-term CV outcome trials. Mipomersen, lomitapide, bempedoic acid, gemcabene, and few other agents have proven to be efficacious in lowering LDL, however, long-term data to show CV benefit are still lacking. Residual risk reduction beyond LDL-C control is a major challenge for the 21st century. Despite evidence-based guidelines recognize and incorporate management strategies for high TG and low HDL, there is still lack of awareness in routine practice. Fibrates in selected patients (high TG and low HDL) and high-dose omega-3 fatty acid (icosapent ethyl) have shown to improve CV outcome. Several upcoming therapies may provide potential, although benefit versus risk analysis, especially in the longer term, requires further consideration.

KEY POINTS

- Low-density lipoprotein lowering has the most robust evidence to lower CV events. Although statin is the first-line drug for LDL lowering, it has efficacy and safety concerns.
- Ezetimibe and PCSK9 inhibitors have considerable efficacy in lowering LDL-C and have shown to reduce CV events in long-term CV outcome trials.
- Mipomersen, lomitapide, bempedoic acid, gemcabene, and few other agents have proven to be efficacious in lowering LDL, however, long-term data to show CV benefit are still lacking.
- High TG and low HDL are being increasingly recognized as lipid-related "residual risk" for CV events after lowering LDL-C.
- Fibrates in selected patients (high TG ≥204 mg/dL and low HDL ≤34 mg/dL) and high-dose omega-3 fatty acid (icosapent ethyl) have shown to improve CV outcome.

REFERENCES

1. Yusuf S, Hawken S, Ôunpuu S, Dans T, Avezum A, Lanas F, et al. Effect of potentially modifiable risk factors associated with myocardial infarction in 52 countries (the INTERHEART study): case-control study. Lancet. 2004;364:937-52.
2. Baigent C, Blackwell L, Emberson J, Holland LE, Reith C, Bhala N, et al. Efficacy and safety of more intensive lowering of LDL cholesterol: a meta-analysis of data from 170,000 participants in 26 randomised trials. Lancet. 2010;376:1670-81.
3. Jacobson TA, Ito MK, Maki KC, Orringer CE, Bays HE, Jones PH, et al. National Lipid Association recommendations for patient-centered management of dyslipidemia: Part 1—executive summary. J Clin Lipidol. 2014;8:473-88.
4. Reiner Z. Managing the residual cardiovascular disease risk associated with HDL-cholesterol and triglycerides in statin-treated patients: a clinical update. Nutr Metab Cardiovasc Dis. 2013;23:799-807.
5. Baigent C, Blackwell L, Emberson J, Holland LE, Reith C, Collaboration CTTC, et al. Efficacy and safety of more intensive lowering of LDL cholesterol: a meta-analysis of data from 170,000 participants in 26 randomised trials. Lancet. 2010;376:1670-81.

6. Stone NJ, Robinson J, Lichtenstein AH, Merz CN, Blum CB, Eckel RH, et al. ACC/AHA guideline on the treatment of blood cholesterol to reduce atherosclerotic cardiovascular risk in adults: a report of the American College of Cardiology/American Heart Association Task Force on Practice Guidelines. Circulation. 2014;129:S1-45.
7. Jellinger PS, Handelsman Y, Rosenblit PD, Bloomgarden ZT, Fonseca VA, Garber AJ, et al. American Association of Clinical Endocrinologists and American College of Endocrinology Guidelines for Management of Dyslipidemia and Prevention of Cardiovascular Disease. Endocr Pract. 2017;23:1-87.
8. Morrone D, Weintraub WS, Toth PP, Hanson ME, Lowe RS, Lin J, et al. Lipid-altering efficacy of ezetimibe plus statin and statin monotherapy and identification of factors associated with treatment response: a pooled analysis of over 21,000 subjects from 27 clinical trials. Atherosclerosis. 2012;223:251-61.
9. Cannon CP, Blazing MA, Giugliano RP, McCagg A, White JA, Theroux P, et al. Ezetimibe added to statin therapy after acute coronary syndromes. N Engl J Med. 2015;372:2387-97.
10. Boden WE, Probstfield JL, Anderson T, Chaitman BR, Desvignes-Nickens P, AIM-HIGH Investigators, et al. Niacin in patients with low HDL cholesterol levels receiving intensive statin therapy. N Engl J Med. 2011;365:2255-67.
11. HPS2-THRIVE Collaborative Group. HPS2-THRIVE randomized placebo-controlled trial in 25,673 high-risk patients of ER niacin/laropiprant: trial design, pre-specified muscle and liver outcomes, and reasons for stopping study treatment. Eur Heart J. 2013;34:1279-91.
12. Gordon DJ, Knoke J, Probstfield JL, Superko R, Tyroler HA. High-density lipoprotein cholesterol and coronary heart disease in hypercholesterolemic men: the Lipid Research Clinics Coronary Primary Prevention Trial. Circulation. 1986;74:1217-25.
13. Giugliano RP, Desai NR, Kohli P, Rogers WJ, Somaratne R, Huang F, et al. Efficacy, safety, and tolerability of a monoclonal antibody to proprotein convertase subtilisin/kexin type 9 in combination with a statin in patients with hypercholesterolemia (LAPLACE-TIMI 57): a randomised, placebo-controlled, dose-ranging, phase 2 study. Lancet. 2012;380:2007-17.
14. Giugliano RP, Pedersen TR, Park JG, De Ferrari GM, Gaciong ZA, Ceska R, et al. Clinical efficacy and safety of achieving very low LDL-cholesterol concentrations with the PCSK9 inhibitor evolocumab: a prespecified secondary analysis of the FOURIER trial. Lancet. 2017;390:1962-71.
15. Steg PG, Szarek M, Bhatt DL, Bittner VA, Brégeault MF, Dalby AJ, et al. Effect of Alirocumab on Mortality After Acute Coronary Syndromes. Circulation. 2019;140:103-12.
16. Ray KK, Landmesser U, Leiter LA, Kallend D, Dufour R, Karakas M, et al. Inclisiran in patients at high cardiovascular risk with elevated LDL cholesterol. N Engl J Med. 2017;376:1430-40.
17. Angeloni E, Paneni F, Landmesser U, Benedetto U, Melina G, Lüscher TF, et al. Lack of protective role of HDL-C in patients with coronary artery disease undergoing elective coronary artery bypass grafting. Eur Heart J. 2013;34:3557-62.
18. Scott R, O'Brien R, Fulcher G, Pardy C, D'Emden M, Tse D, et al. Effects of fenofibrate treatment on cardiovascular disease risk in 9,795 individuals with type 2 diabetes and various components of the metabolic syndrome: the Fenofibrate Intervention and Event Lowering in Diabetes (FIELD) study. Diabetes Care. 2009;32:493-8.
19. Ginsberg HN, Elam MB, Lovato LC, Crouse JR, Leiter LA, ACCORD Study Group, et al. Effects of combination lipid therapy in type 2 diabetes mellitus. N Engl J Med. 2010;362:1563-74.
20. Keech AC, Mitchell P, Summanen PA, O'Day J, Davis TM, Moffitt MS, et al. Effect of fenofibrate on the need for laser treatment for diabetic retinopathy (FIELD study): a randomised controlled trial. Lancet. 2007;370:1687-97.
21. Yokoyama M, Origasa H, Matsuzaki M, Matsuzawa Y, Saito Y, Ishikawa Y, et al. Effects of eicosapentaenoic acid on major coronary events in hypercholesterolemic patients (JELIS): a randomized open-label, blinded endpoint analysis. Lancet. 2007;369:1090-8.
22. Bhatt DL, Steg PG, Miller M, Brinton EA, Jacobson TA, Ketchum SB, et al. Cardiovascular risk reduction with icosapent ethyl for hypertriglyceridemia. N Engl J Med. 2019;380:11-22.
23. ASCEND Study Collaborative Group. Effects of n-3 fatty acid supplements in diabetes mellitus. N Engl J Med. 2018;379:1540-50.
24. Barter PJ, Caulfield M, Eriksson M, Grundy SM, Kastelein JJ, Komajda M, et al. Effects of torcetrapib in patients at high risk for coronary events. N Engl J Med. 2007;357:2109-22.
25. Pai V, Paneerselvam A, Mukhopadhyay S, Bhansali A, Kamath D, Shankar V, et al. A Multicenter, Prospective, Randomized, Double-blind Study to Evaluate the Safety and Efficacy of Saroglitazar 2 and 4 mg Compared to Pioglitazone 45 mg in Diabetic Dyslipidemia (PRESS V). J Diabetes Sci Technol. 2014;8:132-41.

CHAPTER 18

Role of Triglyceride Lowering Drugs on Cardiovascular Outcomes: Focus On Omega-3 Fatty Acid

Rajan Palui, Amarta Shankar Chowdhury

ABSTRACT

Despite improvement in the outcome of cardiovascular disease with optimum statin therapy, a substantial residual risk remains. Epidemiological studies have clearly indicated that elevated triglyceride (TG) levels can predispose to atherosclerotic cardiovascular disease (ASCVD). Three important drugs that primarily reduce TG levels are fibrates, niacin, and omega-3 fatty acids (OM3FAs). Results of outcome studies with fibrates and extended release niacin with or without statin therapy have been inconsistent in lowering the risk of cardiovascular events. OM3FA, particularly, eicosapentaenoic acid (EPA), not only lower TG levels but also have beneficial pleiotropic effects. EPA has been shown to stabilize atherosclerotic plaques and to reduce inflammation, oxidation of atherogenic lipoproteins, cholesterol crystal formation, and endothelial dysfunction. In a recently published large randomized trial, icosapent ethyl, a highly purified ethyl ester of EPA, at a dose of 2 g twice daily reduced a composite of cardiovascular events by 25%. High dose purified EPA when used along with statin therapy is likely to produce beneficial effect in ASCVD risk reduction. Results of ongoing trials will further substantiate the role of OM3FAs in decreasing cardiovascular risks in statin treated patients.

INTRODUCTION

In the year 1947, John R Moreton was one of the first researchers to study the role of triglyceride (TG) rich lipoproteins in atherosclerosis.[1,2] A number of studies have further shown that elevated serum TG level can predispose to atherosclerotic cardiovascular disease (ASCVD).[2] A meta-analysis of 29 prospective western studies which included 262,525 participants has revealed that there was >70% greater risk of coronary heart disease (CHD) in subjects having the highest versus the lowest tertile of TGs.[3] Patients with CHD with high TG levels were independently associated with increased risk of all cause mortality in a study of 22 year follow-up registry.[4] A meta-regression analysis of randomized controlled trials (RCT) of more 370,000 patients found that reduction in TG was associated with a decreased risk of major cardiovascular (CV) events, even after adjustment for low-density lipoprotein cholesterol (LDL-C) reduction.[5] However, it has to be noted that the REDUCE-IT (Reduction of Cardiovascular Events with Icosapent Ethyl–Intervention Trial) results significantly influenced the meta-regression.[5] In the light of recent evidences, treatment of hypertriglyceridemia may have a positive role in the outcome of CV diseases. The three important drugs that primarily reduce TG levels are fibrates, niacin and omega-3 fatty acids (OM3FA). Our focus in this chapter will be on the therapeutic effects of OM3FA.

POTENTIAL ROLE OF OM3FA IN ASCVD EVENT REDUCTION

The major marine derivatives of OM3FA include eicosapentaenoic acid (EPA), docosahexaenoic acid (DHA) and docosapentaenoic acid (DPA). The Food and Drug Administration (FDA) has approved three OM3FA-based drugs for the treatment of severe hypertriglyceridemia: (1) OM3FA ethyl esters (containing EPA and DHA); (2) Icosapent ethyl (exclusively containing EPA); (3) Omega-3 carboxylic acids (containing a mixture of EPA, DHA, and DPA).[6] OM3FA not only lower TG levels but also have been postulated to possess additional atheroprotective properties.[7] Both EPA and DHA reduce TG levels, but clinical studies have shown that they act through different antiatherogenic mechanisms.[8,9] Compared to placebo, treatment with only EPA in patients with hypertriglyceridemia has been shown to reduce inflammatory markers such as lipoprotein-associated phospholipase A2, high-sensitivity C-reactive protein, and oxidized LDL-C levels.[10,11] In another study, EPA (but not DHA) was associated with significant reduction of gene expression of interferon pathway as well as down-regulation of hypoxia inducible factor 1 (HIF1) and cAMP responsive element protein 1 (CREB1).[12] All of these may have a beneficial role in reducing the risks of CHD. EPA is incorporated into membrane phospholipids and has a direct beneficial role in plaque development and stability.[7] EPA has also been found to decrease cholesterol crystal formation and as well as reduce endothelial dysfunction.[7] Ongoing trials will further clarify the role of OM3FAs in decreasing residual cardiovascular risks in statin treated patients.[13,14]

RCTS EVALUATING THE ROLE OF FIBRATES AND NIACIN ON TG LEVEL AND ASCVD EVENT

Fibrates

The role of fibrates in TG lowering and reduction of vascular events have been evaluated in a number of studies with inconsistent findings (**Table 1**).[15-20] In the Helsinki Heart Study, men aged 40-55 years with non-HDL-C levels ≥200 mg/dL were given gemfibrozil 1,200 mg/day.[15] Gemfibrozil significantly lowered CV events by 34% after a mean follow-up of 5 years. In the Veterans Affairs High-Density Lipoprotein Cholesterol Intervention Trial (VA-HIT), gemfibrozil reduced cardiac endpoints by 22% compared with placebo (p = 0.006).[16] In the FIELD (Fenofibrate Intervention and Event Lowering in Diabetes) trial, individuals with type 2 diabetes mellitus (T2DM) were prescribed micronized fenofibrate 200 mg/day.[18] The primary endpoint of nonfatal myocardial infarction (MI) and CHD death did not differ significantly between fenofibrate and placebo at the end of median follow-up of 5 years [hazard ratio (HR) 0.89; 95% confidence interval (CI) 0.75–1.05; p = 0.16)]. Compared to placebo, fenofibrate however, showed a tendency to lower cardiac endpoints in patients with TG levels ≥150.6 mg/dL (p = 0.07) in a pre-specified sub-analysis.[18]

TABLE 1: Randomized control trials of fibrate.

Study	Study participants	TG lowering intervention	Primary endpoint	Results
HHS[15]	4,081 healthy middle-aged men (40–55) years with non-HDL-C ≥200 mg/dL	Gemfibrozil 600 mg twice daily	Fatal or nonfatal MI or cardiac death	TG lowered by 43%; reduction in cardiac endpoints by 34.0% (95% CI 8.2–52.6; p < 0.05) after median f/u of 5 years
VA-HIT[16]	2531 men aged <74 years with CHD and HDL-C ≤40 mg/dL, LDL-C ≤140 and TG ≤300 mg/dL	Gemfibrozil 1,200 mg/day	Nonfatal MI or death from CHD	About 22% reduction in primary outcome (95% CI 7–35; p = 0.006) after median f/u of 5.1 years
BIP[17]	3,090 patients aged between 45–74 years with prior MI or stable angina and HDL-C ≤45 mg/dL, LDL-C ≤180 mg/dL TG ≤300 mg/dL	Bezafibrate 400 mg/day	Fatal or nonfatal MI or sudden death	TG lowered by 21%; after a mean f/u of 6.2 years, the reduction in the cumulative probability of the primary outcome was 7.3% (p = 0.24)
FIELD[18]	9,795 T2DM patients aged 50–75 years and not on any statin treatment at entry	Micronized fenofibrate 200 mg once daily. About 2,547 patients received add on statin therapy during the trial	CHD death or non-fatal MI	Primary endpoint did not differ significantly between fenofibrate and placebo at the end of median f/u of 5 years (HR 0.89, 95% CI 0.75–1.05; p = 0.16)
ACCORD[19,20]	5,518 T2DM (HbA1c ≥7.5%); 40–79 years with clinical CVD or 55–79 years with ≥2 CVD risk factors. LDL-C 60–80 mg/dL; TG <750 mg/dL (if patient without lipid therapy) or TG <400 mg/dL (if patient receiving lipid therapy)	Open-label simvastatin followed by either fenofibrate or placebo. Starting dose of fenofibrate was 160 mg/day but 2004 onward dose was adjusted according to eGFR	CV death, nonfatal MI or nonfatal stroke	The annual rate of the primary outcome did not differ significantly between the fenofibrate and placebo groups (HR 0.92, 95% CI 0.79–1.08, p = 0.32)

(ACCORD: Action to Control Cardiovascular Risk in Diabetes study; BIP: Bezafibrate Infarction Prevention study; CHD: coronary heart disease; CI: confidence interval; CV: cardiovascular; CVD: cardiovascular disease; FIELD: Fenofibrate Intervention and Event Lowering in Diabetes study; f/u: follow-up; eGFR: estimated glomerular filtration rate; HDL-C: high-density lipoprotein cholesterol; HHS: Helsinki Heart Study; HR: hazard ratio; LDL-C: low-density lipoprotein cholesterol; MI: myocardial infarction; T2DM: type 2 diabetes mellitus; TG: Triglyceride; VA-HIT: Veterans Affairs High-Density Lipoprotein Cholesterol Intervention Trial)

One of the most important fenofibrate trial is the ACCORD (Action to Control Cardiovascular Risk in Diabetes) trial.[19] The trial studied the effects of intensive blood glucose lowering therapy in patients with T2DM and included an important sub-study called ACCORD-Lipid. The patients in ACCORD-Lipid were treated with open-label simvastatin following which they were randomized to receive either fenofibrate or placebo. At the end of the study there was a small reduction in CV events in the fenofibrate group but it was not statistically significant. However, there was a trend of ASCVD reduction in a pre-specified subgroup analysis of patients with TG level ≥204 mg/dL and HDL-C levels ≤34 mg/dL (p = 0.057).[19,20]

The PROMINENT (Pemafibrate to Reduce Cardiovascular Outcomes by Reducing Triglycerides in Patients with Diabetes) trial is currently evaluating the role of pemafibrate (selective peroxisome proliferator-activated receptor alpha modulator) in reducing cardiovascular events in high-risk patients with T2DM (NCT03071692). This trial will include around 10,000 patients who are already on statin therapy with a TG level of 200 mg/dL or more and an HDL-C level of less than or equal to 40 mg/dL. The result of this trial is expected in the year 2022.

Niacin

Two impactful studies have revealed that niacin, when added to a statin, may not lower CV diseases.[21,22] In the AIM-HIGH (Atherothrombosis Intervention in Metabolic Syndrome with Low HDL/High TGs: Impact on Global Health Outcomes) trial, a high dose of extended release niacin (1,500–2,000 mg/day) along with a standard dose of simvastatin was administered.[21] Patients who had established ASCVD were given simvastatin 40 mg/day and ezetimibe was added infrequently in some patients in order to achieve an LDL-C level of 40–80 mg/dL. In the placebo arm, patients were given simvastatin along with a matching placebo which had 50 mg immediate-release niacin (for blinding). After a mean follow-up of 36 months, the trial was stopped as no clinical benefit was perceived from the addition of niacin to statin therapy. Interestingly, in a posthoc subgroup analysis of the trial, niacin was found to significantly lower primary ASCVD endpoints in subjects with TG ≥200 mg/dL and HDL-C level <32 mg/dL.[23]

The HPS2-THRIVE (Heart Protection Study 2-Treatment of HDL to Reduce the Incidence of Vascular Events) evaluated the role of extended-release niacin and laropiprant (antiflushing agent) in addition to statin treatment, in subjects with established ASCVD or with diabetes and symptomatic coronary artery disease.[22] At the end of the study period, treatment with extended release niacin could not significantly lower the risk of major vascular events [relative risk (RR) 0.96, 95% CI 0.90–1.03; p = 0.29] but it increased the risk of serious adverse effects.[22]

RCTS EVALUATING THE ROLE OF OM3FA ON TG LEVEL AND ASCVD EVENT

Though RCTs using mixed and pure OM3FA has shown to decrease TG level, but their effect on ASCVD risk reduction remain inconclusive.[24-28] In the JELIS (Japan EPA Lipid Intervention Study), a total of 18,645 Japanese hypercholesterolemic patients (14,981 in primary prevention sub-group and 3,664 in secondary prevention sub-group) were randomized to either in EPA group (EPA 1.8 g/day + statin) or control group (statin alone) and followed up for 4.6 years.[24] Overall, the rate of any major coronary event (including sudden cardiac death, fatal and non-fatal MI, and other non-fatal coronary events) was significantly lower in EPA group (HR 0.81, 95% CI 0.69–0.95, p = 0.011). In secondary prevention subgroup, EPA treatment was associated with a significant 19% decrease in major coronary events, but it did not reach statistical significance in primary prevention sub-group. Though the total number of adverse events including gastrointestinal disturbance, skin abnormality, and hemorrhage were significantly higher in EPA group, most of the side effects were mild in severity. However, the habit of increased consumption of fish among Japanese population may be the reason behind the effectiveness of EPA in much lower dose of 1.8 g/day in JELIS trial. Comparable rise of serum EPA was observed in Japanese population who were on low dosage (1.8 g/day) with respect to western population on high dose of EPA (4 g/day).[29] Another limitation of JELIS trial is that it was done in Japanese population whose risk of death due to coronary artery disease is low in comparison to rest of the world.[24]

A recent meta-analysis included 10 large RCTs (total 77,917 high-risk individuals) evaluating the association of OM3FA with major vascular events.[30] The pooled analysis did not show significant association of OM3FA with any major vascular event (RR 0.97, 95% CI 0.93–1.01; p = 0.10). Similarly, OM3FA failed to show any significant risk reduction of non-fatal MI (RR 0.97), CHD death (RR 0.93) or any CHD event (RR 0.96). In this meta-analysis, all the included RCTs used low dose mixed OM3FA (EPA + DPA) except in JELIS trial (only EPA). However, sensitivity analysis excluding the JELIS trial did not show any change in result for major vascular events.

The role of icosapent ethyl in reducing ASCVD was evaluated in a multi-centric, double blind placebo controlled RCT (REDUCE-IT).[26] A total of 8,179 patients from 11 countries were recruited and among them with 70.7% had CV disease and the rest (29.3%) were having diabetes mellitus and another CV risk factor. They were randomized into either of icosapent ethyl (2 g twice daily) or placebo group and followed up for a median of 4.9 years. All the included patients were already on stable statin therapy and their fasting TG levels were in range of 135–499 mg/dL. The median reduction of TG was 19.7% higher in icosapent group. The risk for primary composite endpoint (CVD, nonfatal MI, nonfatal stroke, coronary revascularization, or unstable angina) was significantly lower (HR 0.75, 95% CI 0.68–0.83; p < 0.001) in icosapent group. This beneficial effect of icosapent persisted irrespective of baseline TG values. The risk of key secondary endpoint (CVD, MI, nonfatal stroke) was also significantly less in icosapent arm (HR 0.74; 95% CI 0.65–0.83; p < 0.001). Among the adverse events, the rates of atrial fibrillation and peripheral edema were significantly higher among patients in icosapent arm.

In the CHERRY (Combination Therapy of eicosapentaenoic Acid and Pitavastatin for Coronary Plaque Regression Evaluated by Integrated Backscatter Intravascular Ultrasonography) trial, a total of 193 CHD patients were randomized to pitavastatin (PTV) group (PTV 4 mg/day) or PTV/EPA group (PTV 4 mg/day and EPA 1,800 mg/day).[31] Regression of total atheroma volume was significantly higher in EPA group (81% vs. 61%, p = 0.002) as assessed by intravascular ultrasonography. Moreover, lipid volume was significantly decreased only in the EPA group. These findings might explain the possible pathological basis of EPA associated CV risk reduction.

Two recent major trials evaluated the role of low dose mixed OM3FA in primary prevention of ASCVD.[27,28] In ASCEND (A Study of Cardiovascular Events in Diabetes) trial, 15,480 diabetic patients without ASCVD were randomized to fatty acid group (460 mg of EPA and 380 mg of DHA) or placebo (olive oil).[27] There was no significant difference in serious vascular events (nonfatal MI, nonfatal stroke, transient ischemic attack or vascular death) between fatty acid or placebo groups (RR 0.97; 95% CI 0.87–1.08; p = 0.55) during a mean follow-up of 7.4 years. In the VITAL (Vitamin D and Omega-3 Trial) study, two-by-two factorial design was used to evaluate role of marine OM3FA (1 g/day) and vitamin D in primary prevention of major ASCVD and cancer.[28] A total of 25,871 participants (male >50 years and female >55 years) were randomized to omega-3 group or placebo group and followed up for a median 5.3 years. However, this study also did not show any significant difference in major CV events (MI, stroke or cardiovascular death) in between two groups (HR 0.92; 95% CI 0.80–1.06; p = 0.24). Both of these RCTs failed to show any benefit of low dose murine mixed OM3FA in primary prevention of ASCVD.[27,28]

Thus, the beneficial effect of OM3FA on reduction of risk of ASCVD is predominantly seen with high dose purified EPA use along with statin therapy. Apart from the TG lowering effects, the pleiotropic effects of EPA (described earlier) might be the reason behind its cardio protective effect.

ONGOING RCTS OF OM3FA (EPA OR EPA + DHA)

Multiple large RCTs evaluating the effect of pure or mixed OM3FA on ASCVD risk are currently in progress. The STRENGTH (Statin Residual Risk Reduction with EpaNova in High Cardiovascular Risk) trial is a randomized placebo controlled trial in which 13,086 statin treated patients with hypertriglyceridemia and low HDL level were included.[13] The aim of this study is to assess the role of high dose mixed OM3FA (EPA + DHA 4 g/day) in reducing CV event in both secondary and high-risk primary prevention (diabetes with risk factor) participants. The result of this study is expected shortly. The RESPECT-EPA (Randomized Trial for Evaluation in Secondary Prevention Efficacy of Combination Therapy–Statin and Eicosapentaenoic Acid) is an open level RCT from Japan in which the effect of low dose purified OM3FA EPA (1.8 g/day) is being evaluated in a secondary prevention cohort who were already receiving statin therapy.[32] The role of low dose mixed OM3FA (EPA + DHA 1.8 g/day) is currently being assessed in OMEMI (OMega-3 fatty acids in Elderly patients with Myocardial Infarction) trial in Norway.[33] In this prospective, placebo-controlled, double blinded and multicenter RCT, a total of 1,400 high risk elderly patients with MI were included. The potential effect of pure EPA (4 g/day) on atherosclerotic plaque is now being assessed in EVAPORATE (Effect of Vascepa on Improving Coronary Atherosclerosis in People With High Triglycerides Taking Statin Therapy) trial.[14] In this multi-centre trial from North America, multidetector computed tomography angiography imaging technique is being used to evaluate coronary plaque characteristics. The details of major ongoing RCTs evaluating the effects of OM3FA on ASCVD risk is given in **Table 2**. In future, the results of these ongoing large trials will clarify the effect of different types (mixed and pure) and dosages (4 g/day or 1.8 g/day) of OM3FA in ASCVD risk reduction among primary and secondary prevention cohorts.

TABLE 2: Major ongoing trials on omega-3-fatty acids.

Study	Study participants	TG lowering intervention	Primary endpoint
STRENGTH[13]	13,086 participants of age 18–99 years (>40 years in diabetic) and TG 180–499 mg/dL; HDL-C <42 mg/dL (male), <47 mg/dL (female)	Mixed EPA + DHA 4 g/day or matching placebo	Time to first event of any component of the composite of CV death, nonfatal MI, nonfatal stroke, coronary revascularization and hospitalization for unstable angina
RESPECT-EPA[32]	3,900 patients of CAD on statin therapy; Age: 20–79 years	EPA 1.8 g/day	First occurrence of any of the following CV events: CV death, nonfatal MI, nonfatal cerebral infarction, unstable angina requiring emergent hospitalization and coronary revascularization and coronary revascularization based on clinical findings
OMEMI[33]	1,400 post MI patients on statin therapy; age 70–82	Mixed EPA + DHA 1.8 g/day	Composite of total mortality, first nonfatal recurring acute MI, stroke and revascularization
EVAPORATE[14]	80 statin treated patients with coronary atherosclerosis with elevated TG (200 MI 499 mg/dL)	EPA 4 g/day	Change in low attenuation plaque volume measured by multi-detector computed tomography angiography

(CAD: coronary artery disease; CV: cardiovascular; DHA: docosahexaenoic acid; EPA: eicosapentaenoic acid; EVAPORATE: Effect of Vascepa on Improving Coronary Atherosclerosis in People With High Triglycerides Taking Statin Therapy; HDL-C: high-density lipoprotein cholesterol; MI: myocardial infarction; OMEMI: OMega-3 fatty acids in Elderly patients with Myocardial Infarction; RESPECT-EPA: Randomized Trial for Evaluation in Secondary Prevention Efficacy of Combination Therapy–Statin and Eicosapentaenoic Acid; STRENGTH: Statin Residual Risk Reduction with EpaNova in High Cardiovascular Risk; TG: Triglyceride)

CONCLUSION

In the view of the promising result of the recent outcome studies, icosapent ethyl is now recommended for CV reduction in diabetic patients with coronary artery disease when their TG levels are elevated (135–499 mg/dL) despite maximally tolerated statin therapy.[34,35] However, the results of major ongoing trials will be crucial to further substantiate the potential role of OM3FA in reducing risk of major CV events.

> **KEY POINTS**
>
> - Several studies have shown that hypertriglyceridemia can predispose to ASCVD and TG lowering therapy can address this associated CV risk.
> - Fibrates and niacin are effective in decreasing TG level but are inconsistent in lowering risk of ASCVD.
> - Omega-3 fatty acid, especially EPA, not only reduces TG but also have additional pleiotropic effects in reducing atherosclerosis.
> - Results of recent RCT evaluating the role of high dose purified EPA along with statin therapy have been found to be favorable in reducing major vascular events.
> - Ongoing trials will further clarify the role of OM3FAs in decreasing residual cardiovascular risks in statin treated patients.

REFERENCES

1. Moreton JR. Atherosclerosis and alimentary hyperlipemia. Science. 1947;106(2748):190-1.
2. Handelsman Y, Shapiro MD. Triglycerides, atherosclerosis, and cardiovascular outcome studies: focus on omega-3 fatty acids. Endocr Pract. 2017;23(1):100-12.
3. Sarwar N, Danesh J, Eiriksdottir G, Sigurdsson G, Wareham N, Bingham S, et al. Triglycerides and the risk of coronary heart disease: 10,158 incident cases among 262,525 participants in 29 Western prospective studies. Circulation. 2007;115(4):450-8.
4. Klempfner R, Erez A, Sagit B-Z, Goldenberg I, Fisman E, Kopel E, et al. Elevated triglyceride level is independently associated with increased all-cause mortality in patients with established coronary heart disease: Twenty-two-year follow-up of the Bezafibrate Infarction Prevention Study and Registry. Circ Cardiovasc Qual Outcomes. 2016;9(2):100-8.
5. Marston NA, Giugliano RP, Im K, Silverman MG, O'Donoghue ML, Wiviott SD, et al. Association between triglyceride lowering and reduction of cardiovascular risk across multiple lipid-lowering therapeutic classes: A systematic review and meta-regression analysis of randomized controlled trials. Circulation. 2019;140(16):1308-17.
6. Zwol W van, Rimbert A, Kuivenhoven JA. The future of lipid-lowering therapy. J Clin Med. 2019;8(7).
7. Preston Mason R. New insights into mechanisms of action for omega-3 fatty acids in atherothrombotic cardiovascular disease. Curr Atheroscler Rep. 2019;21(1):2.
8. Williams JA, Batten SE, Harris M, Rockett BD, Shaikh SR, Stillwell W, et al. Docosahexaenoic and eicosapentaenoic acids segregate differently between raft and nonraft domains. Biophys J. 2012;103(2):228-37.
9. Mason RP, Jacob RF, Shrivastava S, Sherratt SCR, Chattopadhyay A. Eicosapentaenoic acid reduces membrane fluidity, inhibits cholesterol domain formation, and normalizes bilayer width in atherosclerotic-like model membranes. Biochim Biophys Acta. 2016;1858(12):3131-40.
10. Ballantyne CM, Bays HE, Kastelein JJ, Stein E, Isaacsohn JL, Braeckman RA, et al. Efficacy and safety of eicosapentaenoic acid ethyl ester (AMR101) therapy in statin-treated patients with persistent high triglycerides (from the ANCHOR study). Am J Cardiol. 2012;110(7):984-92.
11. Bays HE, Ballantyne CM, Braeckman RA, Stirtan WG, Soni PN. Icosapent ethyl, a pure ethyl ester of eicosapentaenoic acid: effects on circulating markers of inflammation from the MARINE and ANCHOR studies. Am J Cardiovasc Drugs Drugs Devices Interv. 2013;13(1):37-46.

12. Tsunoda F, Lamon-Fava S, Asztalos BF, Iyer LK, Richardson K, Schaefer EJ. Effects of oral eicosapentaenoic acid versus docosahexaenoic acid on human peripheral blood mononuclear cell gene expression. Atherosclerosis. 2015;241(2):400-8.
13. Nicholls SJ, Lincoff AM, Bash D, Ballantyne CM, Barter PJ, Davidson MH, et al. Assessment of omega-3 carboxylic acids in statin-treated patients with high levels of triglycerides and low levels of high-density lipoprotein cholesterol: Rationale and design of the STRENGTH trial. Clin Cardiol. 2018;41(10):1281-8.
14. Budoff M, Muhlestein JB, Le VT, May HT, Roy S, Nelson JR. Effect of Vascepa (icosapent ethyl) on progression of coronary atherosclerosis in patients with elevated triglycerides (200-499 mg/dL) on statin therapy: Rationale and design of the EVAPORATE study. Clin Cardiol. 2018;41(1):13-9.
15. Frick MH, Elo O, Haapa K, Heinonen OP, Heinsalmi P, Helo P, et al. Helsinki Heart Study: primary-prevention trial with gemfibrozil in middle-aged men with dyslipidemia. Safety of treatment, changes in risk factors, and incidence of coronary heart disease. N Engl J Med. 1987;317(20):1237-45.
16. Rubins HB, Robins SJ, Collins D, Fye CL, Anderson JW, Elam MB, et al. Gemfibrozil for the secondary prevention of coronary heart disease in men with low levels of high-density lipoprotein cholesterol. Veterans Affairs High-Density Lipoprotein Cholesterol Intervention Trial Study Group. N Engl J Med. 1999;341(6):410-8.
17. Bezafibrate Infarction Prevention (BIP) study. Secondary prevention by raising HDL cholesterol and reducing triglycerides in patients with coronary artery disease. Circulation. 2000;102(1):21-7.
18. Keech A, Simes RJ, Barter P, Best J, Scott R, Taskinen MR, et al. Effects of long-term fenofibrate therapy on cardiovascular events in 9795 people with type 2 diabetes mellitus (the FIELD study): randomised controlled trial. Lancet. 2005;366(9500):1849-61.
19. Ginsberg HN, Elam MB, Lovato LC, Crouse JR, Leiter LA, Linz P, et al. Effects of combination lipid therapy in type 2 diabetes mellitus. N Engl J Med. 2010;362(17):1563-74.
20. Ginsberg HN. The ACCORD (Action to Control Cardiovascular Risk in Diabetes) Lipid Trial. Diabetes Care. 2011;34(Suppl 2):S107-8.
21. Boden WE, Probstfield JL, Anderson T, Chaitman BR, Desvignes-Nickens P, Koprowicz K, et al. Niacin in patients with low HDL cholesterol levels receiving intensive statin therapy. N Engl J Med. 2011;365(24):2255-67.
22. Landray MJ, Haynes R, Hopewell JC, Parish S, Aung T, Tomson J, et al. Effects of extended-release niacin with laropiprant in high-risk patients. N Engl J Med. 2014;371(3):203-12.
23. Guyton JR, Slee AE, Anderson T, Fleg JL, Goldberg RB, Kashyap ML, et al. Relationship of lipoproteins to cardiovascular events: the AIM-HIGH Trial (Atherothrombosis Intervention in Metabolic Syndrome With Low HDL/High Triglycerides and Impact on Global Health Outcomes). J Am Coll Cardiol. 2013;62(17):1580-4.
24. Yokoyama M, Origasa H, Matsuzaki M, Matsuzawa Y, Saito Y, Ishikawa Y, et al. Effects of eicosapentaenoic acid on major coronary events in hypercholesterolaemic patients (JELIS): a randomised open-label, blinded endpoint analysis. Lancet Lond Engl. 2007;369(9567):1090-8.
25. Galan P, Briancon S, Blacher J, Czernichow S, Hercberg S. The SU.FOL.OM3 Study: a secondary prevention trial testing the impact of supplementation with folate and B-vitamins and/or Omega-3 PUFA on fatal and non fatal cardiovascular events, design, methods and participants characteristics. Trials. 2008;9:35.
26. Bhatt DL, Steg PG, Miller M, Brinton EA, Jacobson TA, Ketchum SB, et al. Cardiovascular Risk Reduction with Icosapent Ethyl for Hypertriglyceridemia. N Engl J Med. 2019;380(1):11-22.
27. Bowman L, Mafham M, Wallendszus K, Stevens W, Buck G, Barton J, et al. Effects of n-3 Fatty Acid Supplements in Diabetes Mellitus. N Engl J Med. 2018;379(16):1540-50.
28. Manson JE, Cook NR, Lee I-M, Christen W, Bassuk SS, Mora S, et al. Marine n-3 Fatty Acids and Prevention of Cardiovascular Disease and Cancer. N Engl J Med. 2019;380(1):23-32.
29. Bays HE, Ballantyne CM, Doyle RT, Juliano RA, Philip S. Icosapent ethyl: Eicosapentaenoic acid concentration and triglyceride-lowering effects across clinical studies. Prostaglandins Other Lipid Mediat. 2016;125:57-64.
30. Aung T, Halsey J, Kromhout D, Gerstein HC, Marchioli R, Tavazzi L, et al. Associations of Omega-3 Fatty Acid Supplement Use With Cardiovascular Disease Risks: Meta-analysis of 10 Trials Involving 77 917 Individuals. JAMA Cardiol. 2018;3(3):225-34.
31. Watanabe T, Ando K, Daidoji H, Otaki Y, Sugawara S, Matsui M, et al. A randomized controlled trial of eicosapentaenoic acid in patients with coronary heart disease on statins. J Cardiol. 2017;70(6):537-44.
32. UMIN Clinical Trials Registry. Randomized trial for Evaluation in Secondary Prevention Efficacy of Combination Therapy-Statin and Eicosapentaenoic Acid. [online] Available from https://upload.umin.ac.jp/cgi-open-bin/ctr_e/ctr_view.cgi?recptno=R000014051. [Last accessed June, 2020].

33. Laake K, Myhre P, Nordby LM, Seljeflot I, Abdelnoor M, Smith P, et al. Effects of 3 supplementation in elderly patients with acute myocardial infarction: design of a prospective randomized placebo controlled study. BMC Geriatr. 2014;14:74.
34. American Diabetes Association. 10. Cardiovascular Disease and Risk Management: Standards of Medical Care in Diabetes-2020. Diabetes Care. 2020;43(Suppl 1):S111-34.
35. Arnold SV, Bhatt DL, Barsness GW, Beatty AL, Deedwania PC, Inzucchi SE, et al. Clinical Management of Stable Coronary Artery Disease in Patients With Type 2 Diabetes Mellitus: A Scientific Statement From the American Heart Association. Circulation. 2020;141(19).

CHAPTER 19

HDL-targeting Therapies: Progress and Future

Sayantan Ray

ABSTRACT

High-density lipoproteins (HDLs) are thought to have a protecting role against atherosclerosis. Probably, this reflects the basic biological function of HDL, facilitating reverse cholesterol transport from the periphery to the liver. Over the last few decades, substantial interest in HDL as an athero- and cardioprotective particle has resulted in the development of HDL (HDL-C) raising as a therapeutic approach to reduce cardiovascular risk. A number of strategies to increase circulating levels of HDL-C were developed that mainly included the use of niacin and fibrates as effective HDL-C raising agents. In the statin era, inhibition of cholesteryl ester transfer protein (CETP), artificially reconstituted HDL infusion, and administration of apolipoprotein A-I mimetics were evaluated as new approaches to raise HDL-C. However, whether increasing the HDL-C by pharmacological approach is effective for the translation of its atheroprotective actions have been questioned. The inter-relationships among HDL-C concentration, HDL particle number, and levels of various HDL particle subpopulations of specific composition are complex, as are their connections with reverse cholesterol transport and other antiatherogenic activities. Finding new targets for the development of HDL-directed therapies is highly desired.

INTRODUCTION

The epidemiological association between high-density lipoprotein cholesterol (HDL-C) and the development of atherosclerotic cardiovascular disease (ASCVD) is strong, graded, and consistent.[1,2] To elucidate this relationship, the HDL hypothesis was proposed by Miller as early as 1975, which stated that "a reduction of plasma concentration of HDL might accelerate the development of atherosclerosis and, therefore, ischemic heart disease by impairing the clearance of cholesterol from the arterial wall".[3] The concept of the key HDL function, which involves reverse cholesterol transport (RCT) from the periphery to the liver, was indirectly present in this hypothesis. Further studies that detailed the notion of HDL functionality revealed that antiatherogenic functions of HDL consist of cellular cholesterol efflux capacity together with pleiotropic effects.[4]

Consequently, doctors and researchers presumed a causal relation linking HDL-C concentrations and cardiovascular (CV) risk.[5]

An ample interest in HDL as an athero- and cardioprotective particle has resulted in the development of HDL-C raising as a therapeutic approach to decrease CV risk persisting after LDL-C lowering. Several methods to increase circulating levels of HDL-C were developed. However, as a therapeutic target, serum HDL-C levels have undergone a rigorous re-evaluation. Clinical trials aimed at increasing HDL-C with fibrates, niacin, apolipoprotein (Apo) A-I infusions, or cholesteryl ester transfer protein (CETP) inhibitors have failed to diminish ASCVD.[6,7] This apparent paradigm has prompted a re-evaluation of the "HDL hypothesis". More recently, novel strategies to target HDL metabolism, e.g., upregulation of hepatic ApoA-I production, have been added to the HDL therapeutics list.[8] This article summarizes present knowledge of HDL-targeting therapies and potential new targets for future pharmacological interventions are discussed.

THE HDL HYPOTHESIS REVISITED

Pharmacological agents that increase HDL-C differ in the degree of their specificity for the elevation of HDL-C levels plus in the extent to which they influence the metabolism and activity of HDL particles.[4,9] Notably, the pleiotropic effects of HDL are weakly correlated with the cholesterol mass of HDL and are generally dependent on the lipidome and proteome of HDL particles.[10,11] A picture thus appears where HDL exerts CV protective effects in healthy individuals and HDL-C may be a biomarker of CV health. Nevertheless, in metabolic disorders and advanced atherosclerosis, HDL-C is no longer predictive of CV outcomes and HDL particles may become dysfunctional. HDL-C is probably not an appropriate therapeutic target and attention has turned toward its function.[12] The main atheroprotective function of HDL is to remove excess cholesterol accumulated in atherosclerotic plaques. The removal of cholesterol happens during the formation of HDL particles, a process known as "HDL biogenesis". The HDL biogenesis process is an attractive potential therapeutic target; however, none of the presently available HDL-C raising drugs has been shown to increase HDL biogenesis in atherosclerotic plaques.

HDL-TARGETED THERAPEUTICS

High-density lipoprotein-targeted therapeutics that have reached clinical trials can be classified into three major groups—CETP inhibitors, HDL mimetics, and activators of ApoA-I synthesis. **Table 1** summarizes the completed trials of HDL lowering therapeutics.

Cholesteryl Ester Transfer Protein Inhibitors

Over the last 15 years, more targeted approaches to increase HDL levels have been developed. The most advanced of these is CETP inhibition, which aims to retain "nonatherogenic" cholesterol in HDL. CETP is a hydrophobic glycoprotein that facilitates transfer of cholesteryl esters (CEs) from HDLs to ApoB-containing lipoproteins as well as transfer of triglycerides (TGs) in the reverse direction. CETP inhibitors decrease heteroexchange of CE and TGs between lipoproteins, thereby reducing the transfer of CE from HDL to TG-rich particles and increasing HDL-C concentrations in human plasma. The thought that CETP inhibition might protect against coronary heart disease (CHD) emerged following reports of a high frequency of *CETP* gene variations that were

TABLE 1: High-density lipoprotein (HDL)-targeted therapeutics with completed trials.			
Type of treatment	Phase	Condition	Status or results
CETP inhibitors:			
Anacetrapib (REVEAL)	3	Cardiovascular disease	Increased HDL cholesterol concentrations and positive effect on primary endpoint (development program terminated because of small effect size)
Evacetrapib (ACCELERATE)	3	Cardiovascular disease, cerebrovascular disease, and peripheral vascular disease	Terminated (lack of efficacy); reduced LDL cholesterol and increased HDL cholesterol concentrations
Dalcetrapib (dal-OUTCOMES)	3	Cardiovascular disease	Increased HDL cholesterol but had no effect on primary endpoint
Torcetrapib (ILLUMINATE)	3	Cardiovascular disease Type 2 diabetes mellitus (T2DM)	Terminated (safety); increased risk of the primary endpoint
HDL infusion agents:			
MDCO-216 (MILANO-PILOT)	2	Acute coronary syndrome	No effect on primary endpoint
CSL112 (AEGIS-I)	2b	Acute coronary syndrome	Safe and efficacious; raised ApoA-I levels and increased the capacity of serum to efflux cholesterol
CER-001 (CARAT)	2	Acute coronary syndrome	No effect on primary endpoint

(ApoA-I: apolipoprotein A-I; HDL: high-density lipoprotein; LDL: low-density lipoprotein)

accompanied by low CETP activity and elevated HDL-C concentrations in Japanese individuals.[13] The first attempt to evaluate the CETP inhibitors was made by Pfizer in the 2000s by introducing into phase I of clinical trials a new drug torcetrapib.[14] However, the development was stopped in 2006 in the phase III of clinical trials because of increased mortality rates in the treatment arm.[15] The adverse effect of torcetrapib is currently thought to reflect the hypertensive effects of the drug. Large RCTs evaluating two other CETP inhibitors (dalcetrapib and evacetrapib) also failed to show reduction in ASCVD event rate, in spite of increasing HDL-C levels.[7,16] Anacetrapib did not show any adverse effects, including blood pressure elevation,[17] as its predecessors have. In a fourth CETP inhibitor trial with anacetrapib, the effect on the primary endpoint was statistically modest and unrelated to the robust rise in HDL-C level.[18] One important caveat is that all of the CETP trials were performed in statin-treated patients, suggesting that their lack of efficacy might have been related to statin treatment.

High-density Lipoprotein Infusions

Till date, in the HDL infusions class, three different formulations have reached clinical trials. From this group, the first series of clinical studies have been carried out on the efficacy of MDCO-216, a complex of ApoA-I Milano with phosphatidylcholine. Only 5 weekly infusions of this formulation at 15–45 mg/kg induced a 4.2% reduction in plaque volume compared with the baseline in 47 patients with acute coronary syndromes (ACS) as assessed by intravascular ultrasound.[19] Further study, assessing efficacy and safety of single infusions of MDCO-216 in healthy individuals (n = 24) and patients with CHD (n = 24), has shown that MDCO-216 remarkably enhanced ATP-binding cassette subfamily A member 1 (ABCA1)-mediated cholesterol efflux and was well tolerated.[20] However, in a RCT of lipid-poor HDL mimetic MDCO-216, there was no regression of plaque volume compared to placebo in patients with ACS.[21]

Short-term infusions of another rHDL preparation, CSL-111, which consisted of wild-type human ApoA-I and soybean phosphatidylcholine (molar ratio 1:150), considerably impacted on atherosclerosis in ERASE (Effect of rHDL on Atherosclerosis-Safety and Efficacy) trial carried out in patients with coronary artery disease (CAD).[22] The subsequent development of rHDL formulations led to the emergence of CSL-112, which surpassed the CSL-111 drug. Trials assessing safety and efficacy revealed that CSL-112 infusions were well tolerated in patients with myocardial infarction.[23] CSL-112 increased ApoA-I levels immediately and caused a rapid and remarkable increase in the serum cholesterol efflux capacity. RCT of CER-001, another HDL-mimetic agent consists of recombinant ApoA-I, sphingomyelin, and phosphatidylglycerol, did not reduce plaque volume in ACS patients.[24] Another study of CER-001 was performed in patients with homozygous familial hypercholesterolemia (HoFH). In contrast, infusions with ApoA-I-containing HDL-mimetic particles caused a significant reduction in the carotid mean vessel wall area.[25] Differential efficacy of CER-001 in CAD patients compared with FH subjects can be explained by increased sensitivity of cholesterol-loaded plaques toward delipidating effects of rHDL in HoFH, and/or enhanced sensitivity of carotid relative to coronary arteries toward rHDL. Whatever the precise mechanism, rHDL infusions merit further investigation as a potential approach for treatment of atherosclerosis.

Apolipoprotein A-I Transcriptional Upregulators

Another promising concept in terms of clinical efficacy is the approach of direct upregulation of hepatic secretion of ApoA-I by small molecules. One class of ApoA-I transcription activator is RVX-208. It is an orally active small molecule inhibiting bromodomain and extra terminal domain (BET) proteins involved in chromatin remodeling.[26] Recent mechanistic studies point out that RVX-208 is a BET inhibitor that induces *ApoA-I* gene transcription, which is followed by an elevation in circulating levels of ApoA-I protein.[27] In the phase II ASSERT efficacy, safety, and tolerability trial, RVX-208 administration for 12 weeks was associated with increases in ApoA-I, HDL-C, and large HDL particles, congruous with facilitation of cholesterol mobilization.[28] Nevertheless, the follow-up ASSURE study performed in patients with angiographic coronary disease and low HDL-C levels has shown that RVX-208 was no more effective in reducing coronary atherosclerosis progression and increasing ApoA-I or HDL-C levels as compared to placebo.[29] While several other molecular targets for the transcriptional upregulation of ApoA-I have been identified, the development of drugs carrying both an acceptable safety profile and significant efficacy has been proven challenging. An even more difficult problem with ApoA-I-directed therapy is qualitative changes in ApoA-I in ASCVD.[12]

Miscellaneous Agents

In this context, another strategy merit mentioning is the direct infusion of recombinant lecithin–cholesterol acyltransferase (LCAT), which is an enzyme converting free cholesterol into CE; this reaction primarily occurs on HDL and increases plasma HDL-C levels. The thought of using recombinant human LCAT in CAD patients is based on the observation that LCAT deficiency accelerates atherosclerosis in mouse models of this disease.[30] In a phase-I study of 16 patients with low HDL-C and CAD, recombinant LCAT (ACP-501) displayed a satisfactory safety profile following a single intravenous infusion.[31] Moreover, lipid and lipoprotein changes indicated that recombinant human LCAT positively altered HDL metabolism, supporting the use of recombinant LCAT in future clinical trials in patients with CHD. However, it remains to be seen whether this treatment would be primarily for patients with LCAT deficiency or whether its use could ever be extended to other high-risk patients having low HDL-C levels. Other potential therapies, e.g., infusions of autologous delipidated HDL, displayed positive effects on biomarkers of lipoprotein metabolism but their influence on atherosclerosis did not attain significance in humans.[32]

The mechanism of action of HDL-raising therapies is illustrated in **Figure 1**.

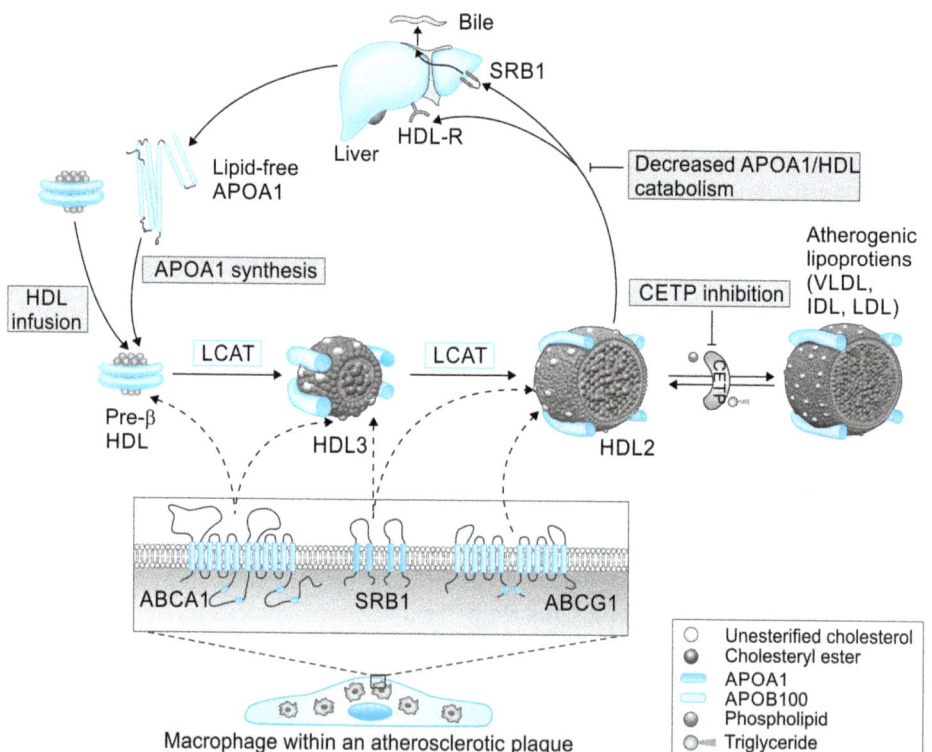

(ApoA1: apolipoprotein A1; CETP: cholesteryl ester transfer protein; IDL: intermediate density lipoprotein; LDL: low-density lipoprotein; VLDL: very low-density lipoprotein; ABCA1: ATP-binding cassette subfamily A member 1; ABCG1: ATP-binding cassette subfamily G member 1)

FIG. 1: Mechanism of action of high-density lipoprotein (HDL)-raising therapies.

Source: Adapted from Kingwell BA, Chapman MJ, Kontush A, Miller NE. HDL-targeted therapies: progress, failures and future. Nat Rev Drug Discov. 2014;13:445-64.

ONGOING TRIALS

Studies of new HDL infusion agents are ongoing to evaluate the possibility to raise serum HDL-C levels and to decrease CV risk by this approach. A Phase III trial is also underway. Among other trials of HDL-targeting agents, the BETonMACE study will determine whether BET inhibition treatment with RVX-208 (Apabetalone) in high-risk patients with type 2 diabetes mellitus (T2DM) and CAD increases the time to major adverse CV events (**Table 2**).

HDL-TARGETED THERAPIES IN PRECLINICAL DEVELOPMENT

Considering the complexity of increasing ApoA-I production, attention has moved toward modulation of ABCA1. Among transcription factors binding to the ABCA1 promoter, liver X receptor (LXR) has been most thoroughly studied as activated LXR induces transcription of ABCA1 as well as other genes functioning in cholesterol removal pathways.[33] Recent efforts are concentrated on developing LXR agonists without lipogenic effects by taking advantage of tissue-specific expression patterns of the LXR subtypes. Numerous other strategies to therapeutically target the metabolism, particle structure, and function of HDL are emerging. These include farnesoid X receptor (FXR) agonists, inhibitors of endothelial lipase (EL), microRNAs (miRNAs) antagonists, and antisense oligonucleotides (ASOs) targeted to genes that are involved in HDL metabolism (**Table 3**). Among miRNA inhibition strategies, antisense oligonucleotides

TABLE 2: Trials of high-density lipoprotein (HDL)-raising agents in progress.

Type of treatment	Agent (study)	Phase	Intervention	Condition	Primary outcome
HDL infusion agents	CSL112 (AEGIS-II)	3	Placebo vs. CSL-112	Acute coronary syndrome	Cardiovascular death, myocardial infarction, or stroke
BET inhibitor	RVX-208 (BETonMACE)	3	Placebo vs. RVX-208 + atorvastatin + rosuvastatin	Type 2 diabetes mellitus and coronary artery disease	Safety and efficacy

(HDL: high-density lipoprotein; BET: bromodomain and extraterminal domain)

TABLE 3: High-density lipoprotein (HDL)-targeted therapies in preclinical development.

Class	Mode of delivery	Mechanism of action
LXR agonists	Oral	Upregulation of ABCA1 and ABCG1 expression
FXR agonists	Oral	Acceleration of cholesterol excretion via HDL
Endothelial lipase inhibitors	Presently unknown	Inhibition of HDL lipid hydrolysis
MicroRNA antagonists	Injection	Upregulation of ABCA1 and ABCG1 expression
ASOs	Subcutaneous injection	Inhibition of *CETP* and *ApoC3* gene expression

(ABCA1: ATP-binding cassette subfamily A member 1; ABCG1: ATP-binding cassette subfamily G member 1; ApoC3: apolipoprotein C3; ASO: antisense oligonucleotide; FXR: farnesoid X receptor; LXR: liver X receptor CETP: cholesteryl ester transfer protein)

Source: Kingwell BA, Chapman MJ, Kontush A, Miller NE. HDL-targeted therapies: progress, failures and future. Nat Rev Drug Discov. 2014;13:445-64.

that can directly target miRNAs have been most encouraging. Modulation of ApoA-I-desmosomal protein desmocollin-1 (DSC1) interactions may represent another novel therapeutic target.[12]

CONCLUSION AND FUTURE PERSPECTIVES

The apparent paradoxical failure of HDL-C-raising therapies to improve CV outcomes has resulted in confusion concerning the exact role of HDL in the CVD pathogenesis. This has led to uncertainty in the clinical arena regarding the potential for future HDL-targeted pharmacotherapies. Different factors have contributed to the slow progress in this field, including the complex metabolism of HDL particles, the dynamic heterogeneity of HDL subpopulations, and the intricacy of their various structure–function features in normolipidemic against dyslipidemia states, plus the lack of drugs that specifically target HDL. Nevertheless, there are several agents in the HDL clinical pipeline that are expected to provide opportunities for more definitive testing of the HDL hypothesis. Existing data indicate that HDL still remains a promising therapeutic target to decrease CV risk. However, it is necessary to resolve the ongoing controversy between the results of epidemiological studies and clinical trials to allow further development of HDL-targeted therapies. Strategies that are focused on the key parameters of HDL biology and structure–function relationships, rather than simple plasma concentration of HDL-C, will be important to the development of future HDL-targeting therapies.

KEY POINTS

- The implications of the HDL hypothesis for therapeutic application remain uncertain.
- The plasma level of HDL-C may be a biomarker of cardiovascular health, but not a good therapeutic target.
- The negative outcomes of studies investigating early HDL cholesterol-raising therapies have increasing pessimism.
- Drug-based future HDL studies will be required to focus on identifying new targets for the promotion of HDL biogenesis.

REFERENCES

1. Gordon DJ, Rifkind BM. High-density lipoprotein—the clinical implications of recent studies. N Engl J Med. 1989;321:1311-6.
2. Gordon DJ, Probstfield JL, Garrison RJ, Neaton JD, Castelli WP, Knoke JD, et al. High-density lipoprotein cholesterol and cardiovascular disease. Four prospective American studies. Circulation. 1989;79:8-15.
3. Miller GJ, Miller NE. Plasma-high-density-lipoprotein concentration and development of ischaemic heart disease. Lancet. 1975;1(7897):16-9.
4. Kontush A, Chapman MJ. High-density Lipoproteins: Structure, Metabolism, Function, and Therapeutics. New York: Wiley & Sons; 2012 p. 648.
5. Van Lenten BJ, Navab M, Shih D, Fogelman AM, Lusis AJ. The role of high-density lipoproteins in oxidation and inflammation. Trends Cardiovasc Med. 2001;11:155-61.
6. Kingwell BA, Chapman MJ, Kontush A, Miller NE. HDL-targeted therapies: progress, failures and future. Nat Rev Drug Discov. 2014;13:445-64.
7. Lincoff AM, Nicholls SJ, Riesmeyer JS, Barter PJ, Brewer HB, Fox KAA, et al. Evacetrapib and cardiovascular outcomes in high-risk vascular disease. N Engl J Med. 2017;376:1933-42.

8. Zakiev E, Feng M, Sukhorukov V, Kontush A. HDL-Targeting Therapeutics: Past, Present and Future. Curr Pharm Des. 2017;23:1-9.
9. Rader DJ, Tall AR. The not-so-simple HDL story: Is it time to revise the HDL cholesterol hypothesis? Nature Med. 2012;18:1344-6.
10. Vaisar T, Tang C, Babenko I, Hutchins P, Wimberger J, Suffredini AF, et al. Inflammatory remodeling of the HDL proteome impairs cholesterol efflux capacity. J Lipid Res. 2015;56:1519-30.
11. Kontush A, Lhomme M, Chapman MJ. Unraveling the complexities of the HDL lipidome. J Lipid Res. 2013;54:2950-63.
12. Genest J, Choi HY. Novel Approaches for HDL-Directed Therapies. Curr Atheroscler Rep. 2017;19:55.
13. Inazu A, Brown ML, Hesler CB, Agellon LB, Koizumi J, Takata K, et al. Increased high-density lipoprotein levels caused by a common cholesteryl-ester transfer protein gene mutation. N Engl J Med. 1990;323:1234-8.
14. Brousseau ME, Schaefer EJ, Wolfe ML, Bloedon LT, Digenio AG, Clark RW, et al. Effects of an inhibitor of cholesteryl ester transfer protein on HDL cholesterol. N Engl J Med. 2004;350:1505-15.
15. Barter PJ, Caulfield M, Eriksson M, Grundy SM, Kastelein JJP, Komajda M, et al. Effects of torcetrapib in patients at high risk for coronary events. N Engl J Med. 2007;357:2109-22.
16. Schwartz GG, Olsson AG, Abt M, Ballantyne CM, Barter PJ, Brumm J, et al. Effects of dalcetrapib in patients with a recent acute coronary syndrome. N Engl J Med. 2012;367:2089-99.
17. Cannon CP, Shah S, Dansky HM, Davidson M, Brinton EA, Gotto AM, et al. Safety of Anacetrapib in Patients with or at High Risk for Coronary Heart Disease. N Engl J Med. 2010;363:2406-15.
18. HPS3/TIMI55–REVEAL Collaborative Group. Effects of anacetrapib in patients with atherosclerotic vascular disease. N Engl J Med. 2017;377:1217-27.
19. Nissen SE, Tsunoda T, Tuzcu EM, Schoenhagen P, Cooper CJ, Yasin M, et al. Effect of recombinant ApoA-I Milano on coronary atherosclerosis in patients with acute coronary syndromes: a randomized controlled trial. JAMA. 2003;290:2292-300.
20. Kallend DG, Reijers JA, Bellibas SE, Bobillier A, Kempen H, Burggraaf J, et al. A single infusion of MDCO-216 (ApoA-1 Milano/POPC) increases ABCA1-mediated cholesterol efflux and pre-beta 1 HDL in healthy volunteers and patients with stable coronary artery disease. Eur Heart J Cardiovas Pharmacother. 2016;2:23-9.
21. Nicholls SJ, Puri R, Ballantyne CM, Jukema JW, Kastelein JJP, Koenig W, et al. Effect of infusion of high-density lipoprotein mimetic containing recombinant apolipoprotein A-I milano on coronary disease in patients with an acute coronary syndrome in the MILANO-PILOT trial: a randomized clinical trial. JAMA Cardiol. 2018;3:806-14.
22. Tardif JC, Gregoire J, L'Allier PL, Ibrahim R, Lespérance J, Heinonen TM, et al. Effects of reconstituted high-density lipoprotein infusions on coronary atherosclerosis: a randomized controlled trial. JAMA. 2007;297:1675-82.
23. Michael Gibson C, Korjian S, Tricoci P, Daaboul Y, Yee M, Jain P, et al. Safety and tolerability of CSL112, a reconstituted, infusible, plasma-derived apolipoprotein A-I, after acute myocardial infarction: the AEGIS-I trial (ApoA-I Event Reducing in Ischemic Syndromes I). Circulation. 2016;134:1918-30.
24. Nicholls SJ, Andrews J, Kastelein JJP, Merkely B, Nissen SE, Ray KK, et al. Effect of serial infusions of CER-001, a pre-β high-density lipoprotein mimetic, on coronary atherosclerosis in patients following acute coronary syndromes in the CER-001 atherosclerosis regression acute coronary syndrome trial: a randomized clinical trial. JAMA Cardiol. 2018;3:815-822.
25. Hovingh GK, Smits LP, Stefanutti C, Soran H, Kwok S, de Graaf J, et al. The effect of an apolipoprotein A-I-containing high-density lipoprotein-mimetic particle (CER-001) on carotid artery wall thickness in patients with homozygous familial hypercholesterolemia: The Modifying Orphan Disease Evaluation (MODE) study. Am Heart J. 2015;169:736-42.
26. Bailey D, Jahagirdar R, Gordon A, Hafiane A, Campbell S, Chatur S, et al. RVX-208: a small molecule that increases apolipoprotein A-I and high-density lipoprotein cholesterol in vitro and in vivo. J Am Coll Cardiol. 2010;55:2580-9.
27. McLure KG, Gesner EM, Tsujikawa L, Kharenko OA, Attwell S, Campeau E, et al. RVX-208, an inducer of ApoA-I in humans, is a BET bromodomain antagonist. PloS One. 2013;8:e83190.
28. Nicholls SJ, Gordon A, Johansson J, Wolski K, Ballantyne CM, Kastelein JJP, et al. Efficacy and safety of a novel oral inducer of apolipoprotein a-I synthesis in statin-treated patients with stable coronary artery disease a randomized controlled trial. J Am College Cardiol. 2011;57:1111-9.

29. Nicholls SJ, Puri R, Wolski K, Ballantyne CM, Barter PJ, Brewer HB, et al. Effect of the BET Protein Inhibitor, RVX-208, on Progression of Coronary Atherosclerosis: Results of the Phase 2b, Randomized, Double-Blind, Multicenter, ASSURE Trial. Am J Cardiovas Drugs. 2016;16:55-65.
30. Furbee JW Jr, Sawyer JK, Parks JS. Lecithin: cholesterol acyltransferase deficiency increases atherosclerosis in the low-density lipoprotein receptor and apolipoprotein E knockout mice. J Biol Chem. 2002;277:3511-9.
31. Shamburek RD, Bakker-Arkema R, Shamburek AM, Freeman LA, Amar MJ, Auerbach B, et al. Safety and Tolerability of ACP-501, a Recombinant Human Lecithin: Cholesterol Acyltransferase, in a Phase 1 Single-Dose Escalation Study. Circulation Res. 2016;118:73-82.
32. Waksman R, Torguson R, Kent KM, Pichard AD, Suddath WO, Satler LF, et al. A first-in-man, randomized, placebo-controlled study to evaluate the safety and feasibility of autologous delipidated high-density lipoprotein plasma infusions in patients with acute coronary syndrome. J Am Coll Cardiol. 2010;55:2727-35.
33. Hong C, Tontonoz P. Liver X receptors in lipid metabolism: opportunities for drug discovery. Nat Rev Drug Discov. 2014;13:433-44.

CHAPTER

20

Statin Intolerance: Impact on Statin Therapy—Assessment and Management

Indira Maisnam

ABSTRACT

Statins are important lipid-lowering agents that provide primary and secondary protection against atherosclerotic cardiovascular diseases (ASCVD). Currently, there is no consensus on the exact diagnostic criteria of statin intolerance. An interplay of modifiable and nonmodifiable factors can cause statin intolerance resulting in statin dose reduction or cessation of use, thereby increasing atherosclerotic risk in such patients. With a systematic approach and correction of or avoidance of confounders, a large number of patients can be successfully rechallenged with statins either with an alternative statin and/or alternative dose. In patients who are completely statin intolerant, despite all measures, alternative lipid-lowering agents like ezetimibe, bile acid sequestrants, and proprotein convertase subtilisin/kexin type 9 (PCSK9) inhibitors can be used.

INTRODUCTION

Statins are important drugs for dyslipidemia and provide primary and secondary protection against atherosclerotic cardiovascular diseases (ASCVD). They are generally well-tolerated drugs, but some individuals develop statin intolerance. Statin discontinuation due to intolerance can increase ASCVD risk.

DEFINITION OF STATIN INTOLERANCE

Statin intolerance is the inability to use statins at recommended dosage for prevention of ASCVD due to the development of symptoms and/or some laboratory abnormalities related to its use. The intolerance to statins may be partial (some statins at some doses) or complete (all statins at any dose). There is no consensus on the definition of statin intolerance.

Statin Intolerance Definitions

The National Lipid Association (NLA) and the International Lipid Expert Panel (ILEP) define statin intolerance as the "inability to tolerate at least two statins: One statin at the lowest starting daily dose and another statin at any daily dose, due to either objectionable symptoms or abnormal laboratory determinations, which are temporally related to statin treatment and reversible upon statin discontinuation".[1,2] The ILEP further adds that "resolution of symptoms or changes in biomarkers or even significant improvement with dose reduction or withdrawal of treatment; symptoms or changes in biomarkers are not attributable to predispositions increasing the risk of statin intolerance".

The EAS (European Atherosclerosis Society) avoids the term statin intolerance because of negative connotation and uses the term statin-associated muscle symptoms (SAMS). The EAS states that "the assessment of SAMS includes the nature of muscle symptoms, increased creatine kinase (CK) levels and their temporal association with initiation of therapy with statin, and statin therapy suspension and rechallenge". The focus of EAS definition is only on muscle symptoms. The intake and intolerance to at least three statins are required for diagnosis of statin intolerance, but no minimum dose is specified.[3]

Statin-associated Muscle Symptoms

The 2014 NLA Statin Muscle Safety Task Force proposed the following terms and definitions of muscle-related symptoms:[4]

- *Myalgia*: A symptom of muscle discomfort, including muscle aches, soreness, stiffness, tenderness, or cramps with or soon after exercise, with a normal CK level.
- *Myopathy*: Muscle weakness (not due to pain), with or without an elevation in CK level.
- *Myositis*: Muscle inflammation.
- *Myonecrosis*: Elevation in muscle enzymes compared with either baseline CK levels (while not on statin therapy) or the upper limit of normal that has been adjusted for age, race, and sex.
    - *Mild*: 3- to 10-fold elevation in CK.
    - *Moderate*: 10- to 50-fold elevation in CK.
    - *Severe*: 50-fold or greater elevation in CK.
- *Clinical rhabdomyolysis*: Myonecrosis with myoglobinuria or acute renal failure.

The SAMSCI (SAMS Clinical Index) score may help identity true statin myalgia and rule out nocebo effect (negative effects of a drug because of side effect warning). A score of <5 had a negative predictive value of 91% in correctly excluding patient with true statin myalgia.

STATINS AND TRANSAMINITIS

Statin-associated transaminitis is rare, mild, and dose related. The transaminitis returns to baseline by 2-4 weeks in most cases. Statin-induced persistent elevation of liver enzymes more than three times upper limit of normal is very rare (<1%) and the number needed to harm for serious liver disease is 1 in million.

EPIDEMIOLOGY OF STATIN INTOLERANCE

Because of differing diagnostic criteria, we may not have the exact prevalence of statin intolerance. Prevalence of statin intolerance was lower in randomized controlled trials (RCTs) (3-5%) but higher in real-world settings.[5] The USAGE (Understanding Statin Use in America and Gaps in Patient Education) study showed that SAMS occurred in 60% of current and 25% of former statin users. SAMS was the main reason for treatment discontinuation.[6] About 10.5% patients had muscle symptoms in the Prediction of Muscular Risk in Observational Conditions (PRIMO) which was conducted in general medicine clinics.[7] The difference between RCTs and real-world data could be due to exclusion of individuals at risk of statin intolerance or their underrepresentation in RCTs.

The Effect of Statins on Skeletal Muscle Function and Performance (STOMP) RCT showed that 9.4% of patients receiving atorvastatin and 4.6% of subjects receiving placebo had myalgia. Thus, incidence of statin-attributable SAMS was <5%.[8] In the ODYSSEY ALTERNATIVE, 46% of patients in the rechallenge arm reported muscle adverse effects and only 22% discontinued taking statin.[9] Thus, patients with a past history of statin intolerance may still tolerate statins in future.

IMPACT OF STATIN INTOLERANCE ON STATIN THERAPY

Due to fear of and development of statin intolerance, patients fail to receive adequate dose of statins or do not receive statins increasing ASCVD morbidity and mortality.

Negative statin-related news was associated with statin discontinuation in a prospective study from Denmark and this was associated with increased cardiovascular (CV) events and deaths.[10] A retrospective analysis of medicare beneficiaries who began statin treatment after hospitalization for myocardial infarction (MI) found that 1.65% of beneficiaries were statin intolerant and had higher rate of recurrent MI and coronary heart disease events compared to those who were statin adherent. However, no increase in all-cause mortality was observed.[11]

In one study, discontinuation of statins after an acute myocardial infarction (AMI) resulted in a four- to seven-fold cumulative cardiac mortality compared to statin adherence.[12]

The excess risk for CV events with statin cessation due to intolerance and the CV protection offered by statins implies that there is urgent need to correctly identify statin intolerance, not overdiagnose it, and manage it appropriately.

PATHOGENESIS AND MECHANISM OF STATIN INTOLERANCE

The mechanisms of statin intolerance are not exactly clear. The proposed mechanisms include alteration in physicochemical properties of membrane lipids with statin use.[13] These changes can alter the functions of membrane ion channels. HMG-CoA reductase inhibition of mevalonate synthesis can impair synthesis of downstream molecules like coenzyme Q and dolichol impairing cellular processes dependent on them like membrane glycoprotein synthesis and adenosine triphosphate (ATP) synthesis.[14,15] The consequent muscle injury releases CK, an important muscle enzyme for ATP generation. ATP depletion can then worsen muscle injury further with necrosis, more release of CK, and ultimately rhabdomyolysis.

RISK FACTORS

Modifiable and nonmodifiable factors determine risk for statin intolerance.

Statin Characteristics and Drug Interaction

Statins differs in toxicity risk and toxicity is dose dependent. Rosuvastatin and pravastatin are hydrophilic statins whereas atorvastatin, fluvastatin, lovastatin, simvastatin, and pitavastatin are lipophilic statins. Lipophilic statins were believed to more myotoxic than hydrophilic statins, but the discovery of organic anion transport polypeptide 2B1 expressed on the sarcolemma which mediates uptake of statins into myocyte suggests that lipophilicity has no role in statin myotoxicity.

Myotoxicity is low with pravastatin, rosuvastatin, pitavastatin, and fluvastatin.[16,17] These are not metabolized by the CYP3A4 and have lower risk for drug interaction. Lovastatin, simvastatin, and to a lesser extent atorvastatin are metabolized by the CYP3A4 and are at higher risk of drug interaction. Fluvastatin, pitavastatin, and rosuvastatin are metabolized by CYP2C9/C8 whereas pravastatin is metabolized by sulfation and conjugation. Pitavastatin and rosuvastatin are also cleared by non-CYP mechanisms and therefore have lesser chance for drug interaction.

Coadministration of simvastatin, lovastatin, and atorvastatin with drugs that inhibit the CYP3A4 (**Table 1**)[18] and CYP3A4 substrates (colchicine, fibrates) increase their toxicity. Cyclosporine also inhibits CYP2C9 and organic anion transporter that regulates hepatic uptake of fluvastatin, rosuvastatin, and pitavastatin. The choice of statins in those on cyclosporine is limited. Fluvastatin can be given up to a dose of 40 mg and reduced dose of rosuvastatin may be given.[19]

Genetic Polymorphisms

A strong association of simvastatin-associated myopathy with the rs4363657 single-nucleotide polymorphism (SNP) within SLCO1B1 was found.[20] Polymorphisms in the *CYP* genes can predispose to statin toxicity. Human leukocyte antigen (HLA) association and statin-associated necrotizing (autoimmune) myopathy have been described.[21]

TABLE 1: Drugs potentially interacting with statins metabolized by CYP3A4.		
Anti-infectives	Calcium channel blockers	Others
• Itraconazole • Ketoconazole • Posaconazole • Erythromycin • Clarithromycin • Telithromycin • Human immunodeficiency virus (HIV) protease inhibitors	• Verapamil • Diltiazem • Amlodipine	• Cyclosporine • Danazol • Amiodarone • Ranolazine • Grapefruit juice • Nefazodone • Gemfibrozil

Preexisting Neuromuscular Disorders

Statins can interact with underlying neuromuscular diseases resulting in their manifestation and worsening. Examples include myasthenia gravis, skeletal muscle glycogen storage diseases, inflammatory myopathies, etc.

Exercise

Vigorous unaccustomed exercise can increase risk of muscle injury with statins.[22] Therefore, exercise should be graded. Patients on statins should continue to exercise. Mild elevation of CK with exercise is mostly subclinical.

Other Factors

Older age, females, Asian ethnicity, low body mass index (BMI), untreated hypothyroidism, vitamin D deficiency, chronic renal failure, and chronic liver disease (CLD) can increase statin-associated myopathy.

ASSESSMENT AND MANAGEMENT

The management of statin intolerance may be approached as follows: (i) Diagnosis, confirmation, and ruling out confounders; (ii) Addressing confounders; (iii) Statin withdrawal or downtitration; (iv) Statin rechallenge; (v) Reassessment after challenge; and (vi) Alternative therapeutic regimens. All patients should follow therapeutic lifestyle changes like dietary modification, weight control, and regular physical activity.

Diagnosis, Confirmation, and Ruling out Confounders

The most common feature of statin intolerance is myalgia; severe side effects like rhabdomyolysis are very rare. Symptoms are often nonspecific, but are typically symmetric, involve proximal muscles and persistent while on therapy.[23] They occur within weeks of starting treatment (usually within 3 months), but can occur anytime. Symptoms abate on cessation of treatment and can reappear on reintroduction.

A detailed history regarding the type and distribution of muscle discomfort, severity, duration, timing in relation to statin intake, the specific statin and dose, relation to exercise, the presence of comorbidities, addiction, and current coprescription including over-the-counter drugs and food supplements should be taken. The possibility of nocebo effect should also be looked into. Blood may be sent for CK, liver function tests, thyroid function tests, 25-hydroxyvitamin D, and baseline reports, if available, may be consulted. Electromyography (EMG) and muscle biopsy are rarely required and may be undertaken when muscle symptoms and CK elevation do not resolve on statin discontinuation.

Addressing Confounders

For statin-intolerant individuals on CYP34A inducers, options include stopping the CYP34A inducers (if feasible) and/or stopping statins. After symptom resolution, shifting to statins that are not metabolized by CYP34A or downtitration of statins are options.

Correction of hypothyroidism and hypovitaminosis D can improve statin-induced muscle symptoms. Dose adjustments in chronic kidney disease (CKD) and CLD, elderly, women, and low BMI are to be done. Avoidance of alcohol and graded exercise help address statin intolerance.

Statin Withdrawal or Downtitration

Once confirmed from history and clinical examination that symptoms are due to statins, based on the patient's clinical condition, statins may be withdrawn or downtitrated. Patients are monitored for resolution of symptoms and normalization of CK. However, in rhabdomyolysis and immune-mediated necrotizing myopathy, additional treatment is required besides immediate stoppage of statins.

Statins should be stopped immediately following rhabdomyolysis and supportive management offered to prevent or treat acute renal failure. Generally, statins are not reintroduced in patients who have had a statin-induced rhabdomyolysis due to risk of recurrence. However, this concept is debated. In the GAUSS-3 Trial, a number of patients who developed rhabdomyolysis were rechallenged with a statin.[24]

An immune-mediated necrotizing myopathy is a very rare condition seen with statins when antibody against HMG-CoA reductase develops. CK levels are markedly elevated and muscle histology shows muscle necrosis without inflammation. The condition is suspected when symptoms and CK do not improve on statin cessation and is to be treated by immunosuppressive therapy.

Statin Rechallenge

Statin rechallenge may be done as follows:
- *Patient on high-intensity statin*: Starting the same statin at lower dose or switching to alternative high-intensity statin.
- *Patient on low-dose statin*: Switching to alternate day statin.
- *Patients need to continue CYP34A inducers*: Switching to statins is not metabolized by CYP34A.

Throughout the process of rechallenge, the physician and the patient should be alert to the possibility of symptom recurrence. If alternate day dosing of statins is used, it is preferable to use long-acting statins like atorvastatin or rosuvastatin.

Reassessment After Challenge

Reassessment should take place throughout the process of initiation of statin rechallenge and at least up to 6 weeks after initiation. The new regimen is to be continued if patient can tolerate it and achieve low-density lipoprotein cholesterol (LDL-C) goals. Additional lipid-lowering therapy will be needed if patient can tolerate the alternative regimen but are not achieving LDL-C goals. Patients are to be offered alternative lipid-lowering therapies, if they are unable to tolerate the rechallenge.

Alternative Therapeutic Regimen

The alternative therapeutic regimen includes ezetimibe, bile acid sequestrants, and proprotein convertase subtilisin/kexin type 9 (PCSK9) inhibitors. These agents have lesser/no myotoxicity and have evidences of CV benefits. There is limited scope of other drugs used in dyslipidemia and limited evidence of nutraceuticals.

Ezetimibe

Ezetimibe is a cholesterol absorption inhibitor. In the IMPROVE-IT (Improved Reduction of Outcomes: Vytorin Efficacy International Trial), addition of ezetimibe to simvastatin 40 mg reduced ASCVD events in post-acute coronary syndrome patients compared to simvastatin alone.[25] Muscle symptoms in GAUSS-3 Trial were similar whether patients were on ezetimibe or evolocumab.[24] However, in ODYSSEY ALTERNATIVE Trial, patients randomized to alirocumab, ezetimibe, and atorvastatin had skeletal muscle adverse effects in 15.9%, 20.2%, and 22.2%, respectively.[9]

Bile Acid Sequestrants

Bile acid sequestrants reduce LDL-C by 15–30%, modestly raise HDL-C but can increase triglycerides level. They should not to be used when triglyceride >300 mg/dL.

The bile acids in clinical use are cholestyramine, colestipol, and the newer colesevelam. Older bile acids produce adverse gastrointestinal symptoms (flatulence, nausea, dyspepsia, and constipation) and interfere with absorption of other drugs. Colesevelam is better tolerated and malabsorption of other drugs is lesser due to its greater affinity for bile acids. Drug malabsorption can be minimized by administering other drugs 1 hour before or 4 hours after the bile acid sequestrants. Bile acid sequestrants reduce CV events with the benefit proportional to degree of LDL-C lowering.

Proprotein Convertase Subtilisin/Kexin Type 9 Inhibitors

Proprotein convertase subtilisin/kexin type 9 inhibitors are new class of effective lipid-lowering drugs. The ODYSSEY ALTERNATIVE and the GAUSS-3 Trials studied PCSK9 inhibitors alirocumab and evolocumab in statin-intolerant individuals. Both agents had no/little evidence of myotoxicity.

The CV benefits of evolocumab and alirocumab were established in the FOURIER Trial and ODYSSEY Outcomes Trial, respectively. In the FOURIER Trial, in patients with ASCVD with LDL-C ≥70 mg/dL on statin therapy, addition of evolocumab compared to placebo significantly reduced the composite of CV death, MI, stroke, hospitalization for unstable angina, or coronary revascularization by 15%, but did not reduce CV or all-cause mortality at 2.2 years.[26] In the ODYSSEY Outcomes Trial, 15% relative reduction in the composite of congenital heart disease (CHD) death, nonfatal MI, ischemic stroke, or unstable angina requiring hospitalization was seen after a median follow-up of 2.8 years in patients who were given alirocumab after an acute coronary event compared to placebo.[27]

Proprotein convertase subtilisin/kexin type 9 inhibitors are alternatives in statin intolerance because of their effectiveness, absence of myotoxicity, and CV benefits. However, they are expensive and the CV benefits in statin-intolerant individuals have not been studied.

CONCLUSION

There is urgent need for a consensus on the definition of statin intolerance. A stepwise approach and identification of modifiable and nonmodifiable risk factors are needed to address statin intolerance. Correction of certain modifiable factors can sometimes help reinitiate statins in statin intolerance. Changes in statin type, dose modulation, and alternative agents are all approaches to statin intolerance. It is important that LDL-C targets are reached for maximum protection.

> **KEY POINTS**
> - Statin intolerance occurs due to modifiable and nonmodifiable factors.
> - Statin nonadherence due to intolerance is a risk for future CV events.
> - Identification and addressing modifiable factors, putting patients on alternative statin, and/or alternative dosing regimen help many patients tolerate statin rechallenge.
> - Alternative agents with documented CV protection are ezetimibe, bile acid sequestrants, and PCSK9 inhibitors.

REFERENCES

1. Jacobson TA, Ito MK, Maki KC, Orringer CE, Bays HE, Jones PH, et al. National Lipid Association recommendations for patient-centered management of dyslipidemia: Part 1-executive summary. J Clin Lipidol. 2014;8:473-88.
2. Banach M, Rizzo M, Toth PP, Farnier M, Davidson MH, Al-Rasadi K, et al. Statin intolerance-an attempt at a unified definition. Position paper from an International Lipid Expert Panel. Expert Opin Drug Saf. 2015;14:935-55.
3. Stroes ES, Thompson PD, Corsini A, Vladutiu GD, Raal FJ, Ray KK, et al. Statin-associated muscle symptoms: impact on statin therapy-European Atherosclerosis Society Consensus Panel Statement on Assessment, Aetiology and Management. Eur Heart J. 2015;36:1012-22.
4. Rosenson RS, Baker SK, Jacobson TA, Kopecky SL, Parker BA, The National Lipid Association's Muscle Safety Expert Panel. An assessment by the Statin Muscle Safety Task Force: 2014 Update. J Clin Lipidol. 2014;8:S58-71.
5. Bays H. Statin safety: an overview and assessment of the data—2005. Am J Cardiol. 2006;97:6C-26.
6. Cohen JD, Brinton EA, Ito MK, Jacobson TA. Understanding Statin Use in America and Gaps in Patient Education (USAGE): an internet-based survey of 10,138 current and former statin users. J Clin Lipidol. 2012;6:208-15.
7. Bruckert E, Hayem G, Dejager S, Yau C, Bégaud B. Mild to moderate muscular symptoms with high-dosage statin therapy in hyperlipidemic patients—the PRIMO study. Cardiovasc Drugs Ther. 2005;19:403-14.
8. Parker BA, Capizzi JA, Grimaldi AS, Clarkson PM, Cole SM, Keadle J, et al. Effect of Statins on Skeletal Muscle Function. Circulation. 2013;127:96-103.
9. Moriarty PM, Thompson PD, Cannon CP, Guyton JR, Bergeron J, Zieve FJ, et al. Efficacy and safety of alirocumab vs ezetimibe in statin-intolerant patients, with a statin rechallenge arm: the ODYSSEY ALTERNATIVE randomized trial. J Clin Lipidol. 2015;9:758-69.
10. Nielsen SF, Nordestgaard BG. Negative statin-related news stories decrease statin persistence and increase myocardial infarction and cardiovascular mortality: a nationwide prospective cohort study. Eur Heart J. 2016;37:908-16.
11. Serban MC, Colantonio LD, Manthripragada AD, Monda KL, Bittner VA, Banach M, et al. Statin Intolerance and Risk of Coronary Heart Events and All-Cause Mortality Following Myocardial Infarction. J Am Coll Cardiol. 2017;11:1386-95.
12. Kim MC, Cho JY, Jeong HC, Lee KH, Park KH, Sim DS, et al. Impact of postdischarge statin withdrawal on long-term outcomes in patients with acute myocardial infarction. Am J Cardiol. 2015;115:1-7.
13. Morita I, Sato I, Ma L, Murota S. Enhancement of membrane fluidity in cholesterol-poor endothelial cells pre-treated with simvastatin. Endothelium. 1997;5:107-13.
14. DiMauro S, Bonilla E, Davidson M, Hirano M, Schon EA. Mitochondria in neuromuscular disorders. Biochem Biophys Acta. 1998;1366:199-210.
15. Laaksonen R, Jokelainen K, Sahi T, Tikkanen MJ, Himberg JJ. Decreases in serum ubiquinone concentrations do not result in reduced levels in muscle tissue during short-term simvastatin treatment in humans. Clin Pharmacol Ther. 1995;57:62-6.
16. Wiggins BS, Saseen JJ, Page RL, Reed BN, Sneed K, Kostis JB, et al. Recommendations for management of clinically significant drug-drug interactions with statins and select agents used in patients with cardiovascular disease: a scientific statement from the American Heart Association. Circulation. 2016;134:e468-95.
17. Catapano AL. Statin-induced myotoxicity: pharmacokinetic differences among statins and the risk of rhabdomyolysis, with particular reference to pitavastatin. Curr Vasc Pharmacol. 2012;10:257-67.
18. Mach F, Baigent C, Catapano AL, Koskinas KC, Casula M, Badimon L, et al. 2019 ESC/EAS guidelines for the management of dyslipidaemias: lipid modification to reduce cardiovascular risk. Eur Heart J. 2020;41:111-88.

19. Holdaas H, Hagen E, Asberg A, Lund K, Hartman A, Vaidyanathan S, et al. Evaluation of the pharmacokinetic interaction between fluvastatin XL and cyclosporine in renal transplant recipients. Int J Clin Pharmacol Ther. 2006;44:163-71.
20. SEARCH Collaborative Group, Link E, Parish S, Armitage J, Bowman L, Heath S, et al. SLCO1B1 variants and statin-induced myopathy—a genomewide study. N Engl J Med. 2008;359:789-99.
21. Mammen AL. Statin-Associated Autoimmune Myopathy. N Engl J Med. 2016;374:664-9.
22. Franc S, Dejager S, Bruckert E, Chauvenet M, Giral P, Turpin G. A comprehensive description of muscle symptoms associated with lipid lowering drugs. Cardiovasc Drugs Ther. 2003;17:459-65.
23. Jacobson TA. Toward "pain-free" statin prescribing: clinical algorithm for diagnosis and management of myalgia. Mayo Clin Proc. 2008;83:687-700.
24. Nissen SE, Stroes E, Dent-Acosta RE, Rosenson RS, Lehman SJ, Sattar N, et al. Efficacy and tolerability of evolocumab vs ezetimibe in patients with muscle-related statin intolerance: the GAUSS-3 randomized clinical trial. JAMA. 2016;315:1580-90.
25. Cannon CP, Blazing MA, Giugliano RP, McCagg A, White JA, Theroux P, et al. Ezetimibe added to statin therapy after acute coronary syndromes. N Engl J Med. 2015;372:2387-97.
26. Sabatine MS, Giugliano RP, Keech AC, Honarpour N, Wiviott SD, Murphy SA, et al. Evolocumab and clinical outcomes in patients with cardiovascular disease. N Engl J Med. 2017;376:1713-22.
27. Schwartz GG, Steg PG, Szarek M, Bhatt DL, Bittner VA, Diaz R, et al. Alirocumab and cardiovascular outcomes after acute coronary syndrome. N Engl J Med. 2018;379:2097-107.

CHAPTER 21

Diabetic Dyslipidemia: Current Concepts and Guidelines

Amritava Ghosh

ABSTRACT

Dyslipidemia in people with type 2 diabetes mellitus is an important risk factor for atherosclerotic cardiovascular disease. It is atherogenic and defined by the triad of increased proportion of small, dense low-density lipoprotein particles, low high-density lipoprotein cholesterol, and elevated triglyceride levels. Apart from glycemic control, other measures, i.e., therapeutic lifestyle changes and pharmacotherapy are required for comprehensive management of these lipid abnormalities. This chapter deals with details of these measures along with the available evidence and guidelines.

INTRODUCTION

A triad of high triglycerides (TG), low high-density lipoprotein (HDL) cholesterol, and increased proportion of small dense low-density lipoprotein (LDL) particles occurring in a person with type 2 diabetes mellitus (T2DM) is known as atherogenic diabetic dyslipidemia (ADD).

Fatty acids derived from diet are largely incorporated into TG in intestinal mucosal cells and secreted in chylomicrons, which bypass the liver and enter the systemic circulation via intestinal lymph through the thoracic duct. In the peripheral tissues, lipoprotein lipase (LPL) hydrolyzes the TG in chylomicrons to release free fatty acids in the process generating chylomicron remnants.

Nonesterified fatty acids (NEFAs) are released from adipocytes through the action of hormone sensitive lipase (HSL) and adipocyte triglyceride lipase, a process which is inhibited by insulin.[1] Conversely, insulin deficiency and/or resistance is associated with excessive release of NEFA from adipocytes, a major driver of dyslipidemia in diabetes mellitus (DM) and obesity.[2]

The liver derives fatty acids from additional lipolysis, uptake of remnant lipoproteins, de novo hepatic lipogenesis, and uptake of albumin-bound NEFA circulating in plasma. Very low-density lipoproteins (VLDLs) assembled and secreted from the liver undergo lipolysis by LPL in peripheral tissues resulting in release of fatty acids and production of VLDL remnant particles, called intermediate-density lipoproteins (IDL). IDL return to the liver where they are partly internalized and partly processed at the cell surface by hepatic lipase (HL) to become LDL. Insulin promotes production of free fatty acid (FFA) and degradation of apolipoprotein B100 (ApoB-100). ApoB is the primary protein constituent of VLDL, IDL, and LDL. As insulin resistance develops and levels of insulin rise, the pathway for FFA production remains insulin sensitive resulting in increased hepatic production of FFA; while there is reduction of insulin-stimulated intracellular degradation of ApoB. These changes at the level of hepatocytes, together with insulin resistance at the level of adipocytes causing increased FFA efflux in bloodstream, is thought to be central to the pathogenesis of ADD. This results in increased production of VLDL cholesterol from the liver.[3] Insulin stimulates LPL activity. Insulin deficiency and/or resistance also results in suboptimal metabolism of VLDL particles. Thus hyperglycemia leads to excess VLDL production and/or inefficient lipolysis.

Triglyceride-rich lipoproteins (TRLs) are involved in exchange of TG and cholesteryl esters with LDL and HDL facilitated by cholesteryl ester transfer protein (CETP). This leads to TG enrichment of LDL and HDL particles. Through subsequent action of HL, the LDL particles become small, dense, and more atherogenic. The TG-rich HDL undergoes similar hydrolysis by HL or LPL resulting in low HDL; ApoA-I dissociates from the reduced-size HDL, which due to its small size is filtered by the renal glomeruli and degraded in renal tubular cells (**Fig. 1**).[4-6]

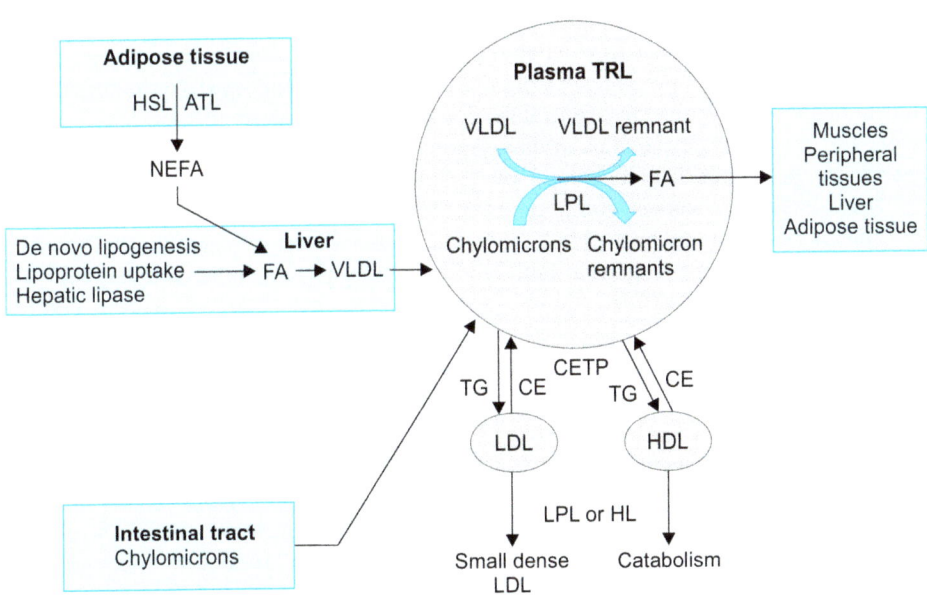

(ATL: adipocyte triglyceride lipase; CE: cholesterol esters; CETP: cholesteryl ester transfer protein; FA: fatty acids; HDL: high-density lipoprotein; HL: hepatic lipase; HSL: hormone-sensitive lipase; LDL: low-density lipoprotein; LPL: lipoprotein lipase; NEFA: nonesterified fatty acids; TG: triglycerides; TRL: triglyceride-rich lipoproteins; VLDL: very low-density lipoprotein)

FIG. 1: Triglyceride-rich lipoprotein metabolism.

APPROACH TO MANAGEMENT OF DIABETIC DYSLIPIDEMIA

In adults with DM, it is advisable to obtain a lipid profile at the time of diagnosis, and at least once in every 5 years thereafter. A lipid profile should also be obtained prior to initiation of statin therapy. LDL levels should be reassessed 4–12 weeks after initiation of statin therapy or- any change in dose of statin. Need for additional testing may also be considered- on an individual basis depending on patient's health status (e.g., to monitor adherence and response to therapy).[7]

Effective management of ADD requires a comprehensive approach including lifestyle measures and pharmacologic intervention.

Lifestyle Modification

Lifestyle changes focused on dietary modification, increase in physical activity and weight loss are the initial steps in the management of ADD. Both exercise and weight loss significantly reduce insulin resistance (IR) and the associated risk factors. Current recommendations for physical activity for adults with DM include at least 150 minutes of moderate- to vigorous-intensity aerobic activity per week, spread over at least 3 days/week with no >2 consecutive days without activity. The exercise regimen should also include 2–3 sessions/week of resistance training on nonconsecutive days.[7,8] A weight loss of as little as 5–10% can produce significant improvements in IR and ADD. Cigarette smoking also contributes to ADD, and cessation of smoking and other tobacco products should be encouraged. Additionally, excessive alcohol intake increases TG concentration. Therefore, moderation (or cessation) of alcohol consumption should be encouraged.[9] The Therapeutic Lifestyle Changes diet has placed less emphasis on substitution of carbohydrate for dietary fat (provided that the diet remains low in saturated fats and trans fats) since a high-carbohydrate diet will raise the TG concentration, exacerbating ADD.[7,9] Nutrition therapy should focus on increasing intake of plant stanols/sterols, ω-3 fatty acids, and viscous fiber (such as in oats, legumes, and citrus).[7] Intake of whole grains, vegetables, fruits, nuts, lean meat, and low-fat dairy products should be encouraged.[10]

Drugs known to worsen components of ADD (e.g., oral estrogens, retinoids, and thiazide diuretics) should be avoided if possible.[9] Achievement of better glycemic control is associated with improvement in lipid levels, particularly in people with very high TG and poor glycemic control.[7]

Drug Therapies for Atherogenic Diabetic Dyslipidemia
Statins

Blocking 3-hydroxy-3-methylglutaryl-coenzyme A (HMG-CoA) reductase results in a reduction in intrahepatic cholesterol, which leads to an increase in LDL receptor turnover and LDL uptake by the liver, thus reducing circulating LDL by 30–50%. LDL plays a fundamental role in atherogenesis, and numerous clinical trials and meta-analyses have demonstrated the beneficial effects of HMG-CoA reductase inhibitors (statins) on atherosclerotic cardiovascular disease (ASCVD) outcomes. Subgroup analyses of participants with DM in larger trials[11-15] and trials in participants with DM[16,17] showed significant primary and secondary prevention of ASCVD events and coronary heart disease (CHD) deaths in participants with DM. In the Heart Protection Study done on participants with T2DM, simvastatin reduced major adverse cardiovascular events (MACE) by 22%

compared to placebo.[12] CARDS (Collaborative Atorvastatin Diabetes Study) showed a 37% reduction in MACE among participants with T2DM and LDL <160 mg/dL randomized to low-dose atorvastatin versus placebo. The mortality end point however, did not reach significance.[17] A meta-analysis involving 14 randomized trials of statin therapy (18,686 participants with DM) demonstrated a 9% proportional reduction in all-cause mortality (related to significant reduction in vascular mortality) and 21% proportional reduction in major vascular events for each mmol/L (39 mg/dL) reduction in LDL cholesterol.[18] The safety and efficacy of statin therapy among patients with DM make it an obvious first-line choice for the treatment of ADD and for cardioprotection.

Ezetimibe

Ezetimibe targets the Niemann-Pick C1-like 1 (NPC1L1) protein resulting in reduced absorption of cholesterol from the intestine,[19] further resulting in the liver using more circulating cholesterol causing decrease in plasma concentration. Ezetimibe reduces LDL cholesterol levels by an additional 23–24%.[20]

IMPROVE-IT (IMProved Reduction of Outcomes: Vytorin Efficacy International Trial), a randomized controlled trial among patients hospitalized with acute coronary syndrome showed that ezetimibe, added to statin therapy, resulted in incremental lowering of LDL cholesterol levels (24% reduction) and reduction in cardiovascular events (absolute risk reduction: 2 percentage points).[21] In the ESD (Ezetimibe and Simvastatin in Dyslipidemia) trial which included participants with DM who had failed to achieve their goal LDL cholesterol level despite statin therapy, the addition of ezetimibe further reduced LDL by 31%, total cholesterol by 22% and ApoB by 20%.[22] Studies in animal models have shown that IR and hyperglycemia seem to drive cholesterol absorption in the setting of statin use, at least in part mediated by upregulation of NPC1L1. Therefore, it is likely that ezetimibe causes greater reduction in LDL and non-HDL cholesterol among people with T2DM.[23,24] Therefore, in addition to its role in patients not at LDL target or those intolerant of statins, there is growing interest in upfront use of ezetimibe as dual therapy (with statins) to mitigate these diabetes-specific pathways.

Proprotein Convertase Subtilisin Kexin Type 9

Proprotein convertase subtilisin kexin type 9 (PCSK9) is a protease that degrades the LDL receptors thereby diverting the receptors from their normal recycling pathway. The liver becomes less capable of clearing LDL particles from the blood and LDL cholesterol levels rise.[25]

Two monoclonal antibodies—alirocumab and evolocumab—are available as therapies targeting PCSK9. Their efficacy at lowering LDL cholesterol and protecting from ASCVD has been corroborated in numerous studies and meta-analyses. LDL reduction has been shown to be identical in people with and without DM.[26]

The FOURIER trial (Further Cardiovascular Outcomes Research with PCSK9 Inhibition in Subjects with Elevated Risk) assessed the efficacy of evolocumab in participants with ASCVD having on statin LDL levels of ≥70 mg/dL. Evolocumab reduced LDL cholesterol levels by 59% to a median of 30 mg/dL. Additionally, evolocumab significantly reduced the primary endpoint of cardiovascular death, myocardial infarction (MI), stroke, hospitalization for unstable angina or coronary revascularization (i.e., 5-point MACE) by 15% and the secondary endpoint of cardiovascular death, MI, and stroke (i.e., 3-point MACE) by 20% compared to placebo.[27]

The Odyssey Outcomes trial assessed alirocumab treatment in patients with a recent history of ACS on maximal statin therapy. Treatment with alirocumab decreased the risk for 5-point primary endpoint (death from CHD, nonfatal MI, fatal or nonfatal ischemic stroke or unstable angina requiring hospitalization) by 15%.[28]

These findings support the efficacy of PCSK9 inhibition in secondary prevention. People with DM are at a higher risk of ASCVD and thus have greater absolute risk reduction (and lower number needed to treat) with PCSK9 inhibition. PSCK9 inhibitors may also confer additional benefit in people with DM on account of its role in reduction of lipoprotein(a) [Lp(a)] levels.[29]

Bile Acid Sequestrants

Bile acid sequestrants or resins act in the lumen of the terminal ileum. They bind to bile acids which are subsequently excreted. This interferes with the enterohepatic circulation of bile acids resulting in depletion of intrahepatic cholesterol. The LDL receptor gene is upregulated leading to increased catabolism of circulating LDL particles and reduction in serum LDL cholesterol levels. Bile acid sequestrants can reduce LDL cholesterol levels by additional 15% compared to statin monotherapy alone.[30,31]

Other LDL Cholesterol-lowering Therapies

Newer drugs directed at LDL cholesterol reduction include the following:
Inclisiran is a chemically synthesized small-interfering RNA which targets the production of PCSK9.[32] It reduces LDL cholesterol by 50–60%. A single subcutaneous dose is capable of achieving sustained reductions of circulating PCSK9 and LDL cholesterol lasting 6–12 months.[33] Inclisiran also decreases total cholesterol, non-HDL cholesterol, ApoB, and Lp(a). A cardiovascular outcomes study is currently ongoing.[34]

Lomitapide is a microsomal triglyceride transfer protein (MTTP) inhibitor which acts by directly inhibiting assembly of ApoB-containing lipoprotein particles in the liver and intestine.[35] It lowers LDL cholesterol and TG each by approximately 40%. However, its use is limited by its high cost and side effects, particularly, an increased risk of fatty liver, and the drug is usually reserved for patient with homozygous familial hypercholesterolemia (FH).[36,37]

Mipomersen is an antisense oligonucleotide therapy that binds to the ApoB mRNA, and blocks its translation. Impaired ApoB synthesis prevents the assembly of LDL particle precursors resulting in reduction in LDL cholesterol levels.[35,38] It is an injectable agent that lowers LDL cholesterol by approximately 30% in patients with homozygous FH.[39] However, it is associated with significant side effects—flu-like symptoms, hepatosteatosis, injection site reactions and chances for its future use are very remote.

Bempedoic acid, an ATP-citrate lyase (an enzyme upstream of HMG-CoA reductase in the cholesterol biosynthesis pathway) inhibitor.[40] In a recent randomized trial of patients with ASCVD, FH, or both with LDL ≥70 mg/dL while on statin (with or without other lipid-lowering medication), bempedoic acid reduced LDL cholesterol by 16.5%, with no difference between the subgroups with and without DM.[41] A cardiovascular outcomes study is currently under way.

Finally, IONIS-APO(a)-LRx, a small-inhibiting RNA molecule, targets Lp(a), and reduces its levels by up to 90%[42] with no effect on other lipids. This agent could prove to be helpful for people with DM with coexistent elevated Lp(a) levels.

Omega-3 Fatty Acids

Omega-3 fatty acids reduce plasma TG through suppression of hepatic VLDL-TG production.[43-45] Doses of 3-4 g/day of eicosapentaenoic acid (EPA) ± docosahexaenoic acid (DHA) reduce TG by up to 45%.[46,47] About 95% icosapent ethyl is an ethyl ester of EPA which has similar efficacy as EPA/DHA formulation in reducing TG. However, while EPA/DHA formulation may raise LDL modestly, icosapent ethyl does not have significant effect on LDL.[46,47] In the recently published REDUCE IT trial (Reduction of Cardiovascular Events with Icosapent Ethyl Intervention Trial), addition of 4 g/day of icosapent ethyl to statins was associated with a significant reduction of 25% in MACE compared to statins alone.[48]

Niacin

Extended release preparation of niacin (2 g/day) has been proven to decrease TG by 25% and increase HDL by 20% in addition to lipid lowering by statin.[49] However, use of niacin has been limited by significant side effects such as severe facial flushing and cutaneous rash. Also, there is lack of evidence of improved outcomes with niacin on background statin therapy.[50,51]

Fibrates

Peroxisome proliferator-activated receptors (PPAR)-α agonists (fibrates) act by activating PPAR-α, which in turn enhances expression of LPL and other lipid-modifying genes.[52] Fibrates are proven to reduce TG by 30-50%[53] but their role in ADD is extensively debated. Meta-analysis of several trials involving fibrates [ACCORD (Action to Control Cardiovascular Risk in Diabetes Lipid) trial, FIELD (Fenofibrate Intervention and Event Lowering in Diabetes) trial, BIP (Bezafibrate Infarction Prevention) study, HHS (Helsinki Heart Study), VA-HIT (Veterans Affairs High-Density Lipoprotein Cholesterol) Intervention Trial] by Sacks et al. showed that fibrates are very likely to be beneficial as an add-on therapy in subgroup of participants characterized by high TG and low HDL cholesterol despite statin usage.[54] Similar systematic review on usefulness of fibrate therapy by Jun et al. demonstrated that participants with a high-mean TG concentration (>204 mg/dL) obtained greater reduction in cardiovascular outcomes.[55]

Glitazars

Combined PPAR-α/ϒ agonists were developed with expectation that they will lower TG not only directly (PPAR-α activity), but also by improving insulin sensitivity and glycemic control (PPAR-ϒ activity). While these agents have been shown to favorably affect lipid and glycemic parameters[56] robust evidence of cardiovascular benefit is lacking, and several tested molecules have been withdrawn on account of safety concerns.[57] However, the adverse events reported appear to be compound specific and of diverse origin, raising hopes of a potential drug that could mitigate these side effects and still have a positive influence on insulin sensitivity and dyslipidemia dominated by high TG, increased proportion of small dense LDL and low HDL cholesterol.

Saroglitazar is a dual PPAR-α/γ agonist with predominant PPAR-α and moderate PPAR-γ agonistic activities developed indigenously in India.[58] Saroglitazar is the first approved agent in the Glitazar class for the management of diabetic dyslipidemia, and has been approved in India.[59] Although, it has shown positive results in improving glycemic and lipid parameters in participants with ADD in short term, a longer-term clinical trial proving its efficacy in improving cardiovascular morbidity and mortality, currently lacking, is much desired. Saroglitazar appears to be safe and well tolerated and has not shown any of the adverse effects that are commonly recognized with its class of molecules.[60] But given the historical concerns, proof of long-term safety is required.

Triglyceride-lowering Therapies in Development

Apolipoprotein (Apo) CIII Inhibitors: ApoCIII increases TG by inhibiting LPL and reducing hepatic uptake of remnant lipoproteins.[61] Volanesorsen is an antisense oligonucleotide inhibitor of ApoCIII mRNA which decreases ApoCIII expression thereby reducing TG levels. In a phase II trial involving participants with T2DM and TG levels of 201–499 mg/dL, volanesorsen reduced TG by 69%, and raised HDL by 42% compared to placebo. It also improved insulin sensitivity and reduced glycated hemoglobin (−0.44%).[62] It appears as a promising agent; however, thrombocytopenia and serious bleeding are concerning side effects.[63]

Angiopoietin-like 3 protein (ANGPTL3) inhibitors: ANGPTL3 is an endogenous inhibitor of LPL. Evinacumab, a monoclonal antibody against ANGPTL3, has been shown to lower fasting TG by up to 76% and LDL by 23% in a dose-dependent manner.[64] An antisense oligonucleotide against ANGPTL3 has also demonstrated efficacy in reduction of atherogenic lipoproteins and appears to be promising.[65] So far, ANGPTL3 appears to be a promising therapeutic target.

Low-density lipoprotein is an established risk factor and reduction of LDL with statins is the first-line therapy for cardiovascular (CV) risk reduction among people with DM. But, despite statin therapy and achievement of LDL goals, a significant residual risk remains which is potentially attributable to increased TG concentration and low HDL cholesterol, a characteristic hallmark of ADD. This has led to renewed interest in treatments that could selectively target high TGs and low HDL cholesterol thus, further reducing the (CV) risk. Omega-3 fatty acids, niacin, fibrates and recently approved dual PPAR agonist saroglitazar are currently available drugs used to target such dyslipidemias. LDL particle number (LDL-P) has also been suggested as a target for treatment in patients with DM. Suggested targets are <1,200 for high-risk and <1,000 for very high-risk patients. However, there is absence of robust clinical trial data to support a wider use of this marker as a target for treatment of ADD. Moreover, there has been a lack of agreement on the target levels among various risk groups, particularly among extreme-risk patients.[66,67]

Guidelines on Management of Diabetic Dyslipidemia

The NCEP ATP III (National Cholesterol Education Program Adult Treatment Panel-III) guidelines had recommended various LDL cholesterol (primary target), non-HDL cholesterol and ApoB targets (secondary targets) for people with DM based on risk status.[9]

The ACC/AHA (American College of Cardiology and the American Heart Association) published a new set of guidelines for the treatment of blood cholesterol in 2013. These guidelines were coauthored by many of the authors associated with the NCEP ATP-III guidelines, and replaced what would have been the NCEP IV guidelines. These newer guidelines did not recommend any specific LDL goals for patients. The reason for not targeting specific LDL levels in the ACC/AHA guidelines was that the data for clinical outcomes are largely derived from fixed-dose studies and that there have not been adequate studies that used a strategy of titrating to an LDL goal which might justify particular clinical targets.[68]

Taking the 2013 ACC/AHA guidelines into consideration, the American Diabetes Association (ADA) also revised its lipid management guidelines in 2015 to recommend treatment initiation (and initial statin dose) driven primarily by risk status rather than LDL cholesterol level. This approach has been endorsed by several subsequent guidelines.

"The 2018 AHA/ACC/AACVPR/AAPA/ABC/ACPM/ADA/AGS/APhA/ASPC/NLA/PCNA Guideline on the Management of Blood Cholesterol," recommends the use of moderate-intensity statin therapy for primary prevention among individuals with T2DM aged between 40 and 75 years. Additionally, it is considered reasonable to assess the 10-year likelihood of development of ASCVD (using race and gender-adjusted risk calculators) among individuals in this age group having LDL between 70 and 189 mg/dL. High-intensity statin therapy should be considered for all individuals with DM and multiple CVD risk factors. Individual risk benefit ratio needs to be considered while continuing or starting statin therapy among people older than 75 years of age. In adults 20–39 years of age with DM that is either of long duration [≥10 years of T2DM, ≥20 years of type 1 diabetes mellitus (T1DM)], or associated with albuminuria (≥30 µg of albumin/mg creatinine), estimated glomerular filtration rate (eGFR) <60 mL/min/1.73 m^2, retinopathy, neuropathy, or ankle-brachial index (ABI; <0.9), it may be reasonable to initiate statin therapy. The guidelines also suggest consideration for ezetimibe (added to maximally tolerated statin therapy) to achieve ≥50% reduction in LDL cholesterol in adults with DM and 10-year ASCVD risk of ≥20%.[69]

The guidelines from the ADA released in 2020 are similar with slight variations. For patients with DM aged 20–39 years with additional ASCVD risk factors (LDL cholesterol ≥100 mg/dL, high blood pressure, smoking, chronic kidney disease, albuminuria, or family history of premature ASCVD), it is considered reasonable to initiate statin therapy. Use of moderate-intensity statin therapy has been recommended for people with DM aged 40–75 years without ASCVD. For those with higher risk in this age group, especially those with multiple ASCVD risk factors or those aged 50–70 years, use of high-intensity statin therapy is considered reasonable. Ezetimibe (added to maximally tolerated statin therapy) to achieve ≥50% reduction in LDL cholesterol has been considered reasonable in adults with DM and 10-year ASCVD risk of ≥20%. Among patients >75 years of age, it may be reasonable to continue or start statins if the physician feels that the risk–benefit ratio warrants treatment. For secondary prevention, the ADA recommends high-intensity statin therapy for patients of all ages with DM and ASCVD. Additionally, among those considered very high risk, if LDL cholesterol is ≥70 mg/dL on maximally tolerated statin dose, adding additional LDL-lowering therapy (such as ezetimibe or PCSK9 inhibitor) should be considered. **Table 1**[69] summarizes the various statins and their doses that qualify for high-intensity or moderate-intensity therapy. The guidelines have suggested consideration for use of medical therapy directed at reduction of TG for patients with fasting TG levels ≥500 mg/dL to reduce the risk of pancreatitis. Additionally, addition of icosapent ethyl has been suggested for patients with ASCVD or other CV risk factors on a statin with controlled LDL cholesterol but elevated TGs (135–499 mg/dL). The ADA guidelines actively discourage the use of combination therapy with niacin or fibrates to target HDL cholesterol or TG.[7]

TABLE 1: High-intensity and moderate-intensity statin therapy.*	
High-intensity statin therapy (lowers LDL cholesterol by ≥50%)	**Moderate-intensity statin therapy** (lowers LDL cholesterol by 30–49%)
• Atorvastatin 40–80 mg • Rosuvastatin 20–40 mg	• Atorvastatin 10–20 mg • Rosuvastatin 5–10 mg • Simvastatin 20–40 mg • Pravastatin 40–80 mg • Lovastatin 40–80 mg • Fluvastatin XL 80 mg • Pitavastatin 1–4 mg

*Once-daily dosing.
(XL: extended release)

The decision to remove LDL targets, however, remains controversial and many clinicians still favor an LDL-guided approach when treating patients with DM.

In fact, the 2017 Guidelines for the Management of Dyslipidemia published by the American Association of Clinical Endocrinologists (AACE) and American College of Endocrinology (ACE) have, like the NCEP III guidelines, emphasized on LDL and non-HDL-C targets. These guidelines have classified patients with DM, as high risk (DM with no other risk factors), very high risk (DM plus 1 or more additional risk factors—cigarette smoking, hypertension, HDL <40 mg/dL, family history of CHD, age ≥45 years for men or ≥55 years for women), or extreme risk (DM and a prior ASCVD event or chronic kidney disease stage 3 or 4). The recommended targets of LDL cholesterol, non-HDL cholesterol and ApoB are <100 mg/dL, <130 mg/dL, and <90 mg/dL, respectively for high-risk individuals, <70 mg/dL, <100 mg/dL and <80 mg/dL, respectively for very high-risk individuals, and <55 mg/dL, <80 mg/dL and <70 mg/dL respectively for extreme-risk individuals. Although statins have been recommended in these guidelines as first-line cholesterol-lowering drug therapy, like NCEP III guidelines, combination therapy remain an option.[70]

The 2019 ESC/EAS (European Society of Cardiology/European Atherosclerosis Society) Guidelines for the Management of Dyslipidemias outlines different LDL goals for people with DM, based on CV risk factors and complications. People with DM having high risk or very risk have been recommended LDL reduction of ≥50% from baseline. The recommended LDL goals for those having high risk and very high risk are <70 mg/dL and <55 mg/dL, respectively. Statins have been recommended as first-line treatment for LDL lowering with consideration for combination with ezetimibe if goal LDL is not reached with statin alone. Consideration for statin therapy has also been suggested for people with DM aged ≤30 years having evidence of end-organ damage and/or an LDL level >2.5 mmol/L.[71]

The RSSDI-ESI Clinical Practice Recommendations for the Management of Type 2 Diabetes Mellitus 2020 while concurring with the ADA recommendations for moderate- and high-intensity statin therapy, has also suggested a primary goal of <100 mg/dL for LDL cholesterol. Furthermore, the recommendations have suggested addition of fibrates if TGs remain >200 mg/dL in patients receiving statin therapy.[72]

CONCLUSION

Atherogenic diabetic dyslipidemia is characterized by a triad of high triglycerides, low HDL cholesterol, and increased proportion of small dense LDL. It is an important risk factor for ASCVD among people with T2DM and should be managed aggressively with lifestyle measures and pharmacotherapy. While, statins are first-line choice for treatment of ADD, the role of other agents in the management of diabetic dyslipidemia is increasingly being recognized.

> **KEY POINTS**
>
> - Atherogenic diabetic dyslipidemia consists of a triad of high triglycerides, low HDL cholesterol, and increased proportion of small dense LDL particles.
> - Dyslipidemia in people with DM can be treated with lifestyle modification and pharmacotherapy directed at reduction of different lipid parameters.
> - Although pharmacotherapy is based mostly on statin therapy, the role of other agents in the management of diabetic dyslipidemia is being increasingly recognized.

REFERENCES

1. Lampidonis AD, Rogdakis E, Voutsinas GE, Stravopodis DJ. The resurgence of Hormone-Sensitive Lipase (HSL) in mammalian lipolysis. Gene. 2011;477(1-2):1-11.
2. Ginsberg HN. Insulin resistance and cardiovascular disease. J Clin Invest. 2000;106(4):453-8.
3. Lewis GF. Fatty acid regulation of very low density lipoproteins production. Curr Opin Lipidol. 1997;8(3):146-53.
4. Bruce C, Chouinard RA Jr, Tall AR. Plasma lipid transfer proteins, high-density lipoproteins and reverse cholesterol transport. Annu Rev Nutr. 1998;18:297-330.
5. Austin MA, King MC, Vranizan KM, Krauss RM. Atherogenic lipoprotein phenotype. A proposed genetic marker for coronary heart disease risk. Circulation. 1990;82(2):495-506.
6. Alexopoulos A-S, Qamar A, Hutchins K, Crowley MJ, Batch BC, Guyton JR. Triglycerides: Emerging targets in diabetes care? Review of Moderate Hypertriglyceridemia in Diabetes. Curr Diab Rep. 201926;19(4):13.
7. American Diabetes Association. Cardiovascular Disease and Risk Management: Standards of Medical Care in Diabetes 2020. Diabetes Care. 2020;43(Suppl. 1):S111-34.
8. Hu FB, Manson JE. Walking: The best medicine for diabetes? Arch Intern Med. 2003;163(12):1397-8.
9. National Cholesterol Education Program (NCEP) Expert Panel on Detection, Evaluation, and Treatment of High Blood Cholesterol in Adults (Adult Treatment Panel III). Third Report of the National Cholesterol Education Program Expert Panel on detection, evaluation, and treatment of high blood cholesterol in adults (Adult Treatment Panel III) Final Report. Circulation. 2002;106(25):3143-421.
10. Maki KC. Fibrates for treatment of the metabolic syndrome. Curr Atheroscler Rep. 2004;6(1):45-51.
11. Pyörälä K, Pedersen TR, Kjekshus J, Faergeman O, Olsson AG, Thorgeirsson G. Cholesterol lowering with simvastatin improves prognosis of diabetic patients with coronary heart disease. A subgroup analysis of the Scandinavian Simvastatin Survival Study (4S). Diabetes Care. 1997;20(4):614-20.
12. Collins R, Armitage J, Parish S, Sleigh P, Peto R; Heart Protection Study Collaborative Group. MRC/BHF Heart Protection Study of cholesterol-lowering with simvastatin in 5963 people with diabetes: a randomised placebo-controlled trial. Lancet. 2003;361(9374):2005-16.
13. Goldberg RB, Mellies MJ, Sacks FM, Moyé LA, Howard BV, Howard WJ, et al. Cardiovascular events and their reduction with pravastatin in diabetic and glucose-intolerant myocardial infarction survivors with average cholesterol levels: subgroup analyses in the Cholesterol And Recurrent Events (CARE) trial. Circulation. 1998;98(23):2513-9.
14. Shepherd J, Barter P, Carmena R, Deedwania P, Fruchart J-C, Haffner S, et al. Effect of lowering LDL cholesterol substantially below currently recommended levels in patients with coronary heart disease and diabetes: the Treating to New Targets (TNT) study. Diabetes Care. 2006;29(6):1220-6.

15. Sever PS, Poulter NR, Dahlöf B, Wedel H, Collins R, Beevers G, et al. Reduction in cardiovascular events with atorvastatin in 2,532 patients with type 2 diabetes: Anglo-Scandinavian Cardiac Outcomes Trial–Lipid-Lowering Arm (ASCOT-LLA). Diabetes Care. 2005;28(10):1151-7.
16. Knopp RH, d'Emden M, Smilde JG, Pocock SJ. Efficacy and safety of atorvastatin in the prevention of cardiovascular end points in subjects with type 2 diabetes: the Atorvastatin Study for Prevention of Coronary Heart Disease Endpoints in Non-Insulin-Dependent Diabetes Mellitus (ASPEN). Diabetes Care. 2006;29(7):1478-85.
17. Colhoun HM, Betteridge DJ, Durrington PN, Graham A Hitman, H Andrew W Neil, Shona J Livingstone, et al. Primary prevention of cardiovascular disease with atorvastatin in type 2 diabetes in the Collaborative Atorvastatin Diabetes Study (CARDS): multicentre randomised placebo-controlled trial. Lancet. 2004;364(9435):685-96.
18. Cholesterol Treatment Trialists' (CTT) Collaborators, Kearney PM, Blackwell L, Collins R, Keech A, Simes J, et al. Efficacy of cholesterol-lowering therapy in 18,686 people with diabetes in 14 randomised trials of statins: a meta-analysis. Lancet. 2008;371(9607):117-125.
19. Sudhop T, Lutjohann D, Kodal A, Igel M, Tribble DL, Shah S, et al. Inhibition of intestinal cholesterol absorption by ezetimibe in humans. Circulation. 2002;106(15):1943-8.
20. Morrone D, Weintraub WS, Toth PP, Hanson ME, Lowe RS, Lin J, et al. Lipid-altering efficacy of ezetimibe plus statin and statin monotherapy and identification of factors associated with treatment response: a pooled analysis of over 21,000 subjects from 27 clinical trials. Atherosclerosis. 2012;223(2):251-61.
21. Cannon CP, Blazing MA, Giugliano RP, McCagg A, White JA, Theroux P, et al. Ezetimibe added to statin therapy after acute coronary syndromes. N Engl J Med. 2015;372(25):2387-97.
22. Ruggenenti P, Cattaneo D, Rota S, , Iliev I, Parvanova A, Diadei O, et al. Effects of combined ezetimibe and simvastatin therapy as compared with simvastatin alone in patients with type 2 diabetes: a prospective randomized double-blind clinical trial. Diabetes Care. 2010;33(9):1954-6.
23. Lally S, Owens D, Tomkin GH. Genes that affect cholesterol synthesis, cholesterol absorption, and chylomicron assembly: the relationship between the liver and intestine in control and streptozotosin diabetic rats. Metabolism. 2007;56(3):430-8.
24. Ravid Z, Bendayan M, Delvin E, Sane AT, Elchebly M, Lafond J, et al. Modulation of intestinal cholesterol absorption by high glucose levels: impact on cholesterol transporters, regulatory enzymes, and transcription factors. Am J Physiol Gastrointest Liver Physiol. 2008;295(5):G873-85.
25. Seidah NG, Chretien M, Mbikay M. The ever-expanding saga of the proprotein convertases and their roles in body homeostasis: Emphasis on novel proprotein convertase subtilisin kexin number 9 functions and regulation. Curr Opin Lipidol. 2018;29(2):144-50.
26. Hoe E, Hegele RA. Lipid management in diabetes with a focus on emerging therapies. Can J Diabetes. 2015;39 Suppl 5:S183-90.
27. Sabatine MS, Giugliano RP, Keech AC, Honarpour N, Wiviott SD, Murphy SA, et al. Evolocumab and clinical outcomes in patients with cardiovascular disease. N Engl J Med. 2017;376(18):1713-22.
28. Schwartz GG, Steg PG, Szarek M, Bhatt DL, Bittner VA, Diaz R, et al. Alirocumab and cardiovascular outcomes after acute coronary syndrome. N Engl J Med. 2018;379(22):2097-107.
29. O'Donoghue ML, Fazio S, Giugliano RP, , Stroes ESG, Kanevsky E, Gouni-Berthold I, et al. Lipoprotein(a), PCSK9 inhibition, and cardiovascular risk. Circulation. 2019;139(12):1483-92.
30. Mazidi M, Rezaie P, Karimi E, Kengne AP. The effects of bile acid sequestrants on lipid profile and blood glucose concentrations: A systematic review and meta-analysis of randomized controlled trials. Int J Cardiol. 2017;227:850-7.
31. The Lipid Research Clinics Coronary Primary Prevention Trial results. I. Reduction in incidence of coronary heart disease. JAMA. 1984;251(3):351-64.
32. Leiter LA, Teoh H, Kallend D, Wright RS, Landmesser U, Wijngaard PLJ, et al. Inclisiran lowers LDL-C and PCSK9 irrespective of diabetes status: The ORION-1 randomized clinical trial. Diabetes Care. 2019;42(1):173-6.
33. Ray KK, Landmesser U, Leiter LA, Kallend D, Dufour R, Karakas M, et al. Inclisiran in patients at high cardiovascular risk with elevated LDL cholesterol. N Engl J Med. 2017;376:1430-40.
34. Stoekenbroek RM, Kallend D, Wijngaard PL, Kastelein JJ. Inclisiran for the treatment of cardiovascular disease: The ORION clinical development program. Future Cardiol. 2018;14:433-42.
35. Cupido AJ, Reeskamp LF, Kastelein JJP. Novel lipid modifying drugs to lower LDL cholesterol. Curr Opin Lipidol. 2017;28(4):367-73.
36. Brahm AJ, Hegele RA. Lomitapide for the treatment of hypertriglyceridemia. Expert Opin Investig Drugs. 2016;25(12):1457-63.

37. Berberich AJ, Hegele RA. Lomitapide for the treatment of hypercholesterolemia. Expert Opin Pharmacother. 2017;18(12):1261-8.
38. Gryn SE, Hegele RA. Novel therapeutics in hypertriglyceridemia. Curr Opin Lipidol. 2015;26(6):484-91.
39. Reeskamp LF, Kastelein JJP, Moriarty PM, Duell PB, Catapano AL, Santos RD, et al. Safety and efficacy of mipomersen in patients with heterozygous familial hypercholesterolemia. Atherosclerosis. 2019;280: 109-17.
40. Burke AC, Huff MW. ATP-citrate lyase: Genetics, molecular biology and therapeutic target for dyslipidemia. Curr Opin Lipidol. 2017;28(2):193-200.
41. Ray KK, Bays HE, Catapano AL, Lalwani ND, Bloedon LT, Sterling LR, et al. Safety and efficacy of bempedoic acid to reduce LDL cholesterol. N Engl J Med. 2019;380:1022-32.
42. Hegele RA, Tsimikas S. Lipid-lowering agents. Circ Res. 2019;124(3):386-404.
43. Nestel PJ, Connor WE, Reardon MF, Connor S, Wong S, Boston R. Suppression by diets rich in fish oil of very low density lipoprotein production in man. J Clin Invest. 1984;74(1):82-9.
44. Harris WS, Connor WE, Illingworth DR, Rothrock DW, Foster DM. Effects of fish oil on VLDL triglyceride kinetics in humans. J Lipid Res. 1990;31(9):1549-58.
45. Durrington PN, Bhatnagar D, Mackness MI, Morgan J, Julier K, Khan MA, et al. An omega-3 polyunsaturated fatty acid concentrate administered for one year decreased triglycerides in simvastatin treated patients with coronary heart disease and persisting hypertriglyceridaemia. Heart. 2001;85(5):544-8.
46. Harris WS, Ginsberg HN, Arunakul N, Shachter NS, Windsor SL, Adams M, et al. Safety and efficacy of Omacor in severe hypertriglyceridemia. J Cardiovasc Risk. 1997;4(5-6):385-91.
47. Bays HE, Ballantyne CM, Kastelein JJ, Isaacsohn JL, Braeckman RA, Soni PN. Eicosapentaenoic acid ethyl ester (AMR101) therapy in patients with very high triglyceride levels (from the Multi-center, plAcebo-controlled, Randomized, double-blINd, 12-week study with an open-label Extension [MARINE] trial). Am J Cardiol. 2011;108(5):682-90.
48. Bhatt DL, Steg PG, Miller M, Brinton EA, Jacobson TA, Ketchum SB, et al. Cardiovascular risk reduction with icosapent ethyl for hypertriglyceridemia. N Engl J Med. 2018;380(1):11-22.
49. Maccubbin D, Bays HE, Olsson AG, Elinoff V, Elis A, Mitchel Y, et al. Lipid modifying efficacy and tolerability of extended-release niacin/laropiprant in patients with primary hypercholesterolemia or mixed dyslipidaemia. Int J Clin Pract. 2008;62:1959-70.
50. Investigators A-H, Boden WE, Probstfield JL, Anderson T, Chaitman BR, Desvignes-Nickens P, et al. Niacin in patients with low HDL cholesterol levels receiving intensive statin therapy. N Engl J Med. 2011;365(24):2255-67.
51. Group HTC, Landray MJ, Haynes R, Hopewell JC, Parish S, Aung T, et al. Effects of extended-release niacin with laropiprant in high risk patients. N Engl J Med. 2014;371(3):203-12.
52. Rosenson RS, Davidson MH, Hirsh BJ, Kathiresan S, Gaudet D. Genetics and causality of triglyceride-rich lipoproteins in atherosclerotic cardiovascular disease. J Am Coll Cardiol. 2014;64(23): 2525-40.
53. Shapiro MD, Fazio S. From lipids to inflammation: new approaches to reducing atherosclerotic risk. Circ Res. 2016;118(4):732-49.
54. Sacks FM, Carey VJ, Fruchart JC. Combination lipid therapy in type 2 diabetes. N Engl J Med. 2010;363:692-4.
55. Jun M, Foote C, LV J, Neal B, Patel A, Nicholls SJ, et al. Effects of fibrates on cardiovascular outcomes: A systematic review and meta-analysis. Lancet. 2010;375(9729):1875-84.
56. Henry RR, Lincoff AM, Mudaliar S, Rabbia M, Chognot C, Herz M. Effect of the dual peroxisome proliferator-activated receptor-alpha/gamma agonist aleglitazar on risk of cardiovascular disease in patients with type 2 diabetes (SYNCHRONY): a phase II, randomised, dose-ranging study. Lancet. 2009;374(9684):126-35.
57. Lincoff AM, Tardif JC, Schwartz GG, Nicholls SJ, Ryden L, Neal B, et al. Effect of aleglitazar on cardiovascular outcomes after acute coronary syndrome in patients with type 2 diabetes mellitus: the AleCardio randomized clinical trial. JAMA. 2014;311(15):1515-25.
58. Jani RH, Kansagra K, Jain MR, Patel H. Pharmacokinetics, safety and tolerability of saroglitazar (ZYH1), a predominantly PPAR agonist with moderate PPAR agonist activity in healthy human subjects. Clin Drug Investig. 2013;33(11):800-16.
59. Agrawal R. The first approved agent in the Glitazar's Class: Saroglitazar. Curr Drug Targets. 2014;15:151-5.
60. Jani RH, Pai V, Jha P, Jariwala G, Mukhopadhyay S, Bhansali A, et al. A multicenter, prospective, randomised, double blind study to evaluate the safety and efficacy of saroglitazar 2 and 4 mg compared with placebo in type 2 diabetes mellitus patients having hypertriglyceridemia not controlled on Atorvastatin therapy (PRESS VI). Diabetes Technol Ther. 2014;16(2):63-71.

61. Ooi EM, Barrett PH, Chan DC, Watts GF. Apolipoprotein C-III: understanding an emerging cardiovascular risk factor. Clin Sci (Lond). 2008;114(10):611-24.
62. Digenio A, Dunbar RL, Alexander VJ, Hompesch M, Morrow L, Lee RG, et al. Antisense-mediated lowering of plasma apolipoprotein C-III by volanesorsen improves dyslipidemia and insulin sensitivity in type 2 diabetes. Diabetes Care. 2016;39(8):1408-15.
63. Gaudet D, Alexander V, Arca M, Jones A, Stroes E, Bergeron J, et al. The APPROACH study: a randomized, double-blind, placebo-controlled, phase 3 study of volanesorsen administered subcutaneously to patients with familial chylomicronemia syndrome (FCS). Atherosclerosis. 2017;283:e10.
64. Dewey FE, Gusarova V, Dunbar RL, O'Dushlaine C, Schurmann C, Gottesman O, et al. Genetic and pharmacologic inactivation of ANGPTL3 and cardiovascular disease. N Engl J Med. 2017;377(3):211-21.
65. Graham MJ, Lee RG, Brandt TA, Tai LJ, Fu W, Peralta R, et al. Cardiovascular and metabolic effects of ANGPTL3 antisense oligonucleotides. N Engl J Med. 2017;377(3):222-32.
66. Toth PP, Grabner M, Punekar RS, Quimboc RA, Cziraky MJ, Jacobsond TA, et al. Cardiovascular risk in patients achieving low-density lipoprotein cholesterol and particle targets. Atherosclerosis. 2014;235(2):585-91.
67. Otvos JD, Mora S, Shalaurova I, Greenland P, Mackey RH, Goff Jr DC. Clinical implications of discordance between LDL cholesterol and LDL particle number. J Clin Lipidol. 2011;5(2):105-13.
68. Stone NJ, Robinson JG, Lichtenstein AH, Bairey Merz CN, Blum CB, Eckel RH, et al. 2013 ACC/AHA guideline on the treatment of blood cholesterol to reduce atherosclerotic cardiovascular risk in adults: a report of the American College of Cardiology/American Heart Association Task Force on Practice Guidelines. J Am Coll Cardiol. 2014;63(25 Pt B):2889-934.
69. Grundy SM, Stone NJ, Bailey AL, Beam C, Birtcher KK, Blumenthal RS, et al. 2018 AHA/ACC/AACVPR/AAPA/ABC/ACPM/ADA/AGS/APhA/ASPC/NLA/PCNA guideline on the management of blood cholesterol: a report of the American College of Cardiology/American Heart Association Task Force on Clinical Practice Guidelines. Circulation. 2019;139(25):e1082-143.
70. Jellinger PS, Handelsman Y, Rosenblit PD, Bloomgarden ZT, Fonseca VA, Garber AJ, et al. American Association of Clinical Endocrinologists and American College of Endocrinology Guidelines for Management of Dyslipidemia and Prevention of Atherosclerosis. Endocr Pract. 2017;23(Suppl 2):1-87.
71. Catapano AL, Graham I, De Backer G, Koskinas KC, Casula M, Badimon L, et al. 2019 ESC/EAS guidelines for the management of dyslipidaemias: lipid modification to reduce cardiovascular risk. Eur Heart J. 2020;41(1):111-88.
72. Chawla R, Madhu SV, Makkar BM, Ghosh S, Saboo B, Kalra S. RSSDI-ESI clinical practice recommendations for the management of type 2 diabetes mellitus 2020. Indian J Endocr Metab. 2020;24:1-122.

CHAPTER 22

Comparative Analysis of Major Cholesterol Guidelines

Soumik Goswami, Mounam Chattopadhyay

ABSTRACT
The guidelines laid down for comprehensive management of dyslipidemia have been subject to significant changes over time. This chapter reviews the important international guidelines vis-à-vis Indian guidelines. All these guidelines have similarly extensively reviewed clinical evidence, underscored the importance of statin therapy for primary and secondary prevention of atherosclerotic cardiovascular disease, and given importance to a clinician–patient risk discussion before management. However, there are marked dissimilarities in statin intensities, use of risk estimators, treatment of specific patient subgroups, and consideration of safety concerns. Clinicians should realize these similarities and differences in the recommendations while treating their patients with dyslipidemia.

PART 1: RISK ESTIMATORS AND STATIN THERAPY FOR PRIMARY PREVENTION OF ATHEROSCLEROTIC CARDIOVASCULAR DISEASE

Introduction

Atherosclerotic cardiovascular disease (ASCVD) is on the rise in India and across the globe and is gradually becoming an uphill task for the clinicians to manage. The natural history of ASCVD is somewhat different in Indian individuals—they usually develop the disease at an earlier age, the disease assumes greater severity and has poorer outcome and survival as compared to the Western population. Healthcare access is also inadequate in India, and the treatment of ASCVD continues to be costly.

Dyslipidemia has the greatest population attributable risk for myocardial infarction (MI) given its huge prevalence and direct causal association. So, treatment of this hazard could go a long way in preventing ASCVDs.

Indians are difficult to treat when it comes to dyslipidemia rather than their western counterparts for reasons, e.g., younger age of onset, genetic susceptibility, and interaction among cardiovascular (CV) risk factors. Hence, forming specialized guidelines that suit our conditions is of utmost importance.[1] The dictum "nothing stands still" bodes well with lipid guidelines published and followed by clinicians across the globe as it goes with history.[2]

Epidemiology

Current trends reveal dyslipidemia has become more of a vexing problem in economically backward nations rather than the so-called developed countries. The "Global Burden of Metabolic Risk Factors Study" holds a testimony to this. But unfortunately there are very few studies carried out in India to support this. Hypercholesterolemia prevalence is shown to be 10–15% in village and 25–30% in urban individuals as shown in **Table 1**.[1]

Atherogenic dyslipidemia is exclusively seen in Indian and South Asian population and is characterized by high triglyceride (TG), low high-density lipoprotein cholesterol (HDL-C) and low total cholesterol. This is intrinsically linked to type 2 diabetes mellitus (T2DM), coronary heart disease, and metabolic syndrome.[3] The first phase of the ICMR-INDIAB (Indian Council of Medical Research-India Diabetes) study has shown the following prevalence rates for different forms of dyslipidemia—hypercholesterolemia in 13.9%, high TG in 29.5%, low HDL-C in 72.3% and high low-density lipoprotein cholesterol (LDL-C) in 11.8%. Moreover, 79% of individuals had in at least one aberrant lipid parameter.[4] The INTERHEART study (**Table 2**) in a bid to show link between dyslipidemia and MI found that apolipoprotein A-1 (ApoA-1) was more reliable as a marker of protection [odds ratio (OR) 0.72; 95% confidence interval (CI) 0.66–0.78] than HDL-C (OR 0.97; 95% CI 0.90–1.05) while raised apolipoprotein B (ApoB): ApoA-1 was most useful predictor of risk.[5]

Risk Estimators

All individuals are not equally prone to develop ASCVD and hence need primary prevention measures of varying intensities to bring down the costs as well as to minimize adverse effects of therapies. All such guidelines make use of different risk estimators and recommend statin therapy for primary prevention. These estimators are entirely based on 10-year risk thresholds. The ACC/AHA (American College of Cardiology/American Heart Association) and USPSTF (United States Preventive Services Task Force) utilize the ACC/

TABLE 1: Prevalence of hypercholesterolemia (≥200 mg/dL) in multisite Indian studies.				
Study	Year reported	Sample size	Prevalence (%)	
			Men	Women
Indian Industrial Population Surveillance Study	2006	10,442	25.1	–
Indian Migration Study: Rural	2010	1,983	21.1	27.8
ICMR IDSP: Urban	2010	15,223	31.7	32.8
ICMR IDSP: Rural	2010	13,517	19.5	26.4
ICMR IDSP: Periurban/urban slum	2010	15,751	18.1	23.4
Indian Women's Health Study: Urban	2013	2,008	–	27.7
Indian Women's Health Study: Rural	2013	2,616	–	13.5
Indian Heart Watch	2014	6,123	25.1	24.9
ICMR–INDIAB Study	2014	2,042	13.9	–

(ICMR–INDIAB: Indian Council of Medical Research-India Diabetes; IDSP: Integrated Disease Surveillance Project)

TABLE 2: Relative risk (95% confidence interval) for acute myocardial infarction (MI) in different ethnic groups for one standard deviation change in various lipid measures in the INTERHEART study.

	South Asians	European	Chinese	Latin America	Overall
Total cholesterol	1.23 (1.14–1.31)	1.08 (1.02–1.15)	1.16 (1.09–1.23)	1.05 (0.97–1.14)	1.16 (1.13–1.19)
HDL-C	0.97 (0.90–1.05)	0.78 (0.73–0.83)	0.83 (0.78–0.88)	1.03 (0.94–1.13)	0.85 (0.83–0.88)
Non-HDL-C	1.23 (1.15–1.31)	1.17 (1.10–1.24)	1.24 (1.18–1.31)	1.04 (0.96–1.28)	1.21 (1.17–1.24)
ApoA-1	0.72 (0.66–0.78)	0.70 (0.66–0.75)	0.67 (0.63–0.71)	0.67 (0.61–0.74)	0.67 (0.65–0.70)
ApoB	1.38 (1.29–1.48)	1.24 (1.16–1.32)	1.28 (1.20–1.36)	1.18 (1.09–1.28)	1.32 (1.28–1.36)
Total cholesterol: HDL-C ratio	1.10 (1.01–1.17)	1.31 (1.21–1.42)	1.34 (1.24–1.45)	0.97 (0.90–1.05)	1.17 (1.13–1.20)
ApoB: ApoA-1 ratio	1.53 (1.42–1.64)	1.47 (1.37–1.59)	1.77 (1.63–1.92)	1.27 (1.17–1.38)	1.59 (1.52–1.64)

(Apo: apolipoprotein; HDL-C: high-density lipoprotein cholesterol)

AHA pooled cohort risk equations (PCE), whereas the Canadian Cardiovascular Society (CCS) and American Association of Clinical Endocrinologists (AACE) advocate using the Framingham Risk Score (FRS), and Veterans Affairs/Department of Defense (VA-DoD) suggests either. The European Society for Cardiology/European Atherosclerosis Society (ESC/EAS) propose use of the systemic coronary risk evaluation (SCORE) estimator.

Risk estimators are deduced from large scale studies in the US or Europe. All entail age, sex, total cholesterol, HDL-C, and systolic blood pressure as predictors. However, ethnicity, treatment for hypertension, diabetes, and smoking status are solely taken into consideration in some; thus, patient risk may differ with respect to different estimators as shown in **Figure 1**.[6]

To predict 10-year risk of ASCVD event (defined as coronary heart disease death, nonfatal MI, or stroke) among adults (40–75 years), we need a powerful tool like the ACC/AHA PCE. The problems with this are age being considered inappropriately as a risk factor and heterogeneity in risk patterns among different ethnic groups. There are certain risk enhancing factors vide **Box 1**[7] that sometimes prove more useful than PCE despite not predicting risk independently and only influencing risk discussion between physician and patient.

The ESC/EAS has come up with SCORE risk estimator which happens to be most specific in estimation of 10-year cumulative risk of first atherosclerotic event. This is needed for apparently healthy people who may harbor subtle risk and not in case of documented ASCVD. Total cardiovascular disease (CVD) risk is three times higher than fatal CVD risk and still higher among women but lower in elderly individuals. HDL-C is also included in the electronic version of SCORE, i.e., HeartScore (http://www.heartscore.org/en_GB/).[8]

Outcome for the FRS is the most inclusive among the risk calculators, predicting 10-year risk of coronary heart disease, cerebrovascular events, peripheral artery disease, or heart failure. AACE and CCS guidelines use it. The other risk scoring tools used in AACE guidelines are elaborated in **Table 3**.[9]

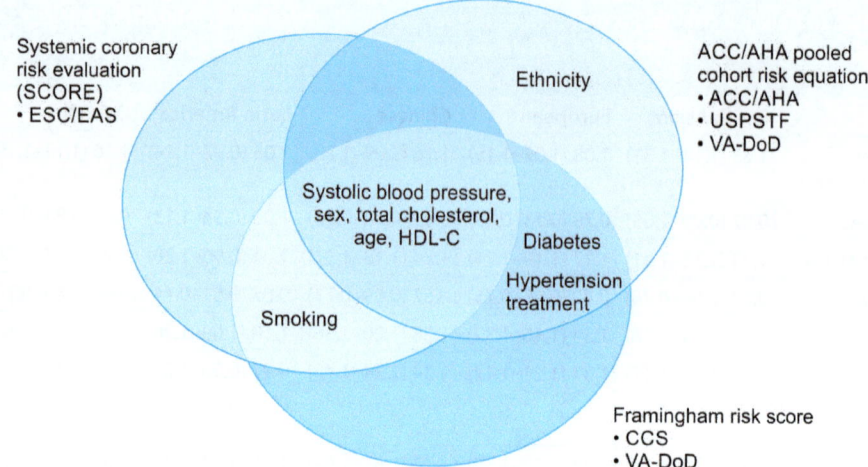

(ACC: American College of Cardiology; AHA: American Heart Association; CCS: Canadian Cardiovascular Society; EAS: European Atherosclerosis Society; ESC: European Society of Cardiology; HDL-C: high-density lipoprotein cholesterol; USPSTF: United States Preventive Services Task Force; VA-DoD: Veterans Affairs/Department of Defense)

FIG. 1: Three chief risk estimators and their respective components.[6]

BOX 1 Risk-enhancing factors.

- Family history of premature ASCVD (males, age <55 years; females, age <65 years)
- Primary hypercholesterolemia [LDL-C: 160–189 mg/dL (4.1–4.8 mmol/L); non-HDL-C: 190–219 mg/dL (4.9–5.6 mmol/L)]
- Metabolic syndrome [increased waist circumference, elevated triglycerides (>175 mg/dL), elevated blood pressure, elevated glucose, and low HDL-C (<40 mg/dL in men; <50 mg/dL in women) are factors; tally of 3 makes the diagnosis]
- Chronic kidney disease (eGFR 15–59 mL/min/1.73 m^2 with or without albuminuria; not treated with dialysis or kidney transplantation)
- Chronic inflammatory conditions such as psoriasis, RA, or HIV/AIDS
- History of premature menopause (before age 40 years) and history of pregnancy-associated conditions that increase later ASCVD risk such as pre-eclampsia
- High-risk race/ethnicities (e.g., South Asian ancestry)
- Lipid/biomarkers: Associated with increased ASCVD risk
Persistently* elevated, primary hypertriglyceridemia (≥175 mg/dL);
If measured:
Elevated high-sensitivity C-reactive protein (≥2.0 mg/L)
Elevated Lp(a): A relative indication for its measurement is family history of premature ASCVD. An Lp(a) ≥50 mg/dL or ≥125 nmol/L constitutes a risk-enhancing factor especially at higher levels of Lp(a)
Elevated ApoB ≥130 mg/dL: A relative indication for its measurement would be triglyceride ≥200 mg/dL. A level ≥130 mg/dL corresponds to an LDL-C ≥160 mg/dL and constitutes a risk-enhancing factor
ABI <0.9

*Optimally, three determinations.
[AIDS: acquired immunodeficiency syndrome; ABI: ankle-brachial index; ApoB: apolipoprotein B; ASCVD: atherosclerotic cardiovascular disease; eGFR: estimated glomerular filtration rate; HDL-C: high-density lipoprotein cholesterol; HIV: human immunodeficiency virus; LDL-C: low-density lipoprotein cholesterol; Lp(a): lipoprotein(a); RA: rheumatoid arthritis]

Comparative Analysis of Major Cholesterol Guidelines

TABLE 3: Key cardiovascular risk scoring tools.[9]

Framingham global risk Risk factors included/questions	Risk group/Framingham global risk (10-year absolute ASCVD risk)	Clinical examples
Risk assessment tool for calculating 10-year risk of having a heart attack for adults of 20 years and older who do not have heart disease or diabetes (using data from the Framingham Heart Study): Age: Sex: Total cholesterol: HDL-C: Smoking: SBP: Antihypertensives:	High >20%	• Established coronary artery disease • Cerebrovascular disease • Peripheral arterial disease • Abdominal aortic aneurysm • Diabetes mellitus • Chronic kidney disease
	Intermediate 10–20%	• Subclinical coronary artery disease • MetS • Multiple risk factors[a] • Markedly elevated levels of a single risk factor[b] • First-degree relative(s) with early onset coronary artery disease
	Lower <10%	• May include women with multiple risk factors, MetS, or 1 or no risk factors
	Optimal <10%	• Optimal levels of risk factors and heart-healthy lifestyle

High-risk: A >20% risk that you will develop a heart attack or die from coronary disease in the next 10 years.
Intermediate risk: A 10–20% risk that you will develop a heart attack or die from coronary disease in the next 10 years.
Low risk: <10% risk that you will develop a heart attack or die from coronary disease in the next 10 years.
[a]Patients with multiple risk factors can fall into any of the three categories by Framingham scoring.
[b]Most women with a single, severe risk factor will have a 10-year risk ≤10%.

Multiethnic study of atherosclerosis (MESA) Risk factors included/questions	Risk calculation outcomes
MESA 10-year ASCVD risk with coronary artery calcification: Sex: Age (45–85 years): Coronary artery calcification: Race/ethnicity: Diabetes: Smoking: Family history of heart attack: Total cholesterol: HDL-C: SBP: Lipid lowering medications: Antihypertensives:	• External validation provided evidence of very good discrimination and calibration • Harrell's C-statistic ranged from 0.779 to 0.816 in validation against existing studies • The difference in estimated 10-year risk between events and nonevents was ~ 8–9%, indicating excellent discrimination • Mean calibration found average predicted 10-year risk within 1/2 of a percent of the observed event rate • The test predicts 10-year risk of an ASCVD event

Continued

Continued

Reynolds risk score Risk factors included/questions	Risk calculation outcomes
Reynolds risk score predicts 10-year risk of heart attack, CVA, or other major heart diseases in healthy people without diabetes. Age: Smoking: Mother or father has heart attack before age 60? Total cholesterol: HDL-C: SBP: hsCRP:	Compared to ATP III/Framingham 10-year risk categorization: • Very little change in categorization of individuals with very low (<5%) risk • About 30% reclassification of those classified as 5% to <10% risk according to ATP III • About 29% reclassification of those classified as 10% to <20% risk according to ATP III • About 25% reclassification of those classified as ≥20% risk according to ATP III Risk is classified as low (<5%), low-to-moderate (5% to <10%), moderate-to-high (10% to <20%), and high (≥20%) ASCVD risk
UKDPS risk score Risk factors included/questions	Risk calculation outcomes

UKPDS risk engine is a model for estimating risk of ASCVD in persons with T2DM (this risk is up to three times greater than for the general population)
Age:
Sex:
Height:
Weight:
Afro-caribbean ethnicity:
Smoking:
Total cholesterol:
HDL-C:
SBP:
HbA1c:
Duration of diabetes:
Regular exercise per week:
	• Survival rates predicted by UKPDS risk score model were similar to rates observed in the UKPDS trial, well within nonparametric confidence intervals
	• Predicted survival rates adjust for HbA1c, blood pressure, and lipid risk factors
	• The UKPDS risk engine provides risk estimates and 95% confidence intervals in individuals with T2DM not known to have heart disease for:
	  ○ Nonfatal and fatal coronary heart disease
	  ○ Fatal coronary heart disease
	  ○ Nonfatal and fatal CVA
	  ○ Fatal CVA

(ATP III: adult treatment panel III; ASCVD: atherosclerotic cardiovascular disease; CVA: cerebrovascular accident; HbA1c: glycosylated hemoglobin; HDL: high-density lipoprotein cholesterol; hsCRP: high-sensitivity C-reactive protein; In: natural logarithm; MetS: metabolic syndrome; MI: myocardial infarction; SBP: systolic blood pressure; T2DM: type 2 diabetes mellitus; UKPDS: United Kingdom Prospective Diabetes Study)

Once risk has been determined, the threshold for a patient to start primary prevention measures also varies across the guidelines. These differences in outcome measures are important when considering differences in treatment thresholds between the guidelines.[10]

Statin Therapy

Multiple reasons make statins the drug of choice among all drugs available presently. They inhibit 3-hydroxy-3-methyl-glutaryl-coenzyme-A reductase, the key enzyme involved in hepatic cholesterol biosynthesis. Besides, they also increase removal of LDL particles from the blood stream by increasing the expression of LDL receptors thereby acting as a powerful LDL-C lowering drug. Moreover, they also reduce very low-density lipoprotein cholesterol (VLDL-C) and increase HDL-C levels. The net effect on lipid profile is a rough reduction of 25–55%, 15–51%, 7–30% and increase of 5–15% in LDL-C, non-HDL-C, TG, and HDL-C, respectively. They also have various pleiotropic effects as described below. The inhibition of the enzyme 3-hydroxy-3-methyl-glutaryl-coenzyme-A reductase also leads to inhibition of the production of isoprenoid intermediates. Isoprenylation of proteins is seen in inflammation and vascular remodeling in disease states, e.g., atherosclerosis and diabetes. Statins block this pathway and provide protection against the progression of atherosclerosis.[11,12] They decrease platelet activation and aggregability in both cholesterol dependent and cholesterol independent mechanisms; curb the proatherogenic effects of oxidized LDL; promote endothelial progenitor cell proliferation, migration and survival; impeding atherogenesis by reducing vascular smooth muscle cell migration and proliferation. The atherosclerotic plaque formed thereby is stabilized by lowering of blood lipids, tissue factors and matrix metalloproteinases. Cardiac remodeling is prevented by reducing fibroblast, collagen, and cardiomyocyte proliferation.[13,14]

Statins play crucial role in reducing CV morbidity and mortality when used for primary and mixed primary and secondary prevention settings as shown in many renowned studies (**Table 4**).[1]

TABLE 4: Major randomized clinical trials with statins.

Trial, year	Statin	Baseline LDL-C level (mg/dL)	Duration; number of patients	Primary findings
WOSCOPS, 1995	Pravastatin	192	4.9 years; 6,595	Nonfatal MI and CHD deaths reduced by 31%
AFCAPS/TexCAPS, 1998	Lovastatin	150	5.2 years; 6,605	About 37% reduction in first acute major coronary event
HPS, 2002	Simvastatin	131	5 years; 20,536	About 13% reduction in all-cause mortality, coronary mortality reduced by 17%
PROSPER, 2002	Pravastatin	147	3.2 years; 5,804	About 15% reduction in the composite of CHD deaths, nonfatal MI and fatal or nonfatal stroke
ASCOT-LLA	Atorvastatin 10 mg/day	131	3.3 years; 10,305	Nonfatal MI and/or CHD deaths reduced by 36%
JUPITER, 2008	Rosuvastatin	108	1.9 years; 17,802	About 22% reduction in the composite of CHD death, nonfatal nonprocedure-related MI, resuscitation after cardiac arrest, or fatal or nonfatal stroke

(AFCAPS/TexCAPS: The Air Force/Texas Coronary Atherosclerosis Prevention Study; ASCOT-LLA: Anglo-Scandinavian Cardiac Outcomes Trial—Lipid Lowering Arm; CHD: coronary heart disease; MI: myocardial infarction; HPS: The Heart Protection Study; JUPITER: The Justification for the Use of statins in Primary prevention: An Intervention Trial Evaluating Rosuvastatin; LDL-C: low-density lipoprotein cholesterol; PROSPER: PROspective Study of Pravastatin in the Elderly at Risk; WOSCOPS: The West Of Scotland COronary Prevention Study)

PART 2: DIFFERENCES IN STATIN INTENSITIES, USE OF RISK ESTIMATORS, TREATMENT OF SPECIFIC PATIENT SUBGROUPS, AND CONSIDERATION OF SAFETY CONCERNS

Difference in Statin Intensities

To bring about a clinically meaningful reduction in LDL-C in most patients, at least moderate- or high-intensity statin therapy is needed. A statin dose that is expected to decrease LDL-C by ~ 30 to <50% is known as moderate-intensity statin therapy whereas high-intensity statin therapy reduces LDL-C by at least 50% from the baseline.[15]

The expected amount of LDL-C reduction with the different dosages of commonly used statins is provided below:

- About ≥50% reduction from baseline (i.e., high-intensity therapy)
    - Rosuvastatin 20–40 mg/day
    - Atorvastatin 40–80 mg/day
- About 30 to <50% reduction from baseline (i.e., moderate-intensity therapy)
    - Rosuvastatin 5–10 mg/day
    - Atorvastatin 10–20 mg/day
    - Simvastatin 20–40 mg/day
    - Pitavastatin 2–4 mg/day
    - Pravastatin 40–80 mg/day
    - Lovastatin 40 mg/day

Clinicians are required to make decisions regarding high- or moderate-intensity therapy given the clinical situations they face. They need to discuss statin therapy highlighting its advantages and disadvantages with the patients as an integral component of good clinical practice. As far as intensity or dose of statins are concerned, acclaimed international guidelines differ. Targeted reduction in LDL-C level is focused by the CCS with no discussion of statin intensity or dose. Similarly, absolute LDL-C levels are used by the ESC/EAS as a treat-to-target goal (**Table 5**).[8]

The ACC/AHA, USPSTF, and VA-DoD advocate statin intensity or dose based on clinical profiles (**Table 6**).

TABLE 5:	Treatment targets and goals for cardiovascular disease (CVD) prevention.
LDL-C	• *Very-high risk in primary or secondary prevention*: A therapeutic regimen that achieves >50% LDL-C reduction from baseline and an LDL-C goal of <1.4 mmol/L (<55 mg/dL)
	• *No current statin use*: This is likely to require high-intensity LDL-lowering therapy
	• *Current LDL-lowering treatment*: An increased treatment intensity is required
	• *High-risk*: A therapeutic regimen that achieves >50% LDL-C reduction from baseline and an LDL-C goal of <1.8 mmol/L (<70 mg/dL)
	• *Moderate risk*: A goal of <2.6 mmol/L (<100 mg/dL)
	• *Low risk*: A goal of <3.0 mmol/L (<116 mg/dL)
Non-HDL-C	• Non-HDL-C secondary goals are <2.2, 2.6, and 3.4 mmol/L (<85, 100, and 130 mg/dL) for very-high-, high-, and moderate-risk people, respectively

(HDL-C: high-density lipoprotein cholesterol; LDL-C: low-density lipoprotein cholesterol)

Comparative Analysis of Major Cholesterol Guidelines

TABLE 6: Guideline recommendations for primary prevention of ASCVD.[10]

	ACC/AHA (2018)	CCS	ESC/EAS (2019)	USPSTF	VA-DoD

Treatment recommendations	• Lifestyle
	• Immediate statin without risk assessment if LDL-C ≥190 mg/dL (high intensity) or diabetes mellitus and age 40–75 years (moderate intensity)
	• Risk assessment for age 40–75 years and LDL-C ≥70 and <190 without DM
	  ○ ≥20% (high-risk) high intensity statin
	  ○ ≥7.5% to <20% 10-year ASCVD risk (intermediate risk): Moderate intensity statin
	  ○ (c) 5%–<7.5% (borderline risk): Moderate intensity, if risk enhancers
	  ○ <5% (low risk) or age <40 years (estimate lifetime risk and FH) or >75 years and LDL-C <190 mg/dL: Considered in select patients
	• Clinician-patient risk discussion prior to statin initiation[7]
		• Lifestyle
		• Target ≥50% reduction in LDL-C or LDL-C <77 mg/dL
		• Clinician-patient risk discussion prior to statin initiation
			• Lifestyle
			• Maximally tolerated dose of statin to achieve target treatment goal to be added as 2nd line in uncontrolled cases of low-risk (<1%) if LDL-C 116 to <190 mg/dL, moderate risk (≥1% to <5%) if LDL-C 100 to <190 mg/dL, high-risk (≥5% to <10%) if LDL-C 70 to <100 mg/dL, very high-risk (>10%) if LDL-C 55 to <70 mg/dL.
			• Statins as 1st line with LSM if low and moderate risk with LDL-C ≥190 mg/dL, high-risk with LDL-C ≥100 mg/dL, very high-risk with LDL-C ≥70 mg/dL
			• Clinician-patient risk discussion prior to statin initiation[8]
				• Lifestyle
				• >10% ASCVD 10-year risk: Low-moderate statin dose
				• 7.5%–10% 10-year risk: Low-moderate dose for select patients
				• Clinician-patient risk discussion prior to statin initiation
					• Lifestyle
					• >12% 10-year ASCVD risk: Moderate-dose statin
					• 6–12% 10-year risk: Moderate-dose statin for select patients
					• Clinician-patient risk discussion prior to statin initiation

(ACC: American College of Cardiology; AHA: American Heart Association; ASCVD: atherosclerotic cardiovascular disease; CCS: Canadian Cardiovascular Society; DM: diabetes mellitus; EAS: European Atherosclerosis Society; ESC: European Society of Cardiology; HDL-C: high-density lipoprotein cholesterol; LDL-C: low-density lipoprotein cholesterol; USPSTF: United States Preventive Services Task Force; VA-DoD: Veterans Affairs/Department of Defense)

Use of Risk Estimators

No risk estimators has been validated in Indian populations. Hence, Lipid Association of India (LAI) has suggested a simpler risk algorithm which stratifies individuals in to various risk categories depending only on the presence or absence of various conventional and nonconventional ASCVD risk factors (**Table 7**).

The above approach should be sufficient in most individuals and yet clinicians might use a more formal risk assessment according to their personal choice and familiarity with the risk scores. Unfortunately, none of the presently available algorithms has been validated in Indian populations, accurate risk assessment in Indians is not feasible. Still the risk calculator proposed by the Joint British Societies Third Iteration (JBS3) appears to provide the most accurate risk estimates in Indians of all the available risk scores.[16]

Those who are considered to have low 10-year ASCVD risk, warrant a lifetime ASCVD risk to be estimated. If the risk is 30–44%, it is taken to be moderately-high and if >45%, it is considered high.[17] JBS3 risk calculator based on QRISK®-lifetime risk model, PCE by ACC/AHA and Lloyd-Jones/Framingham risk algorithm are some of the widely used lifetime risk calculators. Among them JBS3 happens to be the most suited in Indian perspective. The individuals with moderately-high or high lifetime ASCVD risk are recommended to be treated in the same way as those with moderate risk for 10-year ASCVD.

TABLE 7: Recommended approach to ASCVD risk stratification in Indians.

Very high-risk	High-risk	Moderate risk	Low risk
• Pre-existing ASCVD, *or* • Diabetes with ≥2 other major ASCVD risk factors or evidence of target organ damage, *or* • Familial homozygous hypercholesterolemia	• ≥3 major ASCVD risk factors* *or* • ≥1 other high-risk features *or* • ≥1 moderate-risk nonconventional risk factors in moderate risk	• Two major ASCVD risk factors *or* • ≥1 moderate-risk nonconventional risk factors in low risk *or* • ≥30% lifetime ASCVD risk	• 0–1 major ASCVD risk factor
*Major ASCVD risk factors	Other high-risk features		Moderate-risk nonconventional risk factors
• Age ≥45 years in males and ≥55 years in females • Family history of premature ASCVD • Current cigarette smoking or tobacco use • High blood pressure • Low HDL-C	• Diabetes with 0–1 other major ASCVD risk factors and no evidence of target organ damage • CKD stage 3B or 4 • Familial hypercholesterolemia (other than familial homozygous hypercholesterolemia) • Extreme of single risk factor • Coronary calcium score ≥300 • Nonstenotic carotid plaque • Lipoprotein(a) ≥50 mg/dL		• Coronary calcium score 100–299 • Increased carotid IMT or aortic pulse wave velocity • Lipoprotein(a) 20–49 mg/dL • Metabolic syndrome

(ASCVD: atherosclerotic cardiovascular disease; CKD: chronic kidney disease; HDL-C: high-density lipoprotein cholesterol; IMT: intima-media thickness)

Coronary artery calcium (CAC) score is a tool to predict ASCVD events in cases where the traditional algorithms give ambiguous results or statin therapy might prove to be hazardous. It is not reliant on risk factors traditionally considered, e.g., age, sex, and ethnicity. If the score is ≥100 and/or ≥75th percentile, initiation of statin is recommended. Score of 0 means no statin unless there is diabetes, family history of premature CVD or smoking. In between, i.e., 1–99, statin is favored if age is >55 years.[7] Prioritization of patients for statin according to CAC score precludes treatment of all individuals and thereby long-term therapy becomes pocket-friendly. Other alternatives for CAC score are US carotid or femoral plaque burden, carotid intimal medial thickness (inferior to CAC), carotid plaque detection, and coronary CT angiography for luminal stenosis, and plaque composition gives added information over conventional risk algorithms.[8]

The degree of LDL-C reduction that is needed to achieve the target LDL-C in a given patient dictates the type and dose of the statin to be administered. **Table 8** shows the recommended LDL-C targets for various ASCVD risk categories.

The AACE 2017 guidelines also proclaim that treatment goals for dyslipidemia should be personalized according to levels of risk (**Table 9**).[9]

Safety Concerns

There are numerous untoward effects of statin therapy. Some of which are evidence-based, e.g., myopathy, altered liver function, and new onset diabetes. Some do not have no objective evidence of any cause-and-effect relationship. Meticulous observation of 170,000 patients for long-time revealed that statins are no way related to increased incidence of cancer, cerebral hemorrhage, kidney disease, liver disease, dementia, memory impairment, or fatigue.[18] In a meta-analysis of over 70,000 individuals in 18 primary and secondary prevention placebo-controlled trials, a statin was found to prevent 37 CV events and cause 5 adverse events.[19]

Let us briefly discuss how to manage statin intolerance. First we have to encourage lifestyle changes for all, look for reversible cause, e.g., drug interaction, hypothyroidism, vitamin D deficiency, reduce the dose of statin (mild symptoms) or stop it (severe symptoms), resume after they are gone at reduced dose or frequency or combined with ezetimibe or bile acid sequestrants.

TABLE 8: Treatment goals and statin initiation thresholds according to ASCVD risk categories.

Risk category	Treatment goal		Consider drug therapy	
	LDL-C (mg/dL)	Non-HDL-C (mg/dL)	LDL-C (mg/dL)	Non-HDL-C (mg/dL)
Very high-risk	<50	<80	≥50 (preferably in all)	≥80 (preferably in all)
High risk	<70	<100	≥70 (preferably in all)	≥100 (preferably in all)
Moderate risk	<100	<130	≥100	≥130
Low risk	<100	<130	≥130*	≥160*

*After an initial adequate nonpharmacological intervention for at least 3 months.
(ASCVD: atherosclerotic cardiovascular disease; HDL-C: high-density lipoprotein cholesterol; LDL-C: low-density lipoprotein cholesterol)

TABLE 9: Atherosclerotic cardiovascular disease (ASCVD) risk categories and LDL-C treatment goals (AACE 2017).[9]

Risk category	Risk factors[a]/10-year risk[b]	Treatment goals		
		LDL-C (mg/dL)	Non-HDL-C (mg/dL)	ApoB (mg/dL)
Extreme risk	• Progressive ASCVD including unstable angina in patients after achieving an LDL-C <70 mg/dL • Established clinical cardiovascular disease in patients with DM, CKD 3/4, or HeFH • History of premature ASCVD (<55 male, <65 female)	<55	<80	<70
Very high-risk	• Established or recent hospitalization for ACS, coronary, carotid or peripheral vascular disease, 10-year risk >20% • Diabetes or CKD 3/4 with one or more risk factor(s) • HeFH	<70	<100	<80
High-risk	• ≥2 risk factors and 10-year risk 10–20% • Diabetes or CKD 3/4 with no other risk factors	<100	<130	<90
Moderate risk	≤2 risk factors and 10-year risk <10%	<100	<130	<90
Low risk	0 risk factors	<130	<160	NR

[a]Major independent risk factors are high LDL-C, polycystic ovary syndrome, cigarette smoking, hypertension (blood pressure ≥140/90 mm Hg or on hypertensive medication), low HDL-C (<40 mg/dL), family history of coronary artery disease (in male, first-degree relative younger than 55 years; in female, first-degree relative younger than 65 years), chronic renal disease (CKD) stage 3/4, evidence of coronary artery calcification and age (men ≥45 years; women ≥55 years). Subtract one risk factor if the person has high HDL-C.
[b]Framingham risk scoring is applied to determine 10-year risk.
(AACE: American Association of Clinical Endocrinologists; ACS: acute coronary syndrome; ApoB: apolipoprotein B; CKD: chronic kidney disease; DM: diabetes mellitus; HDL-C: high-density lipoprotein cholesterol; HeFH: heterozygous familial hypercholesterolemia; LDL-C: low-density lipoprotein cholesterol; MESA: multi-ethnic study of atherosclerosis; NR: not recommended)

Muscle-related Side-effects

Its incidence is about 10%. No need of routine monitoring. For symptomatic individuals, serum creatine kinase (CK) level has to be checked and drug–drug interactions, vitamin D deficiency, hypothyroidism, and other potential causes of myopathy have to be ruled out. Options to treat include statin treatment in lower doses or frequency or use of alternatives.

Hepatic Effects

An asymptomatic transaminitis (3x ULN) is one of the most common adverse effects of statin therapy although it is still no more than 3%. Hepatitis, cholestasis, and acute liver failure are exceedingly rare. Statins could be safely given in patients with nonalcoholic fatty liver disease/nonalcoholic steatohepatitis (NAFLD/NASH). At times, it is curative. They are strikingly found to be more beneficial for ASCVD events than in those with normal liver function.

New Onset Diabetes

The risk is higher in those already predisposed to develop diabetes but it is significantly outweighed by the benefits. Such candidates need to be screened with fasting plasma glucose or glycosylated hemoglobin (HbA1C), prior to statin therapy. Thereafter, monitoring should be repeated within 1 year and then at intervals no longer than 3 years. Lifestyle modification is the key management.

Neurological Side-effects

Some studies have shown an association with hemorrhagic stroke particularly in poorly controlled hypertensives and patients with diabetes. But risk is outweighed as it is highly effective in preventing strokes. For both cognitive impairment and peripheral neuropathy, there is no definite evidence. But for the latter, in case no definite etiology is found, it is better to stop statin temporarily.

Cancers

- Anecdotal association only and no definite evidence.
- Guideline recommendations for safety concerns for statins are given in **Table 10**.

Treatment of Specific Patient Subgroups

Elderly Patients

Elderly age group is defined as age >75 years or life expectancy <5 years. Guidelines differ in management of this age group.

The ACC/AHA proposes continuing a statin if already tolerating, recommends not starting one for primary prevention, and recommends initiating a moderate intensity statin for secondary prevention. The ESC/EAS considers starting a statin for primary prevention if ASCVD risk is high, albeit at a lower starting dose and gradual titration to reach the target, given the altered pharmacokinetics in the elderly.

TABLE 10: Guideline recommendations for safety concerns.		
	ACC/AHA	**ESC/EAS**
Safety concerns	• Impaired renal or hepatic function • Unexplained ALT elevation ≥3 upper limits of normal • Elderly • Concomitant drugs that alter statin metabolism • Previous statin intolerance or muscle disorders • Asian ancestry	• Impaired renal function • Asian ancestry • Polypharmacy • Multiple comorbidities

(ACC: American College of Cardiology; AHA: American Heart Association; ALT: alanine aminotransferase; EAS: European Atherosclerosis Society; ESC: European Society of Cardiology)

TABLE 11: Statins for dialysis-dependent chronic kidney disease (CKD).		
	Estimated glomerular filtration rate	
Drug (mg/day)	>30 mL/min/1.73 m²	>30 mL/min/1.73 m²
Atorvastatin	10–80	10–80
Simvastatin	20–80	10–40
Rosuvastatin	10–40	5–10
Fluvastatin	20–80	10–80
Pravastatin	20–80	10–20

Note: All adult kidney transplant recipients ought to receive statin therapy.[20]

End-stage Renal Disease

The ACC/AHA has no definite recommendations for or against the initiation or stoppage of statins in end-stage renal disease patients on maintenance hemodialysis. But according to LAI 2017 recommendations—adults between the ages of 18 and 39 years, with chronic kidney disease (CKD) should be treated with statins if they have one of the following, e.g., known coronary artery disease (CAD), stroke or peripheral arterial disease, diabetes mellitus, and lifetime risk of ASCVD >30%. Statins should be continued in those who are already on it and undergoing dialysis. In very high risk patients who are not on statins, it should be started with a low dose and uptitrated cautiously (**Table 11**).[20]

Other Groups

Low-dose initiation and gradual uptitration are the recommendations by the ACC/AHA and ESC/EAS in patients with human immunodeficiency virus infection and solid organ transplantation. They also suggest clinical judgment to start statins with rheumatologic and inflammatory diseases owing to insufficient evidence. ESC/EAS underscores patients with psychiatric disorders as an impediment to medication compliance. Various guidelines treat these special groups with uncertainty as there is no systematic and large-scale data that show significant and unequivocal benefit or harm in statin usage.[10] All are hereby summarized in **Table 12**.

CONCLUSION

Medical science has come a long way in understanding and treating the menace of atherogenic dyslipidemia. Treatment of dyslipidemia as a modality of primary prevention of cardiovascular disease utilizes risk estimators to determine the need for and dose of statin therapy. Statins have pleiotropic benefits beyond lipid lowering and are used in a dose based or a treat to target LDL approach depending on the guideline being followed.

Although there are certain safety issues with the use of statins, these drugs have greater benefits than risks overall in the general population as well as in the elderly and those with renal failure.

TABLE 12: Guideline recommendations for special groups.

	ACC/AHA	CCS	ESC/EAS	USPSTF	VA-DoD
Elderly (age >75 years or life expectancy <5 years)	Continue statin if already tolerating: • *Prevention:* Recommend not starting statins for primary prevention. Statin therapy may be considered in selected individuals. • *Prevention:* Start moderate-intensity statin	If considered high-risk, recommend patient and physician discussion to initiate statin therapy	• *Prevention:* If risk factors for ASCVD, consider starting statin • *Prevention:* Treatment same as younger patients, but start at lower dose	Insufficient evidence to initiate statin for primary prevention	Therapy based on comorbidities, quality of life, and patients' preferences, values, and culture
Diabetes mellitus	Continue or start statin for: • LDL-C 70–189 mg/dL for age 40–75 years • If 10-year ASCVD risk 7.5%, a high intensity statin is reasonable	Statin therapy for: • Age ≥40 years • Age ≥30 years and duration of disease >15 years • Microvascular complications	Statin therapy for: • LDL-C ≥100 mg/dL or LDL-C 70–100 mg/dL and end-organ damage or one additional ASCVD risk factor	• Recommend statin for ≥10% 10-year risk • Consider statin for 7.5–10% 10-year risk	Recommend statin if hypertension and/or smoking present
End-stage renal disease	*Maintenance dialysis:* No recommendation for or against use of statins	Not to initiate therapy in dialysis-dependent patients Continue therapy in those already receiving it at time of dialysis initiation	No recommendations	No recommendations	Therapy based on comorbidities, quality of life, and patients' preferences, values, and culture
Other groups	• *Solid organ transplantation and HIV:* Caution with drug–drug interactions • *Rheumatologic and inflammatory diseases:* Use clinician judgment		• *Solid organ transplantation:* Caution with drug–drug interactions, start at lower dose HIV: Consider as high-risk patients. *Rheumatologic and inflammatory diseases:* Use clinician judgment. • *Mental disorder:* Consider as high-risk patients, attention to lifestyle and medication adherence.		

(ACC: American College of Cardiology; AHA: American Heart Association; ASCVD: atherosclerotic cardiovascular disease; CCS: Canadian Cardiovascular Society; EAS: European Atherosclerosis Society; ESC: European Society of Cardiology; HDL-C: high-density lipoprotein cholesterol; LDL-C: low-density lipoprotein cholesterol; USPSTF: United States Preventive Services Task Force; VA-DoD: Veterans Affairs/Department of Defense)

KEY POINTS

- Dyslipidemia is a growing menace worldwide which leads to ASCVD and its pattern is markedly more atherogenic in the Indian subcontinent.
- Different risk estimators are used to determine future ASCVD risk posed by dyslipidemia in order to institute therapies for primary prevention, e.g., PCE by ACC/AHA, SCORE by ESC/EAS, FRS by AACE, etc.
- When conventional risk assessment is not adequate, newer noninvasive modalities, e.g., CAC score can be helpful.
- Statins are the most widely used agents for managing dyslipidemia as they have several well-validated pleiotropic benefits apart from lipid-lowering effects.
- Some guidelines (ACC/AHA, USPSTF, and VA-DoD) are of the opinion to use statins based on intensity or dosage while others (ESC/EAS and CCS) use absolute LDL-C levels as a treat-to-target goal.
- Guidelines differ considerably in terms of use of risk estimators, statin treatment threshold, intensities, and goals.
- Clinicians should be cautious of numerous safety issues with use of statin, e.g., renal or hepatic impairment, myopathy, drug interaction, and intolerance.
- Special treatment groups, e.g., elderly, end-stage renal disease, and transplant recipients are to be treated with extra precaution.

REFERENCES

1. Iyengar SS, Puri R, Narasingan SN. Lipid association of India expert consensus statement on management of dyslipidemia in Indians 2016: part 1. J Assoc Physicians India. 2016;64:S7–52.
2. Farzadfar F, Finucane MM, Danaei G, Pelizzari PM, Cowan MJ, Paciorek CJ, et al. National, regional, and global trends in serum total cholesterol since 1980: Systematic analysis of health examination surveys and epidemiological studies with 321 country years and 30 million participants. Lancet. 2011;377:578-86.
3. Grundy SM. Atherogenic dyslipidemia: Lipoprotein abnormalities and implications for therapy. Am J Cardiol. 1995;75:45B-52B.
4. Joshi SR, Anjana RM, Deepa M, Pradeepa R, Bhansali A, Dhandania VK, et al. Prevalence of dyslipidemia in urban and rural india: The icmr-indiab study. PLoS One. 2014;9:e96808.
5. Karthikeyan G, Teo KK, Islam S, McQueen MJ, Pais P, Wang X, et al. Lipid profile, plasma apolipoproteins, and risk of a first myocardial infarction among Asians: An analysis from the INTERHEART study. J Am Coll Cardiol. 2009;53:244-53.
6. Lou N. A Visual Guide Across Lipid Guidelines—Five groups have recommendations out; how do they stack up?. MedPage Today. 2018.
7. Grundy SM, Stone NJ, Bailey AL, Beam C, Birtcher KK, Blumenthal RS, et al. 2018 AHA/ACC/AACVPR/AAPA/ABC/ACPM/ADA/AGS/APhA/ASPC/NLA/PCNA Guideline on the Management of Blood Cholesterol: Executive Summary: A Report of the American College of Cardiology/American Heart Association Task Force on Clinical Practice Guidelines. Circulation. 2019;139:e1046-81.
8. Mach F, Baigent C, Catapano AL, Koskinas KC, Casula M, Badimon L, et al. 2019 ESC/EAS Guidelines for the management of dyslipidaemias: lipid modification to reduce cardiovascular risk. Eur Heart J. 2020;41:111188.
9. Jellinger PS. CPG for Managing Dyslidemia and Prevention of CVD, EndocrPract. 2017;23(Suppl 2).
10. Tibrewala A, Jivan A, Oetgen WJ, Stone NJ. A Comparative Analysis of Current Lipid Treatment Guidelines: nothing stands still. J Am soc cardiol. 2018;71(7):794-9.
11. Dichtl W, Dulak J, Frick M, Alber HF, Schwarzacher SP, Ares MPS, et al. HMG-CoA reductase inhibitors regulate inflammatory transcription factors in human endothelial and vascular smooth muscle cells. Arterioscler Thromb Vasc Biol. 2003;23:58-63.

12. Shirai H, Autieri M, Eguchi S. Small gtp-binding proteins and mitogen-activated protein kinases as promising therapeutic targets of vascular remodeling. Curr Opin Nephrol Hypertens. 2007;16:111-5.
13. Nickenig G, Baumer AT, Temur Y, Kebben D, Jockenhovel F, Bohm M. Statin-sensitive dysregulated at receptor function and density in hypercholesterolemic men. Circulation. 1999;100:2131-4.
14. Dechend R, Fiebeler A, Park JK, Muller DN, Theuer J, Mervaala E, et al. Amelioration of angiotensin II-induced cardiac injury by a 3-hydroxy-3-methylglutaryl coenzyme a reductase inhibitor. Circulation. 2001;104:576-81.
15. Stone NJ, Robinson JG, Lichtenstein AH, Noel Bairey Merz C, Blum CB, Eckel RH, et al. 2013 ACC/ AHA guideline on the treatment of blood cholesterol to reduce atherosclerotic cardiovascular risk in adults: A report of the American college of cardiology/American heart association task force on practice guidelines. Circulation. 2014;129:S1-45.
16. Bansal M, Kasliwal RR, Trehan N. Relationship between different cardiovascular risk scores and measures of subclinical atherosclerosis in an Indian population. Indian Heart J. 2015;67:332-40.
17. International Atherosclerotic Society. (2013). International atherosclerotic society position paper: Global recommendations for the management of dyslipidemia 2013. [online] Available from http://www.Athero.Org/iaspositionpaper.Asp. [Last accessed July, 2020].
18. McKinney JS, Kostis WJ. Statin therapy and the risk of intracerebral hemorrhage: A meta-analysis of 31 randomized controlled trials. Stroke. 2012;43:2149-56.
19. Silva MA, Swanson AC, Gandhi PJ, Tataronis GR. Statin-related adverse events: A meta-analysis. Clin Ther. 2006;28:26-35.
20. Iyengar SS, Puri R, Narasingan SN. Lipid association of India expert consensus statement on management of dyslipidemia in Indians 2017: part 2. Clin lipidol. 2017;12(1):56-109.

CHAPTER 23

International Lipid Guidelines: Implications for Our Indian Patients— A Cardiologist's Viewpoint

Soumitra Kumar, Nodee Chowdhury

ABSTRACT

Dyslipidemia is the strongest contributor in the occurrence of acute myocardial infarctions (MIs) in South Asians according to the INTERHEART study published in 2004. Since then, ACC/AHA (American College of Cardiology/American Heart Association) guidelines as well as the ESC/EAS (European Society of Cardiology/European Atherosclerosis Society) guidelines have emerged with equally strong statements for blood cholesterol control in the reduction of atherosclerotic cardiovascular disease (ASCVD), based on the data obtained from randomized control trials. They concentrate more on evidence-based intensity of statin therapy rather than targeting specific goals for low density lipoprotein (LDL) or non-high-density lipoprotein cholesterols (non-HDL-Cs). The data regarding South Asian, especially Indian populations remain incomplete and need further investigation.

INTRODUCTION

Cardiovascular disease (CVD) is still the single largest killer in developed nations. In contrast to the age-adjusted mortality rates demonstrated in developed nations, India shows a younger predilection for cardiovascular (CV) deaths with 52% of CVD deaths under 70 years and 10% under 40 years.[1] Out of the many known risk factors to CVD, dyslipidemia remains the most pertinent cause as per the INTERHEART study, and one of the easily modifiable ones.[2]

The CTTs (Cholesterol Treatment Trial's) collaboration showed a 21% reduction in relative risk of major adverse cardiovascular events (MACE) with reduction of low-density lipoprotein cholesterol (LDL-C) by 39 mg/dL.[3] In another meta-analysis, statin therapy has been documented to reduce the risk of MACE by 19% in females and 24% in males.[4] In fact, primary prevention is the main aim of dyslipidemia control as 90% of the adverse CV events seem to occur due to insufficient primary and primordial prevention.[3]

A 2019 trajectory analysis of the Framingham Heart Study has also pointed to greater risk in the cohort with higher LDL-C, ASCVD rates were 3–4 times that of those with lifelong low LDL-C.[5] Another study published in 2019 from an endocrinologist's perspective, recommends that individuals identified to have metabolic risk should undergo 10-year global risk assessment for ASCVD or coronary heart disease to ascertain the targets of therapy for reduction of apolipoprotein B–containing lipoproteins, along with rigorous lifestyle modification.[6]

The ACC (American College of Cardiology) and the AHA (American Heart Association) have published updated guidelines on blood cholesterol management,[7] followed closely by the joint ESC/EAS (European Society of Cardiology/European Atherosclerosis Society) 2019 guidelines for the same, both of whom stressed highly on lipid lowering as per risk stratification. While these guidelines were based on randomized control trials, there was an under-representation of the Indian population, less focus was laid on younger individuals at risk for CVD and the need for nonstatin drugs was not well addressed.

THE INDIAN SCENARIO

Indians pose a difficulty with regard to management of dyslipidemia due to their phenotypic, genotypic, and metabolic differences from their European counterparts. These anthropometric differences were addressed in the National Obesity and Metabolic Syndrome Summit in compliance with the Adult Treatment Panel (ATP-III) and World Health Organization (WHO) criteria.

The normal values of body mass index (BMI) were given to be 18–22.9 kg/m² with BMI >25 indicating obesity. Waist circumference was deemed abnormal if >90 cm in males and >80 cm in females.[8]

This phenotypic difference has been designated as the "thin-fat" or Asian Indian phenotype, comprising of higher body fat, visceral fat, less skeletal muscle mass, and thinner hips.[9] They are seen to have higher circulating levels of adipokines, which are bioactive molecules secreted from the larger adipocytes that the bearers of this phenotype possess. The presence of a buffalo-hump or double chin are being considered as novel markers of metabolic syndrome and insulin resistance in them.[10]

Certain genetic polymorphisms have also been isolated in the Indian population that may explain the differences in carbohydrate and lipid metabolism. These include: (1) Glucokinase gene polymorphism, (2) Increased levels of plasma cell glycoprotein 1K121Q, (3) Insulin receptor substrate-1 G972A polymorphism,[11] (4) ApoB polymorphisms (Xba I and EcoRI),[12] and (5) *ApoCIII* gene polymorphisms.[13] ApoE3/E3 phenotype is also thought to be related to low high-density lipoprotein (HDL) levels.[10]

Looking at the differences in lipid profile, it can be summarized as a triad of hypertriglyceridemia, low HDL cholesterol (HDL-C), and small dense LDL particles. Triglyceride (TG) levels of >150 mg/dL are found in >50% of Indian men, perhaps secondary to higher glycemic load. This promotes the generation of LDL that is smaller, dense, and more atherogenic due to chylomicrons and very low-density lipoprotein (VLDL) that stay in circulation for longer periods of time. These are used as substrate by hepatic lipase, an enzyme that is upregulated in type 2 diabetes mellitus. A TG/HDL of >3.8 is highly predictive of small dense LDL among Asian Indians.[14]

An ApoB100/ApoA-1 ratio studied in the INTERHEART study as the best indicator of myocardial infarction (MI) risk with 46.8% attributable risk, was also raised in South Asians.[2] About 30–40% of this population also had >20–30 mg/dL of lipoprotein(a) [Lp(a)] that has been found to be higher in young individuals predisposed to MI.[12] These four factors: (1) Elevated TG, 1 (2) low HDL, (3) small dense LDL, and (4) raised LP(a) are labeled as the deadly Indian dyslipidemic quartet.

There are some lifestyle factors that are thought to contribute to these—extensive intake of saturated fats, trans-fats and carbohydrates, low intake of omega 3 fatty acids, lack of physical activity, and poor awareness regarding health. The PURE (Prospective Urban Rural Epidemiological) study also showed the lack of health awareness and poor compliance to medication—only 2.6% of known ASCVD patients are found to be compliant to medications in India.[15] The urban Indian is at a more perilous position, with higher BMI and LDL more than their rural counterparts up to 25 mg/dL.[16] The culture of promoting obesity since childhood has also got a firm foothold in our society. Riddled with myths and poor dietary habits, the young Indian seems to be hurtling toward a future of poor ASCVD-related prognosis.

A consensus statement published in 2016 by an expert committee of the Lipid Association of India along with multidisciplinary support also has highlighted these above issues with respect to international lipid guidelines in the Indian context.[17]

CURRENT LIPID GUIDELINES

There has been a paradigm shift in the way dyslipidemia is treated since the introduction of the 2013 ACC/AHA guidelines. They were updated recently in 2018, and are based on evidence collated from randomized control trials rather than fixed LDL-C or non-HDL-C targets.

These guidelines are briefly summarized as follows:[14]
- *Secondary prevention*: (those already with clinical ASCVD). Firstly, they stratify individuals into those at "Very high-risk" for ASCVD and those not at very high-risk. The group comprising of very high-risk includes:
  - Major ASCVD events:
    - Recent acute coronary syndrome (within the past 12 months)
    - History of MI (other than recent acute coronary syndrome event listed above)
    - History of ischemic stroke
    - Symptomatic peripheral arterial disease (history of claudication with ankle brachial index <0.85, or previous revascularization or amputation)
  - High-risk conditions:
    - Age 65 years
    - Heterozygous familial hypercholesterolemia
    - History of prior coronary artery bypass surgery or percutaneous coronary intervention (PCI) outside of the major ASCVD event(s)
    - Diabetes mellitus
    - Hypertension
    - Chronic kidney disease [estimated glomerular filtration rate (eGFR) 15–59 mL/min/1.73 m^2]
    - Current smoking
    - Persistently elevated LDL-C [LDL-C ≥100 mg/dL (≥2.6 mmol/L)] despite maximally tolerated
    - Statin therapy and ezetimibe
    - History of congestive heart failure

The treatment protocol is thereby divided into two streams:
i. *Very high-risk individuals*: High-intensity or maximal statin use
   - If on maximal statin and LDL-C ≥70 mg/dL (≥1.8 mmol/L), add ezetimibe (Class IIa)
   - Consider PCSK9-I after ezetimibe added to maximal dose statin (Class I)
ii. *Not very high-risk individuals* are further divided into age groups.
   - If <75 years, high intensity statin therapy with goal of ≥50% reduction in LDL-C (Class I)
     If not tolerated, moderate intensity statin; consider adding ezetimibe (Class IIb)
   - If >75 years, moderate or high intensity statin is reasonable (Class IIa)

- Primary prevention (those without ASCVD)
  In this case, some ground rules apply:
  - If LDL-C >190 mg/dL then high intensity statins are to be started without any risk assessment (Class I)
  - In those with coexistent diabetes mellitus and 40–75 years of age, moderate intensity statins are to be started (Class I)
- In those not meeting the above categories, we classify into age groups (**Table 1**):

TABLE 1: Classification of patients as per the age group.		
Age 0–19 years	Age 20–39 years	Age 40–75 years without diabetes

• Lifestyle modification (LSM)
• Diagnosis of familial hypercholesterolemia: Start statin
	• LSM
	• Consider statin if family history, premature ASCVD and LDL-C ≥160 mg/dL
		• Risk stratification as per 10-year ASCVD risk scoring
		  ○ ≤5%: LSM
		  ○ 5–7.5%: Control risk enhancers
		  ○ 7.5–20%: Moderate intensity statin to reduce
		  ○ LDL-C by 30–49% (Class I)
		  ○ ≥20%: Initiate statin to reduce
		  ○ LDL-C ≥50% (Class I)

TABLE 2: Atherosclerotic cardiovascular disease (ASCVD) risk enhancers.	
ASCVD risk enhancers	Lipid/biomarkers

• Family history of premature ASCVD
• Persistently elevated LDL-C ≥160 mg/dL
• Chronic kidney disease
• Metabolic syndrome
• Conditions specific to women (e.g., pre-eclampsia, premature menopause)
• Inflammatory diseases (especially rheumatoid arthritis, psoriasis, HIV)
• Ethnicity factors (e.g., South Asian ancestry)
	• Persistently elevated triglycerides (≥175 mg/mL)
	• In selected individuals, if measured:
	  ○ hs-CRP ≥2.0 mg/L
	  ○ Lp(a) levels >50 mg/dL or >125 nmol/L
	  ○ ApoB ≥130 mg/dL
	  ○ Ankle-brachial index (ABI) <0.9

(hs-CRP: high-sensitivity C-reactive protein; LDL-C: low-density lipoprotein cholesterol)

TABLE 3: Intensity of statins.		
High intensity statin	**Moderate intensity statin**	**Low intensity statin**
Lowers LDL-C by >50%	Lowers LDL-C by 30–49%	Lowers LDL-C by <30%
• Atorvastatin (40–80 mg) • Rosuvastatin (20–40 mg)	• Atorvastatin (10–20 mg) • Rosuvastatin (5–10 mg) • Simvastatin (20–40 mg) • Pravastatin (40–80 mg)	• Simvastatin (10 mg) • Pravastatin (10–20 mg) • Lovastatin (20 mg)

(LDL-C: low-density lipoprotein cholesterol)

The EAS/ESC guidelines came hot on the heels on the ACC/AHA in 2019. They also stand for an intensive approach to lipid lowering for ASCVD risk management (**Table 2**).[18] Some additional points from EAS/ESC guidelines:

- Lifestyle modification with adequate exercise (3.5–7 h moderately vigorous physical activity per week or 30–60 min most days) and heart healthy diet- a DASH type (Dietary Approaches to Stop Hypertension) or Mediterranean diet rich in fruits and vegetables, whole grains and fish is advised. BMI 20–25 kg/m^2, and waist circumference <94 cm (men) and <80 cm (women) and BP <140/90 mm Hg.
- Use of Lp(a) or ApoB for risk stratification in cases where LDL-C levels underestimate the ASCVD risk.

Key differences between ACA/AHA and ESC guidelines:
- ACC/AHA categorizes "Very high-risk" patients in the secondary prevention category, whereas ESC presents them in either primary or secondary prevention group.
- In addition to the above risk enhancers given by ACC/AHA, ESC also adds obesity and depression.
- Apart from reduction in LDL-C by 50%, ECS also mentions an absolute threshold of 55 mg/dL versus 70 mg/dL given by ACC/AHA, also applicable in the addition of nonstatin agents such as ezetimibe or PCSK-9I.
- ESC put together evidence from the Cholesterol Treatment Trialists' (CTT) Collaboration, IMPROVE-IT (IMProved Reduction of Outcomes: Vytorin Efficacy International Trial), FOURIER and ODYSSEY-OUTCOMES trials, to set the above LDL-C goals.

Follow-up:
- Adherence, adverse effects, and response to therapy to be assessed in 4–12 week after induction. Response is designated as >50% reduction in LDL-C for high-intensity and 30–50% reduction for moderate-intensity statins (**Table 3**).
- Monitor alanine aminotransferase (ALT) or creatine kinase (CK) only in symptomatic patients.
- If response is less than anticipated, add a nonstatin agent.
- In intolerant patients the maximally tolerated dose should be used.

HOW THE NEW GUIDELINES AFFECT INDIANS

The new ACC/AHA guidelines, while being comprehensive and taking a pro-active stance on combating dyslipidemia, have some deficiencies from the Indian perspective.

International Lipid Guidelines: Implications for Our Indian Patients—A Cardiologist's Viewpoint

Firstly, there is an under-representation of the Indian population among the various groups upon which the said randomized control trials were conducted, as mainly non-Hispanic world populations were involved.

Second, the primary prevention for patients with LDL-C <190 mg/dL has been limited to the age group of 40–75 years. It is unfortunate for young Indians with predisposition to metabolic syndrome and ASCVD at 10 years before the western average, thus creating an unmet need for treating dyslipidemia in the young.[19]

Next, the division of target groups into those needing high and moderate intensity therapy may not be as beneficial in south Asian populations as they respond quite well to the moderate intensity dosages. In the IRIS study conducted upon an exclusively south Asian population, the use of moderate intensity statins, i.e., atorvastatin 10 or 20 mg and rosuvastatin 10 mg showed a reduction in LDL by 40%, 47%, and 45% respectively.[20] In fact these groups may be exposed to greater chances of adverse effects or statin-induced new onset diabetes mellitus, which has been shown to be a 9% increase compared to placebo.[21] Rather it has been found that the LDL-C requiring intervention in Indians is 30 mg/dL lower than their western counterpart.[22]

Another key point is the high TG, low HDL and high small dense LDL levels found in the Indian phenotype. Neither is there a set target to achieve according to the new guidelines for such patients, nor is there a clear indicator of when to initiate nonstatin drugs and how to monitor therapy. These issues need to be addressed if we are to truly reduce the national ASCVD burden.

An operational hurdle to overcome for cardiology practice in India includes ASCVD risk scoring. Many reputed scales internationally used include the FES (Framingham Risk Score), ACC/AHA Pooled Cohort Equation, WHO/ISH (International Society for Hypertension) Risk Prediction Charts, JBS3 (Joint British Society) score, and the INTERHEART modifiable risk score. However, issues arise when using risk prediction tools derived from a population group which does not include Indian population as a comparison for data. CVD risk among Indians has been estimated to be four times that of Caucasians. Indians are likely to develop CVD at a younger age (~10 years) compared to others. Both the genetic make-up and the early onset of conventional CV risk factors including diabetes may be contributory to this. As a result, these risk calculators, specially FRS, underestimates 10-year CV risk in young individuals, owing to younger age at presentation, due to which many young individuals at a higher CV risk would remain untreated.

Despite their use in clinical practice, many shortcomings have come up during their use in clinical practice. For example, FRS which is commonly used does not include obesity, family history of premature CVD, and exercise as factors leading to atherosclerotic CVD in Indians. Upon comparison in various studies, the JBS3 risk calculator was found to have gathered data on a more comprehensive list of risk factors and also included data on Indians (albeit nonresident Indians).[23] The 2007 WHO/ISH risk prediction charts also take into consideration the contribution of ethnic-geographical factors and may be the next best option. The other possible option is: FRS could be recalibrated by multiplying the calculated risk with a correction factor (1.0 and 0.8 for rural men and women, respectively, 1.81 and 1.54 for urban men and women respectively)[24] or as per the Lipid Association of India, just compile the possible risk factors instead of using any one scoring system.[17]

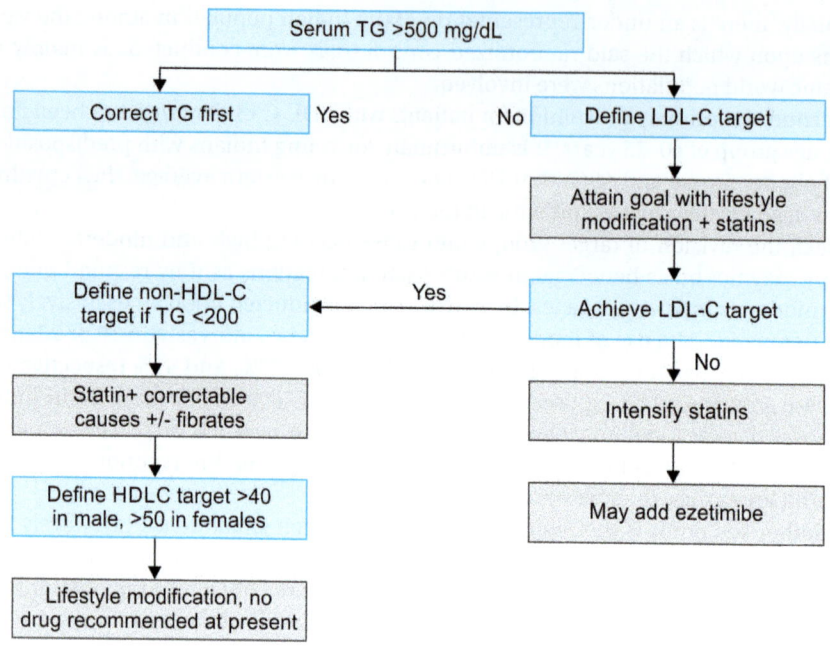

(HDL-C: high-density lipoprotein cholesterol: LDL-C: low-density lipoprotein cholesterol: TG: triglyceride)

FLOWCHART 1: Proposed algorithm for lipid management amongst Indians.[22]

POSSIBLE MODIFICATIONS FOR THE INDIAN POPULATION

A few relevant modifications have been suggested by the Cardiological Society of India in the form of a consensus statement published in the Indian Heart Journal.[25] These can be summarized as follows:

- The NECP-ATP III guidelines recommend obtaining a lipid profile in adults >20 years every 5 years, while the ESC and EAS do so for males >40 years and females >50 years. However, due to financial and logistic constraints, a routine lipid profile is advised above 30 years in the consensus with individualization in younger individuals at high-risk.
- Non-HDL-C may be considered in place of LDL-C as a CV risk marker. ApoB and ApoA-1 are not yet recommended due to issue related to logistics. Standardized reference values and prospective studies with TG as a predictor of CV events are yet to be determined.
- The use of WHO/ISH CV risk predictor score for Indians and the JBS3 risk model might be better applicable in Indians.
- With regard to therapy, the following algorithm may be followed (**Flowchart 1**).

CONCLUSION

The new ACC/AHA guidelines follow evidence from randomised control trials to formulate an intensity-driven statin regimen for the management of dyslipidemia. The applicability of these guidelines in an Indian population is yet to be assessed as fully efficacious due to variations in lipid metabolism and differences in age and anthropometric parameters predisposing to ASCVD. Some areas of dearth in understanding the Indian scenario have been identified and attempts to address them are in progress. Greater population-based studies and randomized control trials with the south Asian phenotype need to emerge before these changes can be proved to be a good fit.

KEY POINTS

- Indians, as an ethnic group, are predisposed to a malignant, often precocious type of ASCVD.
- Till we generate enough data to formulate our own lipid guidelines, we should apply the International guidelines in our practice considering many of our patients to be belonging to high to very high-risk category.

REFERENCES

1. Gupta R. Burden of coronary heart disease in India. Indian Heart J. 2005;57:632-8.
2. Yusuf S, Hawken S, Ounpuu S, Dans T, Avezum A, Lanas F, et al. Effect of potentially modifiable risk factors with myocardial infarction in 52 countries (the INTERHEART study): case-control study. Lancet. 2004;364:937-52.
3. Mihaylova BEJ, Blackwell L, Keech A, Keech A, Simes J, Barnes EH, et al. The effects of lowering LDL cholesterol with statin therapy in people at low risk of vascular disease: meta-analysis of individual data from 27 randomised trials. Lancet. 2012;380:581-90.
4. Dale KM, Colema CI, Henyan NN, Kluger J, White CM. Statins and cancer risk: a meta-analysis. JAMA. 2006;295:74-80.
5. Gidding SS, Norrina BA. Cholesterol and Atherosclerotic Cardiovascular Disease: A Lifelong Problem. J Am Heart Assoc. 2019;8.
6. Rosenzweig JL, Bakris GL, Berglund LF, Hivert M-F, Horton ES, Kalyani RR, et al. Primary Prevention of ASCVD and T2DM in Patients at Metabolic Risk: An Endocrine Society Clinical Practice Guideline. J Clin Endocrinol Metab. 2019;104(9):3939-85.
7. Enas A, Chacko V, Pazhoor SG, Chennikkara H, Devarapalli HP. Dyslipidemia in South Asian Patients. Current Atherosclerosis Reports.2007;9:367-74.
8. Misra A. Revisions of cutoffs of body mass index to define overweight and obesity are needed for Asian-ethnic groups. Int J Obes Relat Metab Disord. 2003;27:1294-6.
9. Enas EA, Mohan V, Deepa M, Farooq S, Pazhoor S, Chennikkara H. The metabolic syndrome and dyslipidemia among Asian Indians: a population with high rates of diabetes and premature coronary artery disease. J Cardiometab Syndr. 2007;2(4):267-75.
10. Misra A, Jaiswal A, Shakti D, Wasir J, Vikram NK, Pandey RM, et al. Novel phenotypic markers and screening score for the metabolic syndrome in adult Asian Indians. Diabetes Res Clin Pract. 2008;79(2):e1-5.
11. Abate N, Carulli L, Cabo-Chan A Jr, Chandalia M, Snell PG, Grundy SM. Genetic polymorphism PC-1 K121Q and ethnic susceptibility to insulin resistance. J Clin Endocrinol Metab. 2003;88:5927-34.
12. Misra A, Luthra K, Vikram NK. Dyslipidemia in Asian Indians: Determinants and Significance. JAPI. 2004; 52:137-42.
13. Miller M, Rhyne J, Chen H, Beach V, Ericson R, Luthra K, et al. APOC3 promoter polymorphisms C-482T and T-455C are associated with the metabolic syndrome. Arch Med Res. 2007;38:444-51.
14. Bhutta ZA, Nundy S, Abbasi K. Twelve years on: a call for papers for another special collection of articles on South Asia. BMJ 2016;353:i3252.
15. Scott MG, Neil JS, Bailey AL, Beam C, Birtcher Kim K, Blumenthal RS et al. 2018 AHA/ACC/AACVPR/AAPA/ABC/ACPM/ADA/AGS/APhA/ASPC/NLA/PCNA Guideline on the Management of Blood Cholesterol. A Report of the American College of Cardiology/American Heart Association Task Force on Clinical Practice Guidelines. J Am Coll Cardiol. 2019;73(24).

16. Iyengar SS, Puri R, Narasingan SN. Lipid Association of India Expert Consensus Statement on Management of Dyslipidemia in Indians 2016: Part 1 - Executive summary. J Clin Prev Cardiol 2016;5:51-61.
17. Sharma M, Ganguly NK. Premature Coronary Artery Disease in Indians and its Associated Risk Factors. Vasc Health Risk Manag. 2005;1(3):217-25.
18. Mach F, Baigent C, Catapano AL, Koskinas KC, Casula M, Badimon L, et al. ESC Scientific Document Group, 2019 ESC/EAS Guidelines for the management of dyslipidaemias: lipid modification to reduce cardiovascular risk: The Task Force for the management of dyslipidaemias of the European Society of Cardiology (ESC) and European Atherosclerosis Society (EAS), Eur Heart J. 2020;41(1):111-88.
19. Sharma M, Ganguly NK. Premature Coronary Artery Disease in Indians and its Associated Risk Factors. Vasc Health Risk Manag. 2005;1(3):217-25.
20. Deedwania PC, Gupta M, Stein M, Ycas J. Comparison of rosuvastatin versus atorvastatin in South-Asian patients at risk of coronary heart disease (from the IRIS Trial). Am J Cardiol. 2007;99(11):1538-43.
21. Sattar N, Preiss D, Murray HM, Welsh P, Buckley BM, de Craen AJM, et al. Statins and risk of incident diabetes: a collaborative meta-analysis of randomised statin trials. Lancet. 2010;375(9716):735-42.
22. Enas EA, Singh V, Munjal YP, Gupta R, Patel KC, Bhandari S, et al. Recommendations of the second Indo-U.S. health summit on prevention and control of cardiovascular disease among Asian Indians. Indian Heart J. 2009;61(3):265-74.
23. Kasliwal R, Mahansaria K, Bansal, M. Cardiovascular risk algorithms and their applicability to Indians. Gurgaon: Elsevier (RELX India Pvt Ltd); 2017. pp. 75-84.
24. Chow CK, Joshi R, Celermajer DS, Patel A, Neal BC. Recalibration of a Framingham risk equation for a rural population in India. J Epidemiol Community Health. 2009;63(5):379-85.
25. Chandra KS, Bansal M, Nair T, Iyengar S, Gupta R, Subhash C, et al. Consensus statement on management of dyslipidemia in Indian subjects. Indian Heart J. 2014;66(Suppl 3):S1-51.

CHAPTER 24

Lipoprotein Apheresis

Satyam Chakraborty, Kaushik Biswas

ABSTRACT

Lipoprotein apheresis (LA), an extracorporeal separation of various lipid components, in patients with abnormally elevated levels of atherogenic lipoproteins primarily low-density lipoproteins (LDLs), initially recommended for patients of familial hypercholesterolemia (FH), has currently been used in other dyslipidemic states. The procedure has undergone enormous modifications since its inception. Often regarded as life-saving procedure in selected patients, its widespread use has been under close scrutiny by the world bodies primarily because of the complexities and the expenses involved. Several gray zones exist regarding the indications of the procedure, courtesy lack of robust data and strong prospective trials. In this review we make an effort to discuss the basic principles, the existing and the recent indications, the beneficial effects, the adverse effects, and the future of the therapy in modern lipidology.

INTRODUCTION AND HISTORY

Apheresis is a term used for extracorporeal separation and then utilization or removal of blood components. After the first unsuccessful attempt in 1964 by Myant, in a child with homozygous familial hypercholesterolemia (FH), lipoprotein apheresis (LA), the removal of cholesterol-rich lipoproteins, was successfully attempted by De Gennes et al. in Paris, in the following year resulting in a 40% reduction in serum cholesterol, in a rather too tedious process which was finally abandoned. Continuous-flow blood cell separator revolutionized the entire concept of separation of blood components and Turnberg et al. used it to clear out a hypercholesterolemic patient in a patient of primary biliary cirrhosis in 1972. Plasma exchange was subsequently used in the western world primarily in treating homozygotic FH patients. Serial venesection followed by ex-vivo mixing of heparin-based agarose-beads and finally reinfusing the resected blood was also attempted, although the process proved more cumbersome than the plasma exchange. The term "low-density lipoprotein (LDL) apheresis" later modified to "Lipoprotein Apheresis" was coined after Stoffel et al. combined cell

Manual of Lipidology

separator to perfuse plasma through a separate column containing antibodies that bound LDL in 1981.[1] Since then the process has undergone several modifications. In this review, we discuss the basic principles and methods of LA, the current indications, the documented adverse effects and its place as a therapeutic option in the presence of recent advancements in pharmacotherapy in modern day lipidology.

CURRENTLY USED METHODS

Most of the data on LA is obtained from Germany (cost 50,000 Euro per year per patient) and Japan as they have reimbursement policies for patients on apheresis.[2] LA is currently carried out on a weekly or bi-weekly schedule. The process primarily involves withdrawing patient's blood through peripheral access through a series of infusion pumps, plasma separators, adequate anti-coagulation, replacement fluids, passage through the liposorber columns and finally reinfusion as depicted in the schematic diagram (**Fig. 1**).[3] As far as anticoagulation is concerned, heparin is the preferred anticoagulant. Citrate is not preferred as it interferes with the electrostatic interaction of the apolipoprotein molecules with the liposorber. In case of heparin-induced thrombocytopenia, lepirudin has been advocated.[4] The liposorber or adsorption column is the heart of the entire system. Various liposorber systems, which are commonly used, have been illustrated in **Table 1**. LA-15 and liposorber D use dextran sulfate as the medium to bind opposite charged lipoprotein molecule with the latter extracting significantly more high-density lipoprotein (HDL) from the plasma than the former.[4] A Canadian comparison between dextran sulfate adsorption and heparin-induced extracorporeal LDL cholesterol precipitation (HELP) apheresis in FH patients exhibited a marginally increased clearance of LDL in the former

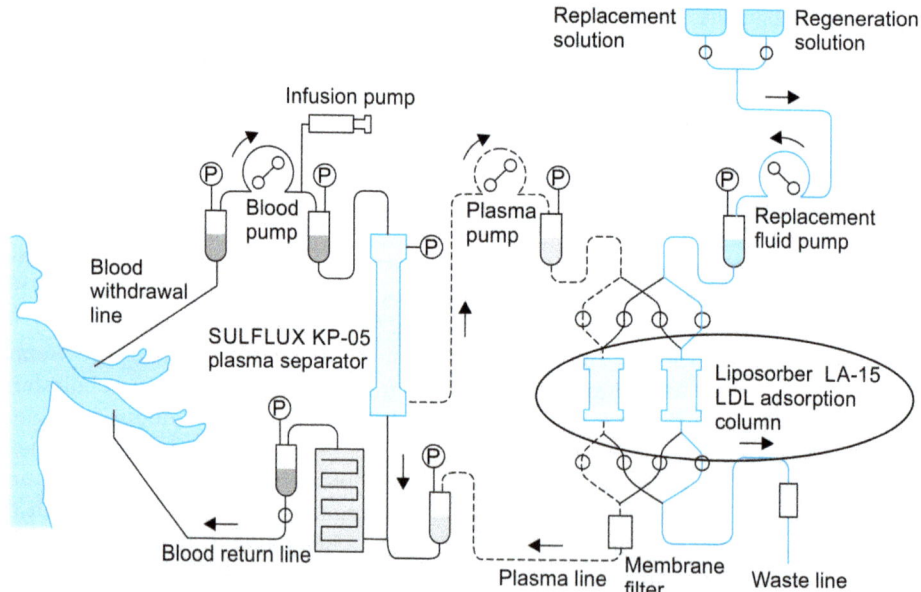

(LA: lipoprotein apheresis; LDL: low-density lipoprotein)

FIG. 1: Liposorber system.

Source: http://www.accessdata.fda.gov/cdrh_docs/pdf12/H120005b.

Lipoprotein Apheresis

TABLE 1: Various Liposorber Systems.

Basic principles	Mediator	Instrument	Moieties separated	Advantages	Disadvantages
Binding of LDL on the basis of electrical charge	Dextran sulfate	Dextran sulfate liposorber (LA-15), liposorber D	LDL: 56–65% HDL: 9–30% TG: 34–40% Lp(a): 52–61%	Column can be regenerated in LA-15	High hematocrit might interfere with plasma separation
	Polyacrylate-coated polyacrylamide beads (DALI)	Lipoprotein hemoperfusion (DALI)	LDL: 61% HDL: 30% TG: 42% Lp(a): 64%	Liposorber D and DALI does not require plasma separation so shorter procedure time	Columns cannot be regenerated in DALI and liposorber D
Mechanical separation based on size by passing through a second filter	Double filtration plasmapheresis		LDL: 56% HDL: 25% TG: 49% Lp(a): 53%	Fibrinogen removed	High hematocrit interference. Additional loss of albumin, HDL and IgG
Apolipoprotein B100 mediated immobilization of LDL moieties	Immobilized sheep apolipoprotein B-100 antibodies	Plasma select	LDL: 64% HDL: 14% TG: 42% Lp(a): 64%	Reusable	System exposure to animal proteins
Precipitation by running through acidic pH	Heparin	Plasmat-secura and plasmat-futura (HELP)	LDL: 67% HDL: 15% TG: 41% Lp(a): 62%	Fibrinogen removed	High hematocrit interference. Complicated system

[DALI: direct adsorption of lipids; HDL: high-density lipoprotein; HELP: heparin-induced extracorporeal; LDL cholesterol precipitation; IgG: immunoglobulin G; LDL: low-density lipoprotein; Lp(a): lipoprotein(a); TG: triglyceride]

Source: Adapted from Winters JL. In: McLeod BC, Weinstein R, Winters JL, Szczepiorkowski ZM (Eds). Apheresis Principles and Practice, 3rd edition. Baltimore: AABB Press; 2010.

(70% vs. 63.5% p = 0.02) probably because a larger amount of plasma being cleared in a single sitting.[5] Dextran sulfate might also clear out increased number of triglycerides as compared to HELP.[4]

INDICATIONS

Most of the lipid guidelines place LA as the last resort, courtesy the lack of convincing outcome trials as most of the data is either derived from retrospective analysis or extrapolation from interventional studies from lipid lowering agents. Since apheresis removes LDL, lipoprotein(a), and triglyceride-rich lipoproteins from plasma, it can well be hypothesized that apheresis might play a crucial role in certain dyslipidemic states. The currently approved indications are discussed here.

Familial Hypercholesterolemia

Autosomal dominant disorder of lipoprotein metabolism is mostly caused by a mutation in the LDL receptor (LDL-R) that results in elevated levels of LDL cholesterol (LDL-C), thereby increasing the risk of premature cardiovascular consequences. Concrete epidemiological data on the prevalence of FH in Indian population is yet to be available. Western data reveals that the entity is far from being rare. Approximately 1 in 500 Americans are heterozygotes, and 1 in 1,000,000 homozygotes. Typically, FH homozygotes have cholesterol in the range of 700–1,200 mg/dL, and heterozygotes have cholesterol in the range of 350–500 mg/dL.[4] Discussion of the pathophysiology is beyond the scope of the article. If untreated tendinous xanthoma, premature cardiovascular disease, and mortality are certain probabilities. Lifestyle modifications and statins remain the mainstay of treatment. Noteworthy, the efficacy of the pharmacologic agents depends on the presence of functional LDL-R. In total absence of the aforementioned receptors in states of complete penetrance, even the PCSK-9 inhibitors are totally ineffective.[6] Apheresis is indicated when LDL remains high even after optimum dietary modifications and maximum tolerated dose of statins for at least 6 months. Current FDA approved indications remain:
- Functional homozygotes with LDL-C >500 mg/dL.
- Functional heterozygotes with LDL-C ≥300 mg/dL.
- Functional heterozygotes with LDL ≥160 mg/dL and documented cardiovascular disease.

In countries such as Germany and Japan the guidelines are a bit more liberal. In Germany, LA is considered for primary prevention in patients of FH with LDL >160 mg/dL with history of cardiovascular events in close relatives. Secondary prevention with LA is offered to patients with progressive cardiovascular disease and serum LDL (S.LDL) >120–130 mg/dL. Japan approves apheresis in patients with coronary artery disease and S.LDL >250 mg/dL.[3]

In children with homozygous FH, it is recommended that LA should begin by 7 years of age to prevent the development of aortic stenosis.[4] Long-term outcomes studies have demonstrated significant reduction in the number of cardiovascular events as well as stabilization or regression of coronary stenosis.

Greater than 60% reduction from baseline LDL-C and reduction in time average total cholesterol levels is contemplated.

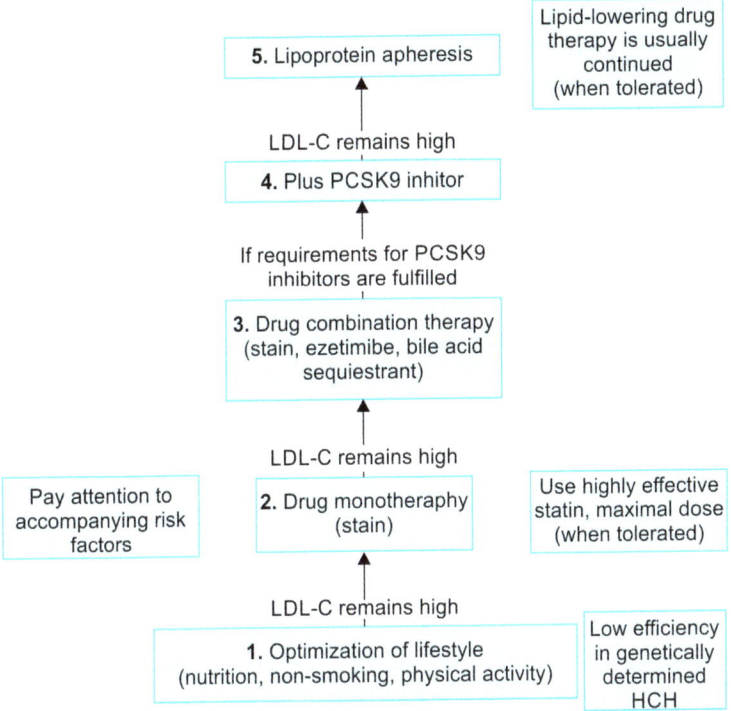

(HCH: hypercholesterolemia; LDL-C: low-density lipoprotein cholesterol)

FLOWCHART 1: Basic protocol for treating high-risk patients with lipoprotein apheresis.

Source: Adapted from Julius U. Current role of lipoprotein apheresis in the treatment of high-risk patients. J Cardiovasc Dev Dis. 2018.

Although not FDA approved, clinical conditions with high LDL-C or lipoprotein(a) [Lp(a)] not responsive to conventional therapy can also be candidates for apheresis. The basic protocol for treating such patients has been illustrated (**Flowchart 1**), target being to bring down the LDL-C levels to 69.6 mg/dL.[6]

Isolated Elevation of Lipoprotein(a)

Since 2008, isolated Lp(a) elevation has been approved by European Joint Federal Committee as an indication of LA after it was recognized an atherogenic marker. Extracorporeal therapy is partially indicated in patients with Lp(a) levels >60 mg/dL with clinical or imaging evidence of atherosclerotic vascular disease.[6] Data on whether reduction in Lp(a) levels reciprocates to clinically significant reductions in cardiovascular event rates is still controversial.[3]

Focal Segmental Glomerulosclerosis

Patients of refractory focal segmental glomerulosclerosis (FSGS) not responding to conventional therapies with steroids and calcineurin inhibitors have been recently approved for LA after convincing evidence from Japan, in adjunction with prednisone in few studies in children of nephrotic syndrome who achieved remission and in prevention of relapse after renal transplantation.[7,8]

Refsum's Disease

It is an autosomal recessive disease resulting from the inability to metabolize phytanic acid due to a defect in the phytanoyl-CoA hydrolase. The accumulation of this branched chain fatty acid may manifest clinically as retinitis pigmentosa, peripheral neuropathy, cerebellar ataxia, sensorineural deafness, and anosmia. Maximum levels of phytanic acid before commencing chronic LA were >300 mg/L. During steady state with lipid apheresis, mean phytanic acid before treatments was 87 mg/L and was reduced to 36 mg/L. Mean reduction rate was 59% per treatment. Symptoms of peripheral neuropathy significantly improved in the four patients tried in the German cohort who were tried apheresis for 5–13 years duration.[9]

Peripheral Vascular Disease

"Prospective Lipoprotein(a) Apheresis Registry Pro(a)Life" study from Germany followed up 170 patients with cardio-vascular disease with or without peripheral vascular disease, either due to high apolipoprotein A or LDL-C. Around 30 peripheral vascular events were recorded in the preceding 2 years prior to apheresis which came down to 11 in the following 2 years. Another monocentric longitudinal cohort of 35 patients with Lp(a) elevation followed up 6.8 ± 5.6 years before and 6.8 ± 4.9 years after initiation of chronic LA found a significant reduction in peripheral vascular events from 11 prior to intervention to 2 cases post-intervention.[10]

Refractory Angina

A single-blinded RCT conducted on 20 patients of refractory angina in patients with isolated Lp(a) excess, not controlled on conventional antianginal rendered significant reduction not only in clinical parameters but also in objective parameters assessed by total carotid wall volume.[11]

Senile Sensory-neural Hearing Loss

About 230 patients from Italy with senile sensory-neural hearing loss (SSNHL) were subjected to HELP apheresis with significant improvement in hearing ability as measured not only by pure tone recovery but also speech perception at 10 and 24 hours post-therapy and tinnitus scores.[12]

PLEIOTROPIC EFFECTS

Several *Pleiotropic* effects of LA have been documented along with lowering of atherogenic lipoproteins.
- Decrease in C-reactive protein, Serum amyloid A, and several other inflammatory markers.
- Reduction in fibrinogen and other coagulation markers.
- Decrease in plasminogen and other fibrinolytic proteins.
- Decrease in blood viscosity.
- Reduced complement levels.[3]

CONTRAINDICATIONS

Being a life-saving procedure, the contraindications are limited and relative. Decompensated heart failure, malignant cardiac arrhythmias, therapy resistant hypotension, underlying malignancies, severe psychiatric disorder, extreme physical or intellectual disability, nonavailability of peripheral access and foreseeable short life-expectancy are usual deterrents for undertaking the procedure.[2]

ADVERSE EFFECTS

An overall adverse effect rate of around 5–11%, without much difference among the various LA systems, has been documented. Vasovagal symptoms are the most common with weakness, malaise, and transient hypotension.[13] Puncture problems (24%), hypotension (20%), bleeding and hematoma (9%), hypertension (3%), anginal pain (4%), edema (3%), nausea, vomiting, and vertiginous sensations have been described.[2] Anaphylaxis has been a recognized complication in patients with angiotensin-converting enzyme (ACE) inhibitors. Although the exact reason is not clear, bradykinin accumulation occurs in the presence of heparin as anticoagulant. The aforementioned levels are clinically nonsignificant but if the patient is on ACE-inhibitors, the Bradykinin levels might be elevated to clinically significant levels causing anaphylaxis. ACE-inhibitors should therefore, be stopped 24 hours prior to the procedure.[14] Angiotensin receptor blockers can, however, be safely used in these patients.[3] Hypocalcemia has been observed in few cases in LA-15 systems where citrate-dextrose (ACD-A) was used as anticoagulant.[14]

CONCLUSION AND FUTURE

A life-saving therapy in homozygote FH patients in whom pharmacological interventions have limited role and can be used safely in children unlike many of its pharmacologic counterparts. In polygenic hypercholesterolemia, the therapy is usually the last resort after lifestyle intervention and lipid lowering therapies (including PCSK-9 inhibitors) fail. While its role in treating refractory dyslipidemic states might gradually become limited with the advent of newer pharmacological therapies, newer avenues open up in various disease states where the therapy might play a key role. More robust data is the need of the hour. The expense and the complicacy of the procedure still remain obvious limitations for widespread use in developing countries.

> **KEY POINTS**
> - Lipoprotein apheresis means extracorporeal removal of lipoproteins in various pathological states.
> - Familial Hypercholesterolemia, particularly the homozygous variant remains the primary indication.
> - An interventional procedure with scope across all age spectrums, even where the pharmacological agents are not indicated.
> - More robust prospective data required for widespread and cost-effective use.

REFERENCES

1. Stefanutti C, Thompson GR. Lipoprotein Apheresis in the Management of Familial Hypercholesterolemia: Historical Perspective and Recent Advances. Curr Atheroscler Rep. 2015;17:465.
2. Julius U. Lipoprotein Apheresis in the management of severe hypercholesterolemia and of elevation of lipoprotein(a):- Ulrich Julius, Current perspectives and patient selection. Med Devices (Auckl). 2016;9:349-60.
3. Feingold KR, Grunfeld C. Lipoprotein Apheresis. In: Feingold KR, Anawalt B, Boyce A, et al. (Eds). Endotext [Internet]. South Dartmouth (MA): MDText.com Inc.; 2000.
4. Paroder-Belenitsky M, Pham HP, LDL Apheresis. In: Transfusion Medicine and Hemostasis: Clinical and Laboratory Aspects; 2019. pp. 493-6.
5. Drouin-Chartier JP, Tremblay AJ, Bergeron J, Pelletier M, Laflamme N, Lamarche B, et al. Comparison of two low-density lipoprotein apheresis systems in patients with homozygous familial hypercholesterolemia. J Clin Apher. 2016;31:359-67.
6. Ulrich J. Current Role of Lipoprotein Apheresis in the Treatment of High-Risk Patients. J Cardiovasc Dev Dis. 2018;5(2):27.
7. Hattori M, Chikamoto H, Akioka Y, Nakakura H, Ogino D, Matsunaga A, et al. A combined low-density lipoprotein apheresis and prednisone therapy for steroid-resistant primary focal segmental glomerulosclerosis in children. Am J Kidney Dis. 2003;42(6):1121-30.
8. Sannomiya A, Murakami T, Koyama I, Nitta K, Nakajima I, Fuchinoue S. Preoperative Low-Density Lipoprotein Apheresis for Preventing Recurrence of Focal Segmental Glomerulosclerosis after Kidney Transplantation. Hindawi Journal of Transplantation. 2018.
9. Zolotov D, Wagner S, Kalb K, Bunia J, Heibges A, Klingel R. Long-Term Strategies for the Treatment of Refsum's Disease Using Therapeutic Apheresis.J Clin Apher. 2012;27:99-105.
10. Weiss N, Julius U. Lipoprotein(a) apheresis in patients with peripheral arterial disease: rationale and clinical results. Clin Res Cardiol Suppl. 2019;14(Suppl 1):39-44.
11. Khan TZ, Hsu L-Y, Arai AE, Rhodes S, Pottle A, Wage R, et al. Apheresis as novel treatment for refractory angina with raised lipoprotein(a): a randomized controlled cross-over trial. Eur Heart J. 2017;38:1561-69.
12. Bianchin G, Russi1 G, Romano1 N, Fioravanti P. Role of H.E.L.P.-apheresis in the treatment of sudden sensorineural hearing loss in a group of 230 patients. Acta Otorhinolaryngol Ital. 2011;31:395-8.
13. Raina R, Young C, Krishnappa V, Chanchlani Rahul. Role of Lipoprotein Apheresis in Cardiovascular Disease Risk Reduction. Blood Purif. 2019;47:301-16.
14. Bambauer R, Bambauer C, Lehmann B, Latza R, Schiel R. LDL-Apheresis: Technical and Clinical Aspects. The Scientific World Journal.

CHAPTER 25

Evolving Targets of Therapies: PCSK9 and Beyond

Subhodip Pramanik

ABSTRACT

Current lipid-lowering therapy to reduce cardiovascular morbidity and mortality focuses mostly on low-density lipoprotein (LDL)-lowering agents, and statins are most studied among them. However, the residual risk remains, despite using full-dose statins and many patients cannot tolerate statin, mandating discovery of other effective and safe LDL-lowering agents. Proprotein convertase subtilisin/kexin type 9 (PCSK9) inhibitors (evolocumab and alirocumab) have shown LDL reduction in the tune of 60% on top of statin and cardiovascular outcome trials have shown a significant reduction in cardiovascular events and mortality. Other LDL-lowering therapies [inclisiran, lomitapide, apolipoprotein B (ApoB) inhibitor, gene-based therapies, LDL apheresis, bempedoic acid, and gemcabene] have shown to reduce LDL significantly but long-term data showing cardiovascular benefit are lacking. Eicosapentaenoic acid, an agent that lowers triglyceride (TG), have recently shown to improve cardiovascular outcome. However, other agents that lower TG (gene therapy, pradigastat, volanesorsen, and evinacumab) are still under trial. Whether elevation of high-density lipoprotein (HDL) or reduction of lipoprotein(a) [Lp(a)] are beneficial for cardiovascular risk reduction is not known, the newer promising agents [Lp(a) lowering—IONIS-APO(a)-LRx and HDL increasing—anacetrapib and ApoA-1 peptide infusions] may answer this question in future. These evolving lipid-lowering drugs have opened up a new horizon in cardiovascular therapeutics.

INTRODUCTION

Lowering low-density lipoprotein cholesterol (LDL-C) has become a major target for primary and secondary prevention of cardiovascular (CV) diseases. Reduction in CV events with statins has been well established during the past two decades in numerous large, randomized, controlled trials (RCT).[1] Recent studies have suggested that reducing LDL-C below current targets may improve CV outcomes further.[1] However, other agents such as ezetimibe, extended-release (ER) niacin, fibrates, and bile acid sequestrants reduce LDL-C only by 12–18%, but may be associated with significant adverse events, and in recent CV outcome trials have not been shown to further reduce CHD events when added to statins.[2] So extensive research is ongoing in search of safe, effective and clinically useful agents to lower LDL-C and triglyceride (TG). In this chapter, we will discuss few novel agents which are recently approved or in the development process with some initial promising results.

Unmet Needs of Current Lipid-lowering Therapies

Although current lipid-lowering therapies such as statins and fibrates meet the need of most patients with dyslipidemia, newer lipid-lowering therapies are needed for individuals with: (1) Severe dyslipidemias that are refractory to current treatments, (2) Intolerable or harmful adverse effects from existing therapies, or (3) Residual atherosclerotic cardiovascular disease (ASCVD) risk, despite having an acceptable response to the current standard of care. Again, newer evidence suggests lowering TG with icosapent ethyl acid reduces ischemic events including CV death.[3] This has raised enthusiasm to evaluate TG-directed therapies for the reduction of ASCVD as well.

CLASSIFICATION OF NOVEL LIPID-LOWERING THERAPIES

The novel lipid-lowering therapies can be classified according to the component of lipid it predominantly affects. Broadly they are classified as: (1) LDL-lowering agent, (2) TG-lowering agents, (3) HDL-increasing agent, and (4) Lp(a)-lowering agents. Except for LDL lowering, the other three interventions are yet not proven undoubtedly to reduce CV events. There has been a search for an effective and safe drug for increasing HDL and lowering Lp(a). We will briefly discuss the mechanism of action, efficacy in terms of lipid-lowering capacity, clinical evidence and few clinical pearls of all the drugs in tabular form (**Table 1**).[4-16] The discussion about PCSK9 inhibitors is done separately.

PCSK9 Inhibitors

History

Proprotein convertase subtilisin/kexin type 9 (PCSK9), the ninth member of the K family of subtilisin-related serine endoproteases, was first described by Abifadel M and his colleagues in 2003 while searching for the cause of *autosomal dominant* hypercholesterolemia (ADH) which were found not to be due to mutation of genes encoding for low-density lipoprotein receptor (LDLR) or apolipoprotein (Apo) B.[17] Further studies in humans by Cohen and colleagues[18] confirmed that few individuals with low plasma levels of LDL-C had "loss-of-function" (LOF) mutations in PCSK9. These families were found to have a significant reduction in lifetime risk of atherosclerotic heart disease, and no adverse health effects.

Structure, Physiology, Function, and Role of PCSK9 in LDL Degradation

Proprotein convertase subtilisin/kexin type 9 is synthesized predominantly in the liver and intestines, and to a lesser extent in the kidneys, as a 692 amino acid, 74 kDa, soluble precursor, consisting of a signal peptide, followed by a prodomain, a catalytic domain, and a C-terminal domain. Autocatalytic cleavage occurs inside endoplasmic reticulum and a

Evolving Targets of Therapies: PCSK9 and Beyond

TABLE 1: Classification, mechanism of action, efficacy, side effect profile, and clinical evidence of novel therapy for dyslipidemia (beyond PCSK9 inhibitors).

Primary target	Agents	Mechanism of action	Efficacy	Side effects	Evidence	Comments
LDL lowering	MTP inhibitors (Lomitapide)	It inhibits microsomal TG transfer protein, which acts to assemble ApoB-containing lipoproteins	LDL-C was reduced at 26 weeks by 50% with lomitapide 40 mg daily	High prevalence of gastrointestinal distress, fatty liver, and transaminitis	Improved quality of life, by reducing the frequency of lipid apheresis treatments[4]	Because of the benefit-to-risk ratio, lomitapide is currently an adjunctive orphan treatment for patients with HoFH
	ApoB inhibitor (Mipomersen)	An antisense oligonucleotide (ASO) inhibitor of ApoB-100 mRNA	A meta-analysis of eight RCTs of mipomersen showed LDL-C was reduced by 36%	High rates of injection-site reactions, transaminase elevations, hepatic steatosis, and flu-like symptoms	Post-hoc analysis of 3 RCTs and an open-label extension trial showed >80% reduced ASCVD endpoints concurrent with mipomersen use[5]	Primarily indicated for the restricted indication of HoFH
	Gene-based therapies (AAV8 vector carrying an LDLR transgene)	Deliver a functional copy of the *LDLR* gene to liver cells				Under trial[6]

Continued

TABLE 1: Classification, mechanism of action, efficacy, side effect profile, and clinical evidence of novel therapy for dyslipidemia (beyond PCSK9 inhibitors).

Primary target	Agents	Mechanism of action	Efficacy	Side effects	Evidence	Comments
LDL lowering	Bempedoic acid	Inhibition of ACL (ATP citrate lyase) and activation of AMPK (adenosine phosphate-activated protein kinase)	When added to background statin therapy, LDL-C was reduced by 24%.	Headache was commonly reported, whereas muscle and liver adverse effects were uncommon	CV outcome trial on heterozygous FH and ASCVD is underway (www.clinicaltrials.gov: NCT02993406)[7]	Promising oral agent
	Gemcabene	Both reduced synthesis and increased clearance of hepatic TG-rich lipoproteins	Reduced LDL-C by 23% and 28%, respectively, over background statin therapy	The trial on pediatric patients with NAFLD was terminated due to unanticipated problems[8]	Trial on adults with metabolic disorders continues (www.clinicaltrials.gov: NCT03508687)	Promising add on therapy
	LDL apheresis	Extracorporeal removal of circulating ApoB-containing lipoproteins, including LDL, lipoprotein(a) and VLDL	Lowers LDL-C acutely by 50–76% but around 30–40% in the long run	Hypotension, anemia, nausea, flushing, and headache	No study has demonstrated significantly improved survival with LDL apheresis[9]	Indicated in established CVD with LDL >200 mg/dL or any patient with LDL >300 mg/dL
	Partial ileal bypass surgery	Prevent enterohepatic circulation	30–40% reduction of LDL-C	Diarrhea, kidney stones, gallstones, and intestinal obstruction	A significant 35% reduction in the combined endpoint of death due to coronary heart disease and confirmed nonfatal myocardial infarction[10]	Not recommended

Continued

TABLE 1: Classification, mechanism of action, efficacy, side effect profile, and clinical evidence of novel therapy for dyslipidemia (beyond PCSK9 inhibitors).

Primary target	Agents	Mechanism of action	Efficacy	Side effects	Evidence	Comments
Triglyceride lowering	Eicosapentaenoic acid	A form of omega-3 fatty acid		Good safety profile	A significant 25% reduction in major ASCVD events	1st TG lowering agent to improve CV outcome
	LPL gene therapy	Expression of the missing gene in liver	Very effective but effect ill sustained			The manufacturer decided not to pursue renewal of the marketing authorization[11]
	DGAT Inhibition[12] (pradigastat)	Inhibit intestinal fat absorption	Fasting TG was reduced in by ≤70%	Gastrointestinal adverse events, such as diarrhea and nausea		Promising agent
	Volanesorsen[13]	Targeted reduction of ApoC-III	52 weeks' sustained TG reduction in the tune of 77% in patients with familial chylomicronemia syndrome	Injection site reaction and thrombocytopenia	Reduces incidences of pancreatitis	Awaiting FDA approval

Continued

TABLE 1: Classification, mechanism of action, efficacy, side effect profile, and clinical evidence of novel therapy for dyslipidemia (beyond PCSK9 inhibitors).

Primary target	Agents	Mechanism of action	Efficacy	Side effects	Evidence	Comments
	Evinacumab	Binding ANGPTL3 and creating immune complexes; thus potentiation of plasma lipase activity	Lowers TG in the tune of 50–70%[14]		Only phase 1 and two trials are done to date	
Drugs increasing HDL	• Torcetrapib • Anacetrapib[15]	Small-molecule inhibitors of CETP	Mild-to-moderate HDL elevation		Modest 9% relative risk reduction of ASCVD events with anacetrapib	Questionable benefit as many confounding factors were present in those trials
	ApoA-1 peptide infusions[16]	Promoting cholesterol efflux from cells thus expected to reduce atherosclerotic plaque size	Not much effect of blood HDL level		Intravascular ultrasound of the coronary arteries showed no regression of plaque volume compared with placebo	Larger trials with clinical endpoints are ongoing
Lp(a) lowering therapy	IONIS-APO(a)-LRx	Antisense oligonucleotide against kringle-IV type 2 which is present in nucleotide level of all isoform	Resulted in mean Lp(a) reductions of 66%, 80%, and 92% in the 10-, 20-, and 40-mg groups, respectively		No clinical data yet	Its yet not known whether lowering Lp(a) is beneficial for cardiovascular diseases

(AAV8: adeno-associated virus 8; ANGPTL3: angiopoietin-like protein 3; ASCVD: atherosclerotic cardiovascular disease; CETP: cholesteryl ester transfer protein; DGAT: diacylglycerol acyltransferase; FDA: Food and Drug Administration; HoFH: homozygous familial hypercholesterolemia; LDL: low-density lipoprotein; LPL: lipoprotein lipase; MTP: microsomal triglyceride transfer protein; NAFLD: nonalcoholic fatty liver disease, PCSK9: proprotein convertase subtilisin/kexin type 9; VLDL: very low-density lipoprotein)

Continued

Evolving Targets of Therapies: PCSK9 and Beyond

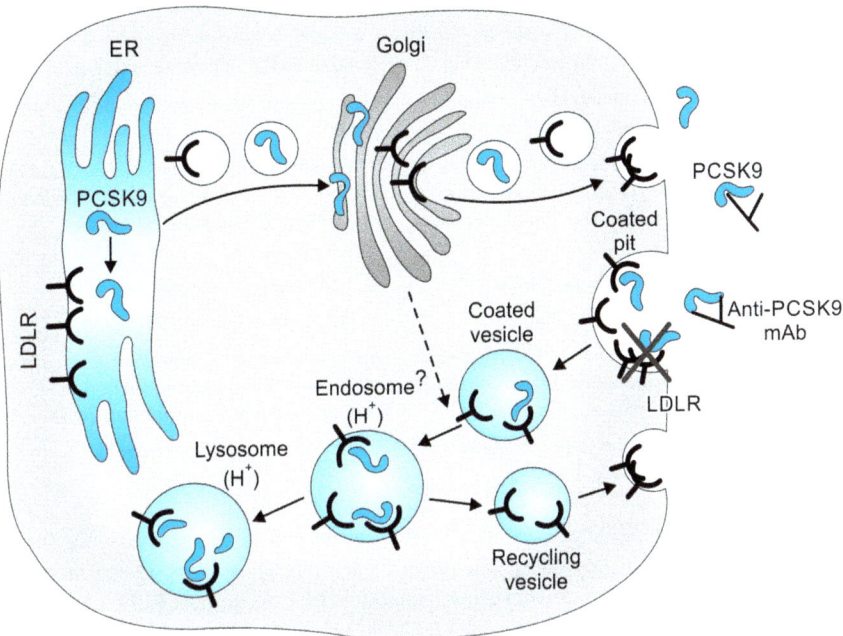

(LDLR: low-density lipoprotein receptor; mAB: monoclonal antibodies; PCSK9: proprotein convertase subtilisin/kexin type 9)

FIG. 1: Physiology of PCSK9 and mechanism of action of PCSK9 inhibitors.

60 kDa catalytic fragment is secreted into blood circulation as guided by the N-terminal domain. It has a short half-life of 5 minutes and is cleared rapidly. Circulating PCSK9 levels vary throughout the day and are synchronous with cholesterol synthesis, with lower PCSK9 levels in the fasting state, a nadir in the evening, and peak level at night. PCSK9 plays a role in the regulation of cholesterol homeostasis via accelerating the degradation of the LDLR. PCSK9 binds to the LDLR on the surface of cells and the complex is internalized and degraded in phagosome. Thus, PCSK9 changes the itinerary of the LDLR, diverting internalized LDLR to degradation in lysosomes and thus preventing them from recycling to the cell surface[19] (**Fig. 1**).

Various PCSK9-targeted Therapy

The discovery that PCSK9 is synthesized in the liver, secreted extracellularly and interact with the LDL receptor has helped to raise following therapeutic approaches to reduce PCSK9 activity.[20]

- *Gene-silencing techniques (inclisiran)*: This targets intracellular production of PCSK9 which leads to reduced extracellular levels of PCSK9 to bind the LDLR. Two different types of silencing techniques are used (**Table 2**). Among them, one small-interfering ribonucleic acid (siRNA) inhibitor of PCSK9 synthesis (inclisiran) has been studied until phase 3. After 6 months of therapy, it reduced LDL-C 52.6% from baseline in a phase 2 study.[21] Recently, a phase 3 study revealed in people with heterozygous familial hypercholesterolemia, inclisiran reduced LDL-C significantly as compared with placebo, with an infrequent dosing regimen and an acceptable safety profile.[22]

TABLE 2: Current status of gene silencing techniques for PCSK9 inhibition.		
Parameters	Antisense oligonucleotide (ASO)	Small-interfering ribonucleic acid (siRNA)
Structure	Short (<25 nucleotides) single-stranded DNA	Short (15–30 nucleotides) double-stranded RNA
Mechanism	This sequence is complementary to mRNA target and blocks translation, leading to degradation of mRNA by mRNAase	After unwinding, antisense siRNA binds to target mRNA in a sequence-specific manner, resulting in degradation of mRNA
Animal trial	• mRNA reduction of 92% • LDL-C reduction of 32%	• mRNA reduction 60–70% • LDL-C reduction of 30%
Human trial	Terminated due to unknown reason (ISIS pharmaceuticals and Santaris)	Inclisiran has been shown to significantly decrease the hepatic production of PCSK9 and cause a marked reduction in LDL-C levels

(LDL-C: low-density lipoprotein cholesterol; PCSK9: proprotein convertase subtilisin/kexin type 9)

- *Small molecule inhibitors and small peptide mimetics*: Small molecule inhibitors interfere with PCSK9 processing (cleavage or secretion) and peptide mimetics block protein function. LDLR epidermal growth factor-A (EGF-A) domain is the site for binding of the PCSK9 catalytic domain to the LDLR, so interference at the EGF-A binding site provides a potentially attractive target. Currently available small molecule inhibitors and small peptide mimetics have a short half-life, necessitating more frequent dosing. They have a less effective binding affinity for LDLR than monoclonal antibodies (mABs).
- *Immunization*: There is only one animal study by Fattori et al. where a humoral response was produced against PCSK9, which produced polyclonal response with a high-titer antibody against PCSK9 resulting in a 66% reduction in circulating PCSK9 and 60% reduction of plasma LDL-C. Human trials are needed to prove its efficacy and safety in the future.
- *Binding proteins to PCSK9*: They are protein particles that bind PCSK9 in vivo and hinders their binding to LDLR. A fusion protein called adnectin, which has been developed from *Escherichia coli*-based manufacturing process, is currently under trial.
- *Monoclonal antibodies*: The most developed strategy to inhibit circulating PCSK9 activity so far is mABs against them which disrupts its interaction with the LDRs. Currently, mABs against PCSK9 are being tested in many human trials and have shown significant LDL-C reduction and cardiovascular benefit.

Monoclonal Antibodies Against PCSK9 Inhibitors

As described above, these agents seem to be the most promising therapy to reduce LDL-C after statins. Much attention is paid toward its development program and early phase 3 trials of its first two drugs have confirmed its efficacy and safety.[23,24]

Agents

Evolocumab was the first agent to be studied in a phase 3 RCT and proven to be efficacious, safe, and have CV benefits in long term.[21] The next agent showing promising results in phase 3 trial was alirocumab.[22] Both agents have received the Food and Drug

Administration (FDA) approval recently. Bococizumab and RG-7652 are undergoing phase 3 trials and few other molecules such as LY3015014 and ALN-PCS02 are still in their initial stages of research. Our discussion in this review will remain confined within the first two agents, namely evolocumab and alirocumab.

Mechanism of Action

Antibody to PCSK9 binds with its ligand in the systemic circulation and prevents them to get internalized with LDL–LDLR complex in hepatocytes, so further degradation of LDL receptors inside the phagosome is prevented. This facilitates the recycling of LDL receptors to the hepatocyte membrane where they can again act as LDL scavenger from blood, leading to lowering of blood LDL level (**Fig. 1**).

Pharmacokinetics

These agents are available in the form of subcutaneous injections, systemic bioavailability being 70–85% following an injection. The effective elimination half-life of the available PCSK9 inhibitors is 11–20 days. Concomitant statin use increases PCSK9 production, thus theoretically PCSK9 inhibition could be more efficacious in patients using high-dose statin. Clinical trials, however, showed LDL lowering the capacity of PCSK9 inhibitors are same whether patient using statin or not. Evolocumab dose is 140 mg subcutaneously every 2 weeks and alirocumab dose is 75–150 mg subcutaneously every 2 weeks or 300 mg subcutaneously monthly.

Efficacy from Clinical Trials

Phase 1 trials with alirocumab and evolocumab were performed and results came out in 2012 which showed dramatic reductions in LDL-C in different of patient phenotypes, when added to diet alone or background statin therapy. The results of alirocumab and evolocumab phase 2 trials were published in 2012. These studies revealed the pharmacokinetic properties of those molecules and their interaction with statins. All patients with familial hypercholesterolemia had excellent results with both alirocumab and evolocumab. A novel and unexpected finding is an as-yet-unexplained reduction in Lp(a), in the 25–30% range which was consistent in all trials.[25] Both alirocumab and evolocumab mABs are fully humanized, thus minimizes the potential for immune reactions and the development of neutralizing antibodies. The major purpose of phase 3 is to assess the efficacy in terms of clinical outcome and long-term safety in a larger population so that regulatory agencies can determine the risk/benefit ratio in approving an LDL-C-lowering agent for general use. On average, LDL reduction was in the tune of 60–70%, even on top of statin therapy.

Cardiovascular Outcome Trials

A 2.2-year follow-up study with evolocumab (the FOURIER outcome study[26]), did not reveal any mortality benefit; however, the risk for myocardial infarction or stroke was significantly reduced. The ODYSSEY OUTCOMES[27] trial followed up patients after acute coronary syndrome for a median of 2.8 years and demonstrated that alirocumab on top of intensive statin treatment reduced all-cause mortality in long term [hazard ratio (HR) 0.85, 95% CI 0.73–0.98, $p = 0.03$].

Side Effects

Local injection site-related problems are common. A significant number of patients undergoing alirocumab therapy found to have myalgia as compared with placebo but not with evolocumab therapy. Although neurocognitive dysfunction was rare, they occurred with increased frequency in those receiving these drugs. However, in a neurocognitive substudy in the FOURIER trial with 1,974 patients, no significant evidence of neurocognitive impairment was found. There was no elevation in hepatic aminotransferases or creatinine phosphokinase as compared with the placebo. A transient rise in binding antibody to evolocumab was observed In long-term therapy but they were not neutralizing in nature.

CONCLUSION

The discovery of PCSK9 considerably changed the therapeutic reality in the lipid field. PCSK9 inhibitors increase LDLR recycling and decrease LDL-C levels. Fully human mAbs are currently the most studied PCSK9 inhibitors. They result in a consistent decrease in LDL-C levels in the tune 60%, either on top of statins or as a monotherapy. The CV outcome trials have shown impressive results in lowering myocardial infarction, stroke and all-cause mortality. PCSK9 inhibitors will offer a novel and powerful therapeutic option for patients with difficult-to-treat hypercholesterolemia. Inclisiran, a siRNA inhibitor of PCSK9 synthesis, has shown impressive LDL cholesterol lowering with much infrequent dosing. There are multiple other novel therapies in the pipeline which reduce LDL, TGs, and Lp(a), or increases HDL. Long-term randomized controlled clinical trials are underway to assess the safety, tolerability, and efficacy of these novel agents to reduce CV events.

> ### KEY POINTS
>
> - There are unmet needs of current lipid-lowering therapies and sometimes therapeutic doses of statins of fibrates cannot be used due to tolerability issues.
> - PCSK9 inhibitors (evolocumab and alirocumab) are human mABs that have shown to reduce LDL in the tune of 60%, even on top of statin therapy.
> - The CV outcome trials have shown impressive results in lowering myocardial infarction, stroke (evolocumab—FOURIER outcome study), and all-cause mortality (alirocumab—ODYSSEY OUTCOMES trial).
> - Other LDL-lowering therapies [microsomal triglyceride transfer protein (MTP) inhibitors, ApoB inhibitor, gene-based therapies, LDL apheresis, bempedoic acid, and gemcabene] are efficacious but long-term data showing CV benefit are lacking.
> - Eicosapentaenoic acid, an agent that lowers TG, has shown to improve CV outcome. However, other agents that lower TG (*LPL* gene therapy, DGAT Inhibition, volanesorsen, and evinacumab) are still under trial.
> - Whether elevation of HDL or reduction of Lp(a) are beneficial for CV risk reduction is not known, the newer promising agents [Lp(a) lowering- IONIS-APO(a)-LRx and HDL increasing- anacetrapib and ApoA-1 peptide infusions] may answer this question in future.

REFERENCES

1. Cholesterol Treatment Trialists' (CTT) Collaboration; C Baigent, L Blackwell, J Emberson, L E Holland, C Reith, N Bhala, et al. Efficacy and safety of more intensive lowering of LDL cholesterol:a meta-analysis of data from 170,000 participants in 26 randomised trials. Lancet. 2010;376(9753):1670-81.
2. Ezzet F, Wexler D, Statkevich P, Kosoglou T, Patrick J, Lipka L, et al. The plasma concentration and LDL-C relationship in patients receiving ezetimibe. J Clin Pharmacol 2001;41(9):943-9.
3. Bhatt DL, Steg PG, Miller M, Brinton EA, Jacobson TA, Ketchum SB, et al. Cardiovascular risk reduction with icosapent ethyl for hypertriglyceridemia. N Engl J Med. 2019;380(1):11-22.
4. Berberich AJ, Hegele RA. Lomitapide for the treatment of hypercholesterolemia. Expert Opin Pharmacother. 2017;18(12):1261-8.
5. Duell PB, Santos RD, Kirwan BA, Witztum JL, Tsimikas S, Kastelein JJP. Long-term mipomersen treatment is associated with a reduction in cardiovascular events in patients with familial hypercholesterolemia. J Clin Lipidol. 2016;10(4):1011-21.
6. Greig JA, Limberis MP, Bell P, Chen SJ, Calcedo R, Rader DJ, et al. Nonclinical pharmacology/toxicology study of AAV8.TBG.mLDLR and AAV8.TBG.hLDLR in a mouse model of homozygous familial hypercholesterolemia. Hum Gene Ther Clin Dev. 2017;28(1):28-38.
7. Ballantyne CM, McKenney JM, MacDougall DE, Margulies JR, Robinson PL, Hanselman JC, et al. Effect of ETC-1002 on serum low-density lipoprotein cholesterol in hypercholesterolemic patients receiving statin therapy. Am J Cardiol. 2016;117(12):1928-33.
8. Stein E, Bays H, Koren M, Bakker-Arkema R, Bisgaier C. Efficacy and safety of gemcabene as add-on to stable statin therapy in hypercholesterolemic patients. J Clin Lipidol. 2016;10(5):1212-22.
9. Julius U. Current Role of Lipoprotein Apheresis in the Treatment of High-Risk Patients. J Cardiovasc Dev Dis. 2018;5(2):27.
10. Buchwald H, Varco RL, Boen JR, Williams SE, Hansen BJ, Campos CT, et al. Effective lipid modification by partial ileal bypass reduced long-term coronary heart disease mortality and morbidity: five-year posttrial follow-up report from the POSCH. Program on the Surgical Control of the Hyperlipidemias. Arch Intern Med. 1998;158(11):1253-61.
11. Gaudet D, Méthot J, Kastelein J. Gene therapy for lipoprotein lipase deficiency. Curr Opin Lipidol. 2012;23(4):310-20.
12. Meyers CD, Amer A, Majumdar T, Chen J. Pharmacokinetics, pharmacodynamics, safety, and tolerability of pradigastat, a novel diacylglycerol acyltransferase 1 inhibitor in overweight or obese, but otherwise healthy human subjects. J Clin Pharmacol. 2015;55:1031-41.
13. Gaudet D, Digenio A, Alexander VJ, et al. The APPROACH study: a randomized double-blind placebo-controlled, phase 3 study of volanesorsen administered subcutaneously to patients with FCS. Athero Suppl. 2017;263:e10.
14. Gaudet D, Gipe DA, Pordy R, Ahmad Z, Cuchel M, Shah PK, et al. ANGPTL3 inhibition in homozygous familial hypercholesterolemia. N Engl J Med. 2017;377(3):296-7.
15. Bowman L, Hopewell JC, Chen F, Wallendszus K, Stevens W, Collins R, et al. Effects of anacetrapib in patients with atherosclerotic vascular disease. N Engl J Med. 2017;377(13):1217-27.
16. Nicholls SJ, Andrews J, Kastelein JJP, et al. Effect of serial infusions of CER-001, a pre-β high-density lipoprotein mimetic, on coronary atherosclerosis in patients following acute coronary syndromes in the CER-001 atherosclerosis regression acute coronary syndrome trial: a randomized clinical trial. JAMA Cardiol. 2018;3(9):815-22.
17. Abifadel M, Varret M, Rabès JP, Allard D, Ouguerram K, Devillers M, et al. Mutations in PCSK9 cause autosomal dominant hypercholesterolemia. Nat Genet. 2003;34(2):54-6.
18. Cohen JC, Boerwinkle E, Mosley TH Jr, Hobbs HH. Sequence variations in PCSK9, low LDL, and protection against coronary heart disease. N Engl J Med. 2006;354(12):1264-72.
19. Maxwell KN, Fisher EA, Breslow JL. Overexpression of PCSK9 accelerates the degradation of the LDLR in a post-endoplasmic reticulum compartment. Proc Natl Acad Sci USA. 2005;102(6):2069-74.
20. Rhainds D, Arsenault BJ, Tardif J-C. PCSK9 inhibition and LDL cholesterol lowering: the biology of an attractive therapeutic target and critical review of the latest clinical trials. Clinical Lipidology. 2012;7(6):621-40.
21. Ray KK, Landmesser U, Leiter LA, Kallend D, Dufour R, Karakas M, et al. Inclisiran in patients at high cardiovascular risk with elevated LDL cholesterol. N Engl J Med. 2017;376:1430-40.

22. Raal FJ, Kallend D, Ray KK, Turner T, Koenig W, Wright RS, et alJ. Inclisiran for the treatment of heterozygous familial hypercholesterolemia. N Engl J Med. 2020;382(16):1520-30.
23. Sabatine MS, Giugliano RP. Efficacy and safety of evolocumab in reducing lipids and cardiovascular events. N Engl J Med. 2015;372(16):1500-09.
24. Robinson JG, Farnier M. Efficacy and safety of alirocumab in reducing lipids and cardiovascular events. N Engl J Med. 2015;372(16):1489-99.
25. Raal FJ, Giugliano RP, Sabatine MS, et al. Reduction in lipoprotein(a) with the PCSK9 monoclonal antibody evolocumab (AMG 145): a pooled analysis of over 1300 patients in 4 phase 2 trials. J Am Coll Cardiol. 2014;63(13):1278-88.
26. Giugliano RP, Pedersen TR, Park JG, De Ferrari GM, Gaciong ZA, Ceska R, et al. Clinical efficacy and safety of achieving very low LDL-cholesterol concentrations with the PCSK9 inhibitor evolocumab: a prespecified secondary analysis of the FOURIER trial. Lancet. 2017;390(10106):1962-71.
27. Steg PG, Szarek M, Bhatt DL, et al. Effect of Alirocumab on Mortality After Acute Coronary Syndromes. Circulation. 2019;140(2):103.

Section 4

Dyslipidemia in Special Population

CHAPTER 26

Management of Dyslipidemia in Chronic Kidney Disease

Ipsita Ghosh

ABSTRACT
Chronic kidney disease (CKD) has been shown to be associated with cardiovascular disease (CVD) in multiple studies, which is the most common cause of morbidity and mortality across the globe and South Asians seem to have the highest rates of coronary artery disease (CAD). It has been documented in a meta-analysis involving 1.4 million that there is a linear increase in CV mortality with a decreasing estimated glomerular filtration rate (eGFR) <75/mL/min/1.73 m². In patients with CKD stage 3 and CKD stage 4, this increase in mortality is by two times and three times, respectively. They also have a higher incidence of myocardial infarction and chronic heart disease. Dyslipidemia is defined as abnormal levels of lipids in the blood that contributes to the development of atherosclerosis, a precursor for ischemic heart disease (IHD). Based on type of abnormality, it is further subclassified as isolated hypercholesterolemia if there is increase in cholesterol only, isolated hypertriglyceridemia if only triglycerides (TGs) are high, or mixed hyperlipidemias when both cholesterol and TGs are raised. Lipids were further classified into various lipoproteins (LP) based on their composition.

There is another term called "atherogenic dyslipidemia" characterized by elevated serum TG, elevated small low-density lipoprotein (LDL) particles, and reduced serum high-density lipoprotein (HDL) cholesterol. Dyslipidemia may be primary due to genetic causes or secondary to other underlying conditions such as diabetes. It is diagnosed by measuring plasma levels of total cholesterol, TGs, and individual lipoproteins. Various clinical practice guidelines such as National Cholesterol Education Program (NCEP), American Heart Association/American College of Cardiology (AHA/ACC), National Institute for Health and Clinical Excellence (NICE), European Society of Cardiology (ESC) and the European Atherosclerosis Society (EAS) have made recommendations for the treatment of elevated cholesterol levels.

Different populations of patients differ both in their presentation, in terms of their risk factors and in their response to treatment. The purpose of this review is to summarize current knowledge on the therapeutic challenge posed by dyslipidemia in patients suffering from CKD.

INTRODUCTION

Chronic kidney disease (CKD) is defined as a sustained reduction in kidney function or evidence of kidney damage present for 3 months or longer. Depending on kidney function and glomerular filtration rate (GFR), it is classified into five grades (**Table 1**).[1]

As CKD progresses, the risk for associated comorbidities such as hypertension, cardiovascular disease (CVD), dyslipidemia, anemia increases.[2] American Heart Association mentioned that CKD is an independent risk factor for development of CVD and "patients with CKD are more likely to die of CVD than because of kidney failure". Diabetes, high blood pressure, anemia, and inflammation are the risk factors for development of dyslipidemia and CVD in patients with CKD.[3] The causes of dyslipidemia in CKD are complex.[4-7] According to a new theory of prenatal programming, i.e., the intrauterine malnutrition causes obesity, dyslipidemia, type 2 diabetes mellitus (T2DM), and a deficit in nephron numbers, thus increasing in parallel cardiovascular (CV) and renal risks (**Table 2**).[8,9]

TABLE 1: Grades of chronic kidney disease (CKD) on the basis of glomerular filtration rate (GFR) (mL/min/1.73 m²).

Grade	GFR		Description of CKD
G1	≥90	Normal or high	Normal kidney function, but urine findings, structural abnormalities, or genetic trait point to kidney disease
G2	60–89	Mildly decreased	Mildly reduced kidney function and other findings (as for G1) point to kidney disease
G3A	45–59	Mildly to moderately decreased	Mild to moderately reduced kidney function
G3B	30–44	Moderately to severely decreased	Moderate to severely reduced kidney function
G4	15–29	Severely decreased	Severely reduced kidney function
G5	<15	Kidney failure	Very severe or end-stage renal disease (ESRD), sometimes called as dialysis established kidney failure

TABLE 2: Coevolution of cardiovascular and renal risk in individuals with intrauterine malnutrition.[12]

Prenatal programming leads to	
CV risk factors ↓	Nephron underdosing ↓
Target organ damage ↓	Microalbuminuria ↓
CV event, MI, stroke ↓	Macroproteinuria ↓
Congestive heart failure ↓	GFR loss ↓
CV death	End-stage renal disease

(CV: cardiovascular; GFR: glomerular filtration rate; MI: myocardial infarction)

Another contributor is impaired insulin resistance, a known component of metabolic syndrome and T2DM, both of which are associated with dyslipidemia and kidney disease.[10]

LIPID ABNORMALITIES IN CHRONIC KIDNEY DISEASE PATIENTS

Lipids were further classified into various lipoproteins based on their composition (**Table 3**). The main dyslipidemic disturbance in CKD is high triglycerides. It starts developing in early stages of the disease and is present in up to 70% of end-stage renal disease (ESRD) patients. Total cholesterol and low-density lipoprotein (LDL) cholesterol are usually in the normal range or low.[11] High-density lipoprotein (HDL) cholesterol is low in the majority of patients due to low activity of lecithin-cholesterol acyltransferase which normally increases the uptake of esterified cholesterol by HDL-C.[12] Due to diminished activity of lipoprotein-lipase and hepatic-lipase catabolism of lipoproteins and chylomicrons decreases and their intermediates such as chylomicron remnants, very LDL remnants as well as intermediate-density lipoproteins accumulate and their concentrations increase.[13]

EVIDENCE FROM LIPID LOWERING INTERVENTIONS

Cardiovascular disease risk is known to increase in patients with CKD, but its pathology may not always be related to atherothrombotic plaques. It has been found that structural heart disease with vascular calcification and stiffness along with overactive sympathetic nervous system are also common thus questioning the sustained meaningful risk reduction with lipid lowering medications.

Fibrates appear to be the medication of choice for the treatment of dyslipidemia in CKD as hypertriglyceridemia is the most common lipid abnormality. But, because of risk of myopathy and rhabdomyolysis they are not used so frequently.[14] In contrast, results from statins studies are mixed. Two trials named 4D (*Die Deutsche Diabetes Dialyse*) and *AURORA* (*A* study to evaluate the *U*se of *R*osuvastatin in subjects *O*n *R*egular hemodialysis: an *A*ssessment of survival and cardiovascular events) evaluated the benefits of statin therapy in patients with ESRD and concluded that despite a trend toward CV risk reduction, atorvastatin, and rosuvastatin did not have a statistically significant effects on CV death, nonfatal myocardial infarction (MI), and stroke in patients receiving hemodialysis.[15,16] On the other hand, *SHARP* (*S*tudy of *H*eart *A*nd *R*enal *P*rotection), a recent study in patients with moderate-to-severe CKD found that the combination of simvastatin/ezetimibe reduced the incidence of major atherosclerotic events.[17]

TABLE 3: Classification of lipoproteins on the basis of their composition.[3]	
Lipoproteins	Lipid composition
Chylomicrons	Triglycerides
Low-density lipoprotein	Cholesterol >>> Triglycerides
Intermediate-density lipoproteins	Cholesterol > Triglycerides
Very low-density lipoproteins	Triglycerides >>> Cholesterol
High-density lipoproteins	Good cholesterol >>> Triglyceride

GUIDELINES FOR LIPID MANAGEMENT IN PATIENTS WITH CHRONIC KIDNEY DISEASE[18-20]

Attention to lifestyle modifications are of utmost importance, though there is a trend favoring modest benefit with lipid lowering treatment in those without ESRD and not on HD. Important factors being, cessation of smoking, weight reduction, exercise, glucose control especially in patients with impaired renal function and lipid modulation.

NCEP ATP III (National Cholesterol Education Program Adult Treatment Panel III) guidelines: The guidelines opine that chronic renal failure can cause secondary dyslipidemia and elevated triglycerides so treatment should be considered in those with persistent dyslipidemia. It states that dyslipidemia is a common finding with stage 5 CKD, but since there is an increased risk of adverse drug-related reactions such as myopathy in CKD, lipid lowering drugs should be used with caution.[21,22]

KDIGO (Kidney Disease: Improving Global Outcomes) K/DOQI (Kidney Disease Outcomes Quality Initiative) guidelines (**Table 4**): Evidence-based algorithm for the management of dyslipidemia mainly according to CKD stage and it also asserts that all CKD patients should be evaluated for dyslipidemia.[23]

Given the risk of toxicity from high dose statins KDIGO recommends the use of moderate doses specifically in patients with eGFR <60 mL/min/m². The doses being atorvastatin 20 mg, rosuvastatin 10 mg, simvastatin 40 mg, pravastatin 40 mg, pitavastatin 2 mg, and fluvastatin 80 mg.

Other important guidelines are ACC/AHA 2013 (**Table 5**) and National Lipid Association (NLA) (**Table 6**).

TABLE 4: Pharmacological cholesterol-lowering treatment in adults.[23]		
Age	eGFR	Treatment
≥50 years	<60 mL/min/1.73 m² Not on HD or renal transplant	Statin or statin/ezetimibe combination
≥50 years	≥60 mL/min/1.73 m²	Statin
18–49 years	CKD but not treated with chronic dialysis or kidney transplantation	Statin recommended if: • Known coronary disease • DM • Prior ischemic stroke • Estimated 10-year incidence of coronary death or nonfatal MI >10%
Adults	Dialysis-dependent CKD	Statins or statin/ezetimibe
Adults	Dialysis-dependent CKD	Already receiving statins or statin/ezetimibe should be continued
Adults	Kidney transplant recipients	Statin
Adults	CKD but treated with chronic dialysis or kidney transplantation and hypertriglyceridemia	Therapeutic lifestyle modifications

(CKD: chronic kidney disease; DM: diabetes mellitus; K/DOQI: kidney disease outcomes quality initiative; KDIGO: Kidney Disease: Improving Global Outcomes; MI: myocardial infarction)

Management of Dyslipidemia in Chronic Kidney Disease

TABLE 5: American College of Cardiology/American Heart Association (ACC/AHA) 2013 guidelines.		
Adults with clinical ASCVD	1–5 not on HD	High intensity statin, if not then moderate intensity
Adults with LDL >190 mg/dL	1–5 not on HD	High intensity statin, if not then moderate intensity
Adults (40–75 years) with diabetes and LDL 70–189 mg/dL	1–5 not on HD	High intensity statin, if not then moderate intensity
Adults (40–75 years) with estimated 10 years ASCVD risk ≥7.5%	1–5 not on HD	High intensity statin, if not then moderate intensity
Adults with dialysis dependent-CKD	HD or PD	No recommendation

(ASCVD: atherosclerotic cardiovascular disease; CKD: chronic kidney disease; HD: hemodialysis; LDL: low-density lipoprotein; PD: peritoneal dialysis)

TABLE 6: National Lipid Association (NLA) guidelines.		
Adults with advanced CKD	3B or 4	Goal non-HDL <130 mg/dL; LDL <100 mg/dL
Adults with ASCVD or diabetes	1–4	Goal non-HDL <100 mg/dL; LDL <70 mg/dL
Adults with 0–1 major ASCVD risk factors	1–3A	• Start treatment when non-HDL >190 mg/dL; LDL >160 mg/dL • Goal non-HDL <130 mg/dL; LDL <100 mg/dL
Adults with two major ASCVD risk factors	1–3A	• Start treatment when non-HDL >160 mg/dL; LDL >130 mg/dL • Goal non-HDL <130 mg/dL; LDL <100 mg/dL
Adults with ≥3 major ASCVD risk factors	1–3A	• Start treatment when non-HDL > 130 mg/dL; LDL > 100 mg/dL • Goal non-HDL <130 mg/dL; LDL <100 mg/dL
Adults with stage 5 CKD or on HD	5	No recommendation

(ASCVD: atherosclerotic cardiovascular disease; CKD: chronic kidney disease: HD: hemodialysis; HDL: high-density lipoprotein; LDL: low-density lipoprotein)

National Institute for Health and Clinical Excellence 2014: Patients with eGFR <60 mL/min/1.73 m^2 and/or albuminuria are at increased risk of CVD, hence risk assessment tool should not be used. For both primary or secondary prevention of CVD 20 mg atorvastatin can be used and to increase the dose if >40% reduction in non-HDL cholesterol is not achieved with eGFR is 30 mL/min/1.73 m^2 or more and those with eGFR <30 mL/min/1.73 m^2 to consult with a renal specialist.[24]

CONCLUSION

Patients suffering from CKD are substantially at a higher risk of CVD and hence assessment of lipids and its treatment is important. There is a clear benefit of statin treatment in CKD stage 1-4 and renal transplant cases, but not in patients on dialysis.

KEY POINTS

- Inverse graded independent relation between eGFR and CVD risk.
- Dysregulation is mainly increased TG, oxidized LP, and reduced HDL.
- With declining eGFR, there is a small relative risk reduction of CV risk with statin therapy except dialysis patients.
- Declining renal function may lead to more deaths due to heart failure and arrhythmia than due to myocardial infarction—a unique phenotype.
- Novel therapies are expensive and have very little evidence to support its use.

REFERENCES

1. Levey AS, de Jong PE, Coresh J, El Nahas M, Astor BC, Matsushita K, et al. The definition, classification, and prognosis of chronic kidney disease: a KDIGO Controversies Conference report. Kidney Int. 2011;80(1):17-28.
2. K/DOQI Task Force. K/DOQI Clinical Practice Guidelines for Managing Dyslipidemias in Chronic Kidney Disease. Am J Kidney Dis. 2003;41(4 Suppl 3):S1-S92.
3. Sarnak MJ, Levey AS, Schoolwerth AC, Coresh J, Culleton B, Hamm LL, et al. Kidney disease as a risk factor for development of cardiovascular disease: a statement from the American Heart Association Councils on Kidney in Cardiovascular Disease, High Blood Pressure Research, Clinical Cardiology, and Epidemiology and Prevention. Circulation. 2003;108(17):2154-69.
4. Enas EA, Chacko V, Pazhoor SG, Chennikkara H, Devarapalli P. Dyslipidemia in South Asian Patients. Curr Atheroscler Rep. 2007;9(5):367-74.
5. Franssen R, Monajemi H, Stroes ES, Kastelein JJ. Obesity and Dyslipidemia. Med Clin North Am. 2011;95(5):893-902.
6. Fredrickson DS, Lees RS. A system for phenotyping hyperlipoproteinemia. Circulation. 1965;31:321-7.
7. Third Report of the National Cholesterol Education Program (NCEP) Expert Panel on Detection, Evaluation, and Treatment of High Blood Cholesterol in Adults (Adult Treatment Panel III). Circulation. 2002;106(25):3143-421.
8. Barker DJ, Bagby SP, Hanson MA. Mechanisms of disease: in utero programming in the pathogenesis of hypertension. Nat Clin Pract Nephrol. 2006;2:700-7.
9. Vikse BE, Irgens LM, Leivestad T, Hallan S, Iversen BM. Low birth weight increases risk for end-stage renal disease. J Am Soc Nephrol. 2008;19:151-7.
10. Krane V, Wanner C. The metabolic burden of diabetes and dyslipidaemia in chronic kidney disease. Nephrol Dial Transplant. 2002;17 (Suppl 11):23-7.
11. Chmielewski M, Carrero JJ, Nordfors L, Lindholm B, Stenvinkel P. Lipid disorders in chronic kidney disease: reverse epidemiology and therapeutic approach. J Nephrol. 2008;21:635-44.
12. Ikewaki K, Schaefer JR, Frischmann ME, Okubo K, Hosoya T, Mochizuki S, et al. Delayed in vivo catabolism of intermediate-density lipoprotein and low-density lipoprotein in hemodialysis patients as potential cause of premature atherosclerosis. Arterioscler Thromb Vasc Biol. 2005;25: 2615-22.
13. Weintraub M, Burstein A, Rassin T, Liron M, Ringel Y, Cabili S, et al. Severe defect in clearing postprandial chylomicron remnants in dialysis patients. Kidney Int. 1992;42:1247-52.
14. Harper CR, Jacobson TA. Managing dyslipidemia in chronic kidney disease. J Am Coll Cardiol. 2008;51:2375-84.
15. Wanner C, Krane V, März W, Olschewski M, Mann JF, Ruf G, et al. Atorvastatin in patients with type 2 diabetes mellitus undergoing hemodialysis. N Engl J Med. 2005;353(3):238-48.
16. Fellström BC, Jardine AG, Schmieder RE, Holdaas H, Bannister K, Beutler J, et al. Rosuvastatin and cardiovascular events in patients undergoing hemodialysis. N Engl J Med. 2009;360(14):1395-407.
17. Baigent C, Landray MJ, Reith C, et al. The effects of lowering LDL cholesterol with simvastatin plus ezetimibe in patients with chronic kidney disease (Study of Heart and Renal Protection): a randomised placebo-controlled trial. Lancet. 2011;377:2181-92.

18. Grundy SM, Cleeman JI, Merz CN, Brewer HB, Clark LT, Hunninghake DB, et al. Implications of recent clinical trials for the National Cholesterol Education Program Adult Treatment Panel III Guidelines. J Am Coll Cardiol. 2004;44:720-32.
19. National Institute for Health and Clinical Excellence. Clinical guideline 67: Lipid modification. London, 2008.
20. Reiner Z, Catapano AL, De Backer G, Graham I, Taskinen MR, Wiklund O, et al. ESC/EAS Guidelines for the management of dyslipidaemias: the Task Force for the management of dyslipidaemias of the European Society of Cardiology (ESC) and the European Atherosclerosis Society (EAS). Eur Heart J. 2011;32(14):1769-818.
21. National Cholesterol Education Program (NCEP) Expert Panel on Detection, 1. Evaluation and Treatment of High Blood Cholesterol in Adults (Adult Treatment Panel III), Third Report of the National Cholesterol Education Program (NCEP) Expert Panel on Detection, Evaluation, and Treatment of High Blood Cholesterol in Adults (Adult Treatment Panel III) final report. Circulation. 2002;106:3143-421.
22. Expert Panel on Detection, Evaluation, and Treatment of High Blood Cholesterol in Adults. Executive Summary of The Third Report of The National Cholesterol Education Program (NCEP) Expert Panel on Detection, Evaluation, and Treatment of High Blood Cholesterol in Adults (Adult Treatment Panel III). JAMA. 2001;285(19):2486-97.
23. K/DOQI Practice Guideline for Lipid Management in Chronic Kidney Disease. Kidney International Supplements. 2013;3:303-5.
24. National Clinical Guideline Centre (UK). Lipid modification: cardiovascular risk assessment and the modification of blood lipids for the primary and secondary prevention of cardiovascular disease. London: National Institute for Health and Care Excellence (UK); 2014.

CHAPTER 27

Dyslipidemia in Children and Adolescents: Screening, Diagnosis, and Management

Sayantan Ray

ABSTRACT

In the last few decades, it has become apparent that atherosclerosis starts in childhood. While cardiovascular disease (CVD) events are not common in children, studies have found the subclinical disease associated with measurable risk factors during childhood. As a result, the justification for considering measures to prevent CVD in childhood is compelling. In the United States, near 20% of youths (aged 6–19 years) have abnormal levels for at least one lipid parameter. Dyslipidemia in childhood has a large genetic component, but environmental factors such as diet and physical inactivity can also affect the lipid profile. Early detection and appropriate management of these children are essential to decrease CVD morbidity and mortality. Lipid screening during childhood and adolescence has been recommended by the National Heart, Lung, and Blood Institute's (NHLBI) Expert Panel for CVD risk reduction during childhood. In addition to screening children with a family history of hypercholesterolemia and/or premature CVD, the Expert Panel also advocated universal screening for lipid levels in children at 9–11 years of age, and that low-density lipoprotein cholesterol (LDL-C) values should be kept <110 mg/dL in children and adolescents. Statins are considered as first-line therapy because of their well-known efficacy to reduce LDL-C and improve additional lipid parameters in children. A particular concern with children is the safety issue; however, studies done to date have documented a reassuring side-effect profile of statins in children.

INTRODUCTION

In adults, dyslipidemia is a well-known risk factor for cardiovascular disease (CVD) and correction of dyslipidemia reduces the risk of CVD. Multiple studies demonstrate that CVD risk factors track from childhood into adulthood predisposing individuals to early atherosclerotic cardiovascular disease (ASCVD). Since abnormal lipid levels in children and adolescents play a role in the development of atherosclerosis, researchers have focused on whether screening lipid levels in youth could contribute to the prevention of adult CVD.[1,2] In particular, familial hypercholesterolemia (FH) links the effects of long-standing elevation of low-density lipoprotein cholesterol (LDL-C) in childhood to premature ASCVD in young adults. The most recent National Health and Nutrition Examination Surveys (NHANES) data has documented the prevalence of adverse levels of triglycerides (TG), high-density lipoprotein cholesterol (HDL-C), and LDL-C in children as 10.2%, 12.1%, and 6.4%, respectively.[3] In 2011, the National Heart, Lung, and Blood Institute (NHLBI) experts developed guidelines for screening, evaluation, and treatment of dyslipidemia in children and adolescents as part of a process to build-up an evidence-based, integrated set of guidelines that would cover all CVD risk factors in children.[2] The NHLBI guidelines have subsequently been endorsed by the American Academy of Pediatrics (AAP) and the AHA (American Heart Association). Publication of the NHLBI guidelines has resulted in a healthy debate, both pros and cons, on how best to prevent CVD starting in youth. The definition of dyslipidemia in children, when and how to evaluate and treat children with lipid disorders will be discussed here.

DEFINITION OF PEDIATRIC DYSLIPIDEMIA

After birth, total cholesterol (TC), LDL-C, and HDL-C levels increase during the first 2 years of life, followed by a plateau until the puberty. During puberty, both TC and LDL-C levels fall by 10% to 20% or more.[4] The NHLBI experts revised the lipid cut-off values considering US normative data and categorized them as acceptable, borderline, and high (**Table 1**). These definitions are also consistent with the 2008 AAP policy statement.[5] Nevertheless, it is important to note that these cut-offs have not been validated as accurate predictors for accelerated atherosclerosis or CVD events.[6]

In general, there are four classes of pediatric lipid disorders: Genetic lipid disorders, dyslipidemia associated with lifestyle factors, drug-related dyslipidemias, and dyslipidemias secondary to a medical condition such as hypothyroidism, nephrotic syndrome or type 2 diabetes mellitus (T2DM) or obesity. Lifestyle-induced lipid disorders are characterized by the triad of low HDL-C, elevated TG, and mildly raised LDL-C from an increased small, dense LDL particle numbers. Monogenic lipid disorders usually have more narrow abnormalities in lipid profiles.

TABLE 1: Definition of lipid levels in children and adolescents from the 2011 expert panel integrated guidelines.

Category	Acceptable (mg/dL)	Borderline (mg/dL)	Abnormal (mg/dL)
TC	<170	170–199	≥200
LDL-C	<110	110–129	≥130
Non-HDL-C	<120	120–144	≥145
Triglycerides 0–9 years 10–19 years	 <75 <90	 75–99 90–129	 ≥100 ≥130
HDL-C	>45	45–40	<40

Note: The threshold points for high and borderline-high values represent approximately the 95 and 75 percentiles, respectively. Low threshold points for HDL-C represent approximately the 10 percentile.

(HDL-C: high-density lipoprotein cholesterol; LDL-C: low-density lipoprotein cholesterol; TC: total cholesterol)

SCREENING

Childhood screening for lipid disorders is based on the rationale that early identification and management of dyslipidemia will lower the risk and severity of CVD in adult life.[2,7] Practicing a healthy lifestyle at a young age is the best way to avert premature CVD in individuals devoid of mutation in the LDL-C metabolic pathway; however, effective and safe interventions are there for the management of those at moderate to high risk, particularly children with genetic dyslipidemias such as FH. Lipid disorders are usually clinically silent and selective screening alone (i.e., screening children with a positive family history only) fails to identify a large number of children with lipid disorders.[8] Moreover, fixating on excess weight in screening will also miss a substantial proportion with abnormal lipids and specifically genetic lipid disorders. About 10 years of age (range, 9–11 years) has been proposed as a suitable time to undergo lipid checking.

Randomized controlled trials (RCTs) evaluating the long-term effectiveness of screening and treatment in childhood are lacking, and no data are available on the cost-effectiveness of lipid screening methods in the pediatric population. Nevertheless, after reviewing the existing literature, the 2011 United States NHLBI expert panel advocated the universal screening for lipid disorders in children before and after the onset of puberty and also recommended focused screening at other ages.[2] The recommendation for universal screening is based, in part, on the likelihood of identifying and treating the greatest number of individuals with FH, a group at high risk for significant morbidity and early mortality.[5] Nevertheless, universal screening remains controversial owing to a lack of data on whether the early diagnosis reduces future ASCVD events, the postdiagnosis stigmatization, and the cost involved in universal screening programs. Consequently, the 2016 US Preventative Task Force (USPTF) review concluded that there was inadequate evidence to recommend for or against universal lipid screening.[9]

Whom to screen?

Two categories of screening have been identified.[2] In the first category, targeted screening is recommended in any child over 2 years in whom one or both parents are known to have hypercholesterolemia or are taking lipid-lowering medications, who have a family history of premature CVD, and who have a moderate to high-risk for premature CVD. The second category (universal screening) is recommended for all children aged 9–11 years, regardless of general health or the presence/absence of CV risk factors. Screening to be repeated at 17–21 years of age.

What to order?

For screening, either a fasting or nonfasting lipid panel can be advised. If the individual is fasting, a standard lipid panel can be ordered (TC, TG, HDL-C, and LDL-C). The LDL-C is calculated using the Friedewald equation: LDL-C = TC − (HDL-C + TG/5). For individuals who are not fasting, a non-HDL-C (TC-HDL-C) is recommended as the first screening test because it is not affected by food, which often elevates TG or TG dependent values such as a calculated LDL-C. Ingestion of food or beverages has a small effect on directly measured LDL-C and the calculated non-HDL-C. When the non-HDL-C is above 145 mg/dL, two fasting lipid profiles should be obtained and the results averaged before determining the most appropriate intervention. Evaluation of secondary causes of dyslipidemia is also needed. Values of concern and appropriate actions to take are shown in **Table 2**.

TABLE 2: Interpreting the results of lipid screening in children.		
Initial screening	Fasting	Nonfasting
LDL-C, mg/dL	>130	–
Non-HDL-C, mg/dL	>145	>145
HDL-C, mg/dL	<40	<40
Triglycerides, mg/dL <10 years old >10 years old	 ≥100 ≥130	 – –
Action	• Repeat the fasting lipid profile after 2 weeks, but before 3 months • Average results	• Obtain a fasting lipid profile x 2 • Average results

(HDL-C: high-density lipoprotein cholesterol; LDL-C: low-density lipoprotein cholesterol)
Source: Expert Panel on Integrated Guidelines for Cardiovascular Health and Risk Reduction in Children and Adolescents, National Heart, Lung, and Blood Institute. Expert panel on integrated guidelines for cardiovascular health and risk reduction in children and adolescents: summary report. Pediatrics. 2011;128:S213-S256.

INTERVENTIONS

Instructions should be given for an age-appropriate diet and tips for a healthy lifestyle for all children and their families and, in some cases, this will include consultation with a dietitian. For children over 2 years of age, the primary beverage should be fat-free, unflavored milk. Older children should be advised to drink water and avoid sugar-sweetened beverages (SSBs). The NHLBI Cardiovascular Health Integrated Lifestyle Diet (CHILD-1) is suggested for all patients with dyslipidemia. With this diet, the total fat should be limited to 30% of total calories, with <10% saturated fats, up to 20% as mono- and polyunsaturated fats, while avoiding saturated and trans fats.[2] For patients with persistently elevated LDL-C or FH while on the CHILD-1 diet, the CHILD-2 LDL-C diet modification is recommended. The CHILD-2-LDL recommends plant sterol and stanol esters and water-soluble fiber psyllium. Similarly, for patients with persistently elevated TGs, while being on the CHILD-1 diet, the CHILD-2 TG diet modification is recommended which eliminates SSB and added sugars and focuses on lowering the dietary glycemic index. Children and adolescents should be encouraged the moderate and vigorous physical activity everyday. Importantly, lifestyle modifications can improve lipid levels without affecting weight immediately.[10]

Approach to Treatment

An algorithm, comprising the average of the child's two LDL-C values and the presence or absence of risk factors and conditions, is used to select patients who are appropriate for drug therapy (**Table 3** and **Flowchart 1**).[2,11] Adolescents with LDL-C levels >250 mg/dL and/or TGs > 500 mg/dL should be sent to a specialist. Besides, patients with lower LDL-C values with a significant history of premature CVD in family and any child for whom drug treatment may be indicated would likely benefit from referral.[2]

Secondary causes of dyslipidemia should be excluded and, if required, families should be referred to a dietitian for age-appropriate dietary instruction. Depending upon the type and severity of dyslipidemia, selected cases may require specific dietary strategies to reduce LDL-C or TGs.[2]

TABLE 3: Risk-factor definitions for dyslipidemia algorithms.

Parameter	Moderate level risk factors	High-level risk factors
BMI	≥95th to 96th percentile	≥97th percentile
BP	High BP without treatment	High BP with treatment
Smoker	–	Current smoker
HDL	HDL-C <40 mg/dL	–
Risk conditions	• Kawasaki disease with regressed coronary aneurysms • Chronic inflammatory diseases • Human immunodeficiency virus infection • Nephrotic syndrome	• Kawasaki disease with current coronary aneurysms • Types 1 and 2 diabetes mellitus • Postorthotopic heart transplant • Chronic renal disease/end-stage renal disease/postrenal transplant

(BMI: body mass index; BP: blood pressure; HDL-C: high-density lipoprotein cholesterol)

Overweight or obese children should be encouraged to attain and maintain a healthy weight, and all children should be instructed in obesity prevention by adopting a healthy lifestyle, including moderate to strenuous exercise for 60 minutes daily. Therapeutic lifestyle changes are particularly important for children who do not meet diagnostic criteria for being overweight/obese yet but have crossed body mass index (BMI) percentiles.

Drug Therapy for Raised Low-density Lipoprotein Cholesterol

When pharmacological therapy is indicated to lower LDL-C, a statin is the drug of choice.[2] Overall, statin therapy is not recommended below the age of 10 years unless there is a significant family history of premature CVD, the child has ≥1 high-risk conditions, or the child has multiple risk factors. Treatment should start with the lowest available dose of a statin administered once daily. All statins, except pitavastatin, have been approved by the United States Food and Drug Administration (FDA) for use in children with FH; pravastatin at 8 years and older, all others (simvastatin, fluvastatin, lovastatin, atorvastatin, and rosuvastatin) at 10 years and older. An overview of the statins currently approved by the FDA in children is provided in **Table 4**.[12]

In children with high LDL-C, genotype confirmation provides unequivocal evidence of FH and supports the need for early medical treatment.[2] Although high LDL-C values are also found in polygenic dyslipidemias such as familial combined hyperlipidemia (FCH), genotyping usually is not recommended in these cases. Individuals with dyslipidemia secondary to polygenic disorders also can get benefit from early medical treatment. However, pediatricians have been hesitant to initiate statin therapy for lifestyle-related dyslipidemias. Recent publications evaluating pharmacotherapy for heterozygous FH concluded that statin use decreased ASCVD events near that of the general population and the progression of atherosclerotic plaques was retarded.[13,14] The treatment goal is to attain an LDL-C <130 mg/dL, ideally <110 mg/dL.[2] If the target is not achieved, then the current statin dose may be increased by one increment (generally 10 mg), and liver function tests should be repeated in 4 weeks. If the LDL-C level remains above the target, the statin dose may be further increased by one increment. However, it should be kept in mind that

Dyslipidemia in Children and Adolescents: Screening, Diagnosis, and Management

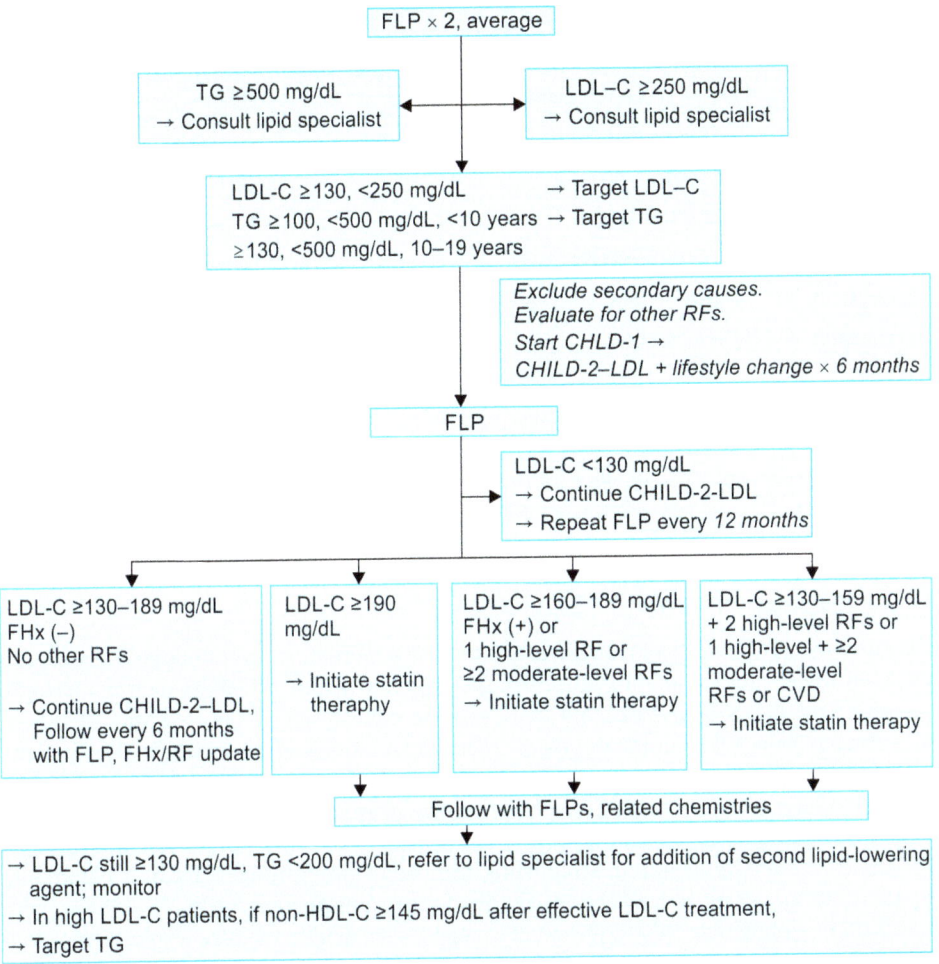

FLOWCHART 1: Dyslipidemia algorithm: target low density lipoprotein cholesterol (LDL-C).

Source: Expert Panel on Integrated Guidelines for Cardiovascular Health and Risk Reduction in Children and Adolescents, National Heart, Lung, and Blood Institute. Expert panel on integrated guidelines for cardiovascular health and risk reduction in children and adolescents: summary report. Pediatrics. 2011;128(Suppl 5):S213-S256.

each doubling of the statin dose only results in approximately a 6% reduction in LDL-C. If lifestyle modifications combined with a statin are unsuccessful, adding other lipid-lowering medications can be considered.[2] Ezetimibe, a cholesterol-absorption inhibitor, and colesevelam, a bile acid sequestrant, are FDA approved to lower LDL-C in children ≥10 years with FH. Proprotein convertase subtilisin/kexin type 9 (PCSK-9) inhibitors have come out as the most promising approach for treating severe hypercholesterolemia, including FH. Despite many published studies with these drugs, there are very little data concerning their effects on pediatric patients with FH.

TABLE 4: Food and Drug Administration (FDA)-approved statins.			
Statin	Licensed indication in children and adolescents	Initial daily dose	Maximum daily dose in children
Atorvastatin	10–17 years with FH	5–10 mg	20 mg/day
Rosuvastatin	10–17 years with FH	5 mg	20 mg/day
Simvastatin	10–17 years with FH	10 mg	40 mg/day
Lovastatin	10–17 years with FH	10 mg	40 mg/day
Fluvastatin	10–16 years with FH	20 mg	80 mg/day
Pravastatin	>8 years with FH	10 mg	• 8–13 years: 20 mg/day • 14–18 years: 40 mg/day
Pitavastatin	Not currently licensed in children or adolescents	NA	

(FH: familial hypercholesterolemia; NA: not applicable)
Source: Eiland LS, Luttrell PK. Use of statins for dyslipidemia in the pediatric population. J Pediatr Pharmacol Ther. 2010;15:160-72.

Drug Therapy for Raised Triglycerides

Children with TG levels >500 mg/dL are at risk for pancreatitis. Even though there are no FDA approved medications to prevent pancreatitis in youth <18 years with a strikingly elevated TG level (>500 mg/dL), the NHLBI Expert Panel suggested fibric acid derivatives, and omega-3 fatty acids could be considered in consultation with a specialist.[2] These agents mainly influence hepatic TG production. The markedly elevated TG levels in familial hyperchylomicronemia, however, are caused primarily by dietary-derived chylomicrons, which cannot be removed from the circulation without effective lipoprotein lipase (LPL) activity.[15] There is information supporting the use of preprandial orlistat, a pancreatic lipase inhibitor, along with a very low-fat diet for preventing pancreatitis in children with familial LPL deficiency.[16] The addition of omega-3 fatty acids to reduce modestly elevated TG levels can also be considered a potential treatment, although it is not yet approved for children. The effective dose has not yet been established.

Drug Therapy for Mixed Dyslipidemia

Although a major focus of the NHLBI guidelines is the detection and treatment of children with elevated LDL-C, especially FH and FCH, the most common dyslipidemia found in clinical practice is an elevated TG combined with a low HDL-C.[2] Even though this pattern of "atherogenic dyslipidemia" can be genetic, it is most commonly seen in children who are obese (BMI ≥95th percentile) and insulin-resistant (clinically manifest by the presence of acanthosis nigricans, prediabetes, hypertension; and in girls, polycystic ovary syndrome). Over the last few decades, dyslipidemia secondary to obesity has been increasing. However, it should be noted that obesity can exacerbate underlying genetic dyslipidemia and thereby accelerate the atherosclerosis process. The safest and most effective treatment for acquired dyslipidemia is lifestyle changes, primarily, encouraging healthy eating habits, moderate to vigorous physical activity, and achieving a healthier body weight (5–10% reduction in body weight). Avoiding the adverse consequences of

Dyslipidemia in Children and Adolescents: Screening, Diagnosis, and Management

smoking is particularly important in youth with CVD risk factors or conditions. Currently, no medical interventions are approved in children with low HDL-C, other than alleviating the typical CVD risk factors with lifestyle changes.

MONITORING

Table 5 lists the recommended monitoring of children following initiation of statin therapy. The child's risk factors and family history need to be updated frequently.

TABLE 5: Recommended monitoring.

Statin therapy	1st year				2nd year	Targets
	Baseline	1 month	2 months	Every 3–4 months	Every 6 months	
Laboratory monitoring						
Lipid profile	√	√	√	√		LDL-C
ALT	√	√	√	√		<3 times ULN
AST	√	√	√	√		<3 times ULN
CK	√	√				<10 times ULN
Physical examination						
Height	√	√	√	√	√	Assess other factors (weight gain, smoking, physical inactivity) regularly
Weight	√	√	√	√	√	
BMI	√	√	√	√	√	
BP	√	√	√	√	√	
Sexual maturation	√	√	√	√	√	
Medication Monitoring						
Compliance	√	√	√	√	√	Advise about potential medication interactions; advise women and girls about concerns about pregnancy and need for appropriate contraception
Tolerance	√	√	√	√	√	
Adverse effects	√	√	√	√	√	

Note: If laboratory abnormalities are noted or symptoms are reported, temporarily withhold medication and repeat blood work in 2 weeks. When abnormalities resolve, medication may be restarted, with close monitoring.

(ALT: alanine aminotransferase; AST: aspartate aminotransferase; BMI: body mass index; BP: blood pressure; CK: creatine kinase; LDL-C: low-density lipoprotein cholesterol; ULN: upper limit of normal)

Source: Expert Panel on Integrated Guidelines for Cardiovascular Health and Risk Reduction in Children and Adolescents, National Heart, Lung, and Blood Institute. Expert panel on integrated guidelines for cardiovascular health and risk reduction in children and adolescents: summary report. Pediatrics. 2011;128(Suppl 5):S213-S256.

CONCERNS AND CONTROVERSY

The modern practice of medicine is directed, whenever possible, by published evidence to ensure efficacy while avoiding the potential for harm. In the absence of published evidence, practitioners, in general, rely on consensus opinions from experts in the field. The NHLBI Expert Panel carefully and systematically examined the available evidence relevant to CV health and risk reduction in children and adolescents. Although RCTs were included in the Expert Panel reviews, particularly absent were the results of long-term RCTs on safety and efficacy. The latter is challenging in children because such trials take a long time to complete. Given the considerable beneficial effects of statin in decreasing CVD-related mortality in adults, withholding effective treatment in moderate- to high-risk children would be concerning to many healthcare providers.[17,18] Knowing these constraints, the "Expert Panel" extensively reviewed epidemiological, observational, and case-control studies in formulating its recommendations for youth. Only time and continued surveillance of the pediatric population at risk will let us know whether, these recommendations are flawed. In view of the serious consequences of unrecognized or insufficiently managed CVD risk factors as well as the acceptable safety profile of statins in children, such recommendations seem reasonable.

BENEFITS AND HARMS

Any screening program should depend on the assessment of benefits and harms and whether there are overall health gains. The screening for childhood dyslipidemia appears justified by the benefits of offsetting premature atherosclerosis[19] and preventing death.[18] This has been well-established for FH and the statin use for FH provides convincing data on the long-term safety of their use in children.[20,21] However, there are limited data on the effect of statins on coronary atherosclerosis in children presently. Although long-term data are absent, cost-effectiveness for FH screening has been documented in the Netherlands,[22] and screening family members of a proband was shown to be effective in the United Kingdom by increasing the detection rate and cost savings in long-term health.[23] The 2016 USPTF report also has acknowledged the importance of early detection of FH and multifactorial dyslipidemias.[9] No psychosocial effects were found in 8-10 years-old children during subsequent dietary treatment for hypercholesterolemia in the well-powered DISC (Dietary Intervention Study in Children).[24] Measures should be taken to ensure a balanced dietary intake.[2,25]

STATIN SAFETY

Cholesterol is an essential component of cell membranes and used as a precursor in the steroid hormones synthesis. Therefore, it is important in the growth and development of children, as well as neurological development.[12] Understandably then, there is a need for assessment of the long-term adverse effects of statins in children and adolescents. No published studies have suggested harm from statin treatment, and there are no data reports regarding adverse events in children.[26] Muscle and liver adverse effects are recognized as potential adverse effects of statin therapy and so need to be monitored along with other recognized side effects. Concerns that the effect of statins on decreasing cholesterol synthesis may affect brain development or function seem to be baseless according to results of the pediatric and adult trials and absence of effects in children with homozygous FH who typically start statin treatment at the age of 2 years. Numerous studies evaluating the efficacy of statins in FH also reported on the effect of statin therapy on sexual maturation. However, these studies failed to find a significant impact of statins on the sexual maturation of adolescents.[27,28] A recent Cochrane review of nine RCTs of statins in children demonstrated no difference between statin and placebo in liver enzyme levels, the occurrence of rhabdomyolysis, or sexual maturation.[29] Long-term follow-up is required to entirely explore any long-term effects on growth and sexual maturation. Drugs metabolized in common with statin pathways are to be avoided to prevent cross-reactions, and prescribing for sexually active girls should be performed along with pregnancy counseling and precautions.

CONCLUSION

Growing evidence suggests that childhood is a critical window in atherosclerosis development. Pediatric dyslipidemia contributes to early atherosclerosis and subsequently significantly increases the risk for ASCVD events earlier in life. Screening helps to detect children with dyslipidemia, enabling them to initiate an effective management program early to avoid the future risk of CVD. The current pediatric dyslipidemia guidelines represent the best evaluation and synthesis of evidence available. Considering the impact of premature CVD on the well-being of future generations and its cost to society, understanding and consistent implementation of such guidelines by primary care providers must be ensured. Doing so will create the opportunity of improving future recommendations. In the absence of high-quality RCT data, clinicians should make the best use of the available evidence to construct reasonable inferences that can help decision making for their patients and their families. Statins have long been used in the management of adult dyslipidemias, and quite a few have been approved by the FDA for use in children with FH. Statin treatment in children seems to be safe in the short-term, but long-term safety remains unknown. The results of trials with PCSK9 inhibitors in pediatric patients are awaited. Because this is such an important area from a clinical standpoint, the evidence continues to build-up and guidelines should undergo a process of re-evaluation.

KEY POINTS

- Dyslipidemia in childhood is one of the key modifiable precursors of adult CVD.
- Many levels of evidence support screening and early treatment of severe hypercholesterolemia, including FH.
- Familiarity with the management of lipid disorders in children has become increasingly necessary.
- The initial step in the management of children with dyslipidemias is targeted lifestyle modifications.
- Briefly, statins are effective in children with no major safety issues.

REFERENCES

1. Kavey RE, Allada V, Daniels SR, Hayman LL, McCrindle BW, Newburger JW, et al. Cardiovascular risk reduction in high-risk pediatric patients: a scientific statement from the American Heart Association Expert Panel on Population and Prevention Science; the Councils on Cardiovascular Disease in the Young, Epidemiology and Prevention, Nutrition, Physical Activity and Metabolism, High Blood Pressure Research, Cardiovascular Nursing, and the Kidney in Heart Disease; and the Interdisciplinary Working Group on Quality of Care and Outcomes Research: endorsed by the American Academy of Pediatrics. Circulation. 2006;114:2710-38.
2. Expert Panel on Integrated Guidelines for Cardiovascular Health and Risk Reduction in Children and Adolescents, National Heart, Lung, and Blood Institute. Expert panel on integrated guidelines for cardiovascular health and risk reduction in children and adolescents: summary report. Pediatrics. 2011;128:S213-S256.
3. Perak AM, NingH, Kit BK, de Ferranti SD, Van Horn LV, Wilkins JT, et al. Trends in levels of lipids and apolipoprotein B in US youths aged 6 to 19 years, 1999–2016. JAMA. 2019;321:1895-905.
4. Olson RE. Atherogenesis in children: implications for the prevention of atherosclerosis. Adv Pediatr. 2000;47:55-78.
5. Daniels SR, Greer FR. Committee on Nutrition. Lipid screening and cardiovascular health in childhood. Pediatrics. 2008;122:198-208.
6. Gidding SS. New cholesterol guidelines for children? Circulation. 2006;114:989-91.
7. Magnussen CG, Koskinen J, Chen W, Thomson R, Schmidt MD, Srinivasan SR, et al. Pediatric metabolic syndrome predicts adulthood metabolic syndrome, subclinical atherosclerosis, and type 2 diabetes mellitus but is no better than body mass index alone: the Bogalusa Heart Study and the Cardiovascular Risk in Young Finns Study. Circulation. 2010;122:1604-11.
8. Ritchie SK, Murphy EC-S, Ice C, Cottrell LA, Minor V, Elliott E, et al. Universal versus targeted blood cholesterol screening among youth: the CARDIAC Project. Pediatrics. 2010;126:260-5.
9. Bibbins-Domingo K, Grossman DC, Curry SJ, Davidson KW, Epling JW, García FAR, et al. Screening for lipid disorders in children and adolescents: US Preventive Services Task Force Recommendation Statement. JAMA. 2016;316:625-33.
10. Zachariah JP, Chan J, Mendelson MM, Regh T, Griggs S, Johnson PK, et al. Adolescent dyslipidemia and standardized lifestyle modification: benchmarking real-world practice. J Am Coll Cardiol. 2016;68:2122-3.
11. Wilson DP, McNeal C, Blackett P. Pediatric dyslipidemia: recommendations for clinical management. South Med J. 2015;108:7-14.
12. Eiland LS, Luttrell PK. Use of statins for dyslipidemia in the pediatric population. J Pediatr Pharmacol Ther. 2010;15:160-72.
13. Luirink IK, Wiegman A, Kusters DM, Hof MH, Groothoff JW, de Groot E, et al. 20-year follow-up of statins in children with familial hypercholesterolemia. N Engl J Med. 2019;381:1547-56.
14. Vallejo-Vaz AJ, De Marco M, Stevens CAT, Akram A, Freiberger T, et al. Overview of the current status of familial hypercholesterolaemia care in over 60 countries—The EAS Familial Hypercholesterolaemia Studies Collaboration (FHSC). Atherosclerosis. 2018;277:234-55.
15. Burnett JR, Hooper AJ, Hegele RA. Familial lipoprotein lipase deficiency. Gene Rev. 1999. [online] Available from https://www.ncbi.nlm.nih.gov/books/NBK1308/ [Last accessed July, 2020].
16. Blackett P, Tryggestad J, Krishnan S, Li S, Xu W, Alaupovic P, et al. Lipoprotein abnormalities in compound heterozygous lipoprotein lipase deficiency after treatment with a low-fat diet and orlistat. J Clin Lipidol. 2013;7:132-9.
17. Neil A, Cooper J, Betteridge J, Capps N, McDowell I, Durrington P, et al. Reductions in all-cause, cancer, and coronary mortality in statin-treated patients with heterozygous familial hypercholesterolaemia: a prospective registry study. Eur Heart J. 2008;29:2625-33.
18. Scientific Steering Committee on behalf of the Simon Broome Register Group. Mortality in treated heterozygous familial hypercholesterolaemia: implications for clinical management. Atherosclerosis. 1999;142:105-12.
19. Wiegman A, Hutten BA, de Groot E, Rodenburg J, Bakker HD, Büller HR, et al. Efficacy and safety of statin therapy in children with familial hypercholesterolemia: a randomized controlled trial. JAMA. 2004;292:331-7.
20. Avis HJ, Vissers MN, Stein EA, Wijburg FA, Trip MD, Kastelein JJP, et al. A systematic review and meta-analysis of statin therapy in children with familial hypercholesterolemia. Arterioscler Thromb Vasc Biol. 2007;27:1803-10.
21. Wiegman A, Gidding SS, Watts GF, Chapman MJ, Ginsberg HN, Cuchel M, et al. . Familial hypercholesterolaemia in children and adolescents: gaining decades of life by optimizing detection and treatment. Eur Heart J. 2015;36:2425-37.

22. Wonderling D, Umans-Eckenhausen MA, Marks D, Defesche JC, Kastelein JJP, Thorogood M. Cost effectiveness analysis of the genetic screening program for familial hypercholesterolemia in the Netherlands. Semin Vasc Med. 2004;4:97-104.
23. Marks D, Wonderling D, Thorogood M, Lambert H, Humphries SE, Neil HAW. Cost-effectiveness analysis of different approaches of screening for familial hypercholesterolaemia. BMJ. 2002;324:1303.
24. Lavigne JV, Brown KM, Gidding S, Evans M, Stevens VJ, von Almen TK, et al. A cholesterol-lowering diet does not produce adverse psychological effects in children: three-year results from the dietary intervention study in children. Health Psychol. 1999;18:604-13.
25. Lifshitz F, Tarim O. Considerations about dietary fat restrictions for children. J Nutr. 1996;126:1031S-1041S.
26. Vuorio A, Kuoppala J, Kovanen PT, Humphries SE, Tonstad S, Wiegman A, Drogari E. Statins for children with familial hypercholesterolemia. Cochrane Database Syst Rev. 2014;7:CD006401.
27. Avis HJ, Hutten BA, Gagné C, Langslet G, McCrindle BW, Wiegman A, et al. Efficacy and safety of rosuvastatin therapy for children with familial hypercholesterolemia. J Am Coll Cardiol. 2010;55:1121-6.
28. Braamskamp MJ, Langslet G, McCrindle BW, Cassiman D, Francis GA, Gagne C, et al. Efficacy and safety of rosuvastatin therapy in children and adolescents with familial hypercholesterolemia: Results from the CHARON study. J Clin Lipidol. 2015;9:741-50.
29. Vuorio A, Kuoppala J, Kovanen PT, Humphries SE, Tonstad S, Wiegman A, et al. Statins for children with familial hypercholesterolemia. Cochrane Database Syst Rev. 2017;7(7):CD006401.

CHAPTER 28

Dyslipidemia in Pregnancy

Sayantan Ray

ABSTRACT

Metabolic adaptation in pregnancy is linked with an increase in blood lipid levels that, to a certain degree, is considered physiological. Lipid parameters increase mostly in mid-gestation, particularly triglycerides (TGs) and cholesterol. In contrast, supraphysiological changes in lipid levels in predisposed women or with familiar forms of hyperlipidemia can carry bring an increased risk for maternal-fetal complications. Additionally, cardiometabolic dysfunction in mother may contribute to the long-term effects on vascular health of the mother and child. Current evidence underscores the importance of preventing or optimizing maternal dyslipidemias for the benefit of the mother and the child. However, presently, no reference standards defined for lipid parameters during pregnancy are available. Limited choices are there for effective lipid-lowering therapy during pregnancy. Increases in the usual age of pregnant women and the frequency of obesity have led to exposure to statins in certain women in the first trimester of pregnancy. The teratogenic risk associated with the statin use is uncertain because of the conflicting data, but statins remain contraindicated in pregnancy.

INTRODUCTION

Traditionally, dyslipidemia during pregnancy has not been thought to be pathophysiologically relevant. Lipids have not been regularly measured at any time point throughout pregnancy, in spite of their role in cardiovascular disease (CVD) or outcomes of pregnancy. Recent evidence relating enhanced fatty streak formation in the aortas of offspring born to hypercholesterolemic mothers[1] and animal studies have challenged the hypothesis that maternal cholesterol does not cross the placental barrier. Fetal exposure to high levels of cholesterol and oxidative byproducts of cholesterol metabolism has been shown to cause the programming of arterial cells with a tendency to atherosclerosis in later life. In parallel with obesity, the occurrence of dyslipidemia and the prescription of statins are increasing in women.[2] The current recommendation is to discontinue cholesterol-lowering treatments before conception. On the other hand, more than half of the pregnancies are unplanned,[3] and the number of women likely to continue statin through the first trimester of pregnancy is rising every year. Additionally, a therapeutic potential of statins in pre-eclampsia is reported in recent publications. Studies addressing the question of the teratogenicity of statins are urgently needed. This review provides an update on this progressively appreciated subject. The new insight that disorder of maternal cholesterol and triglyceride (TG) metabolism is possible to have a long-standing effect on future generations is reviewed. The conjectural benefits and supposed risks of statins in pregnancy are also discussed.

PREVALENCE OF CARDIOVASCULAR DISEASE RISK FACTORS

Prepregnancy cardiometabolic and inflammatory risk factors predict the risk of hypertensive disorders of pregnancy. Obese women have a greater risk of hypertension. Around 50% of pregnancies are unplanned, posing difficulty to identify CVD risk factors in women prior to pregnancy. Recently, a Kaiser Family Foundation national survey found that the CVD screening rate for women aged 18–44 years was much lower compared with women aged 45–64 years.[4] Moreover irrespective of the specialty providing the care, <70% of women underwent lipid screening, and nutrition counseling. In the last few decades, women in their 30s and 40s have experienced the greatest increase in the prevalence of obesity. Given the well-known connection between obesity and dyslipidemia, the implications are deep for this trend. Moreover, around 43% of pregnant women gain more than the recommended weight during pregnancy. For many women, the reproductive years can span two decades, representing an ideal time to reduce CVD risk factors prior to conception, for the benefit of both the mother and the offspring.

LIPIDS AND PREGNANCY: IMPORTANCE OF LIPIDS IN PLACENTAL AND FETAL DEVELOPMENT

In pregnancy, plasma TG levels increase 2- to 4-fold, and cholesterol increases by 25–50%.[5,6] Free fatty acids (FFAs), cholesterol, and cholesterol esters are the major lipids transported by the placenta. Low-density lipoprotein (LDL), which is involved in the plasma transport of TGs, is smaller, denser, and more prone to oxidation during pregnancy.[7] Fatty acids (FAs) function as an important source of energy for the fetus and placenta,[8,9] and are utilized in fetal and placental cell membrane formation and in the growth of fetal nerve cells with a precise balance between saturated, monounsaturated, and polyunsaturated FAs.[10-12] FAs are also used in the synthesis of prostaglandins, which play a vital role in parturition.[13] Cholesterol is essential for fetal and placental cell membrane building and steroid hormone synthesis.[14] Sources of cholesterol in fetus comprise of the maternal circulation, endogenous production, and synthesis within the placenta or yolk sac. The maternal circulation serves as an exogenous source of cholesterol [LDL and high-density lipoprotein (HDL)] for the fetus. Cholesterol from maternal circulation has been shown to enter the fetal circulation through placenta, contributing to a large extent to the fetal cholesterol pool in humans.[15,16] Maternal cholesterol crosses the trophoblast and the fetal vascular endothelium through the process of efflux and secretion. LDL-Rs and scavenger receptors (SCARs) are the two main families of receptors that enable cholesterol to enter the trophoblast cells.

The impact of the maternal cholesterol level on fetoplacental metabolism is not fully understood. In women with familial hypercholesterolemia (FH), hypercholesterolemia during pregnancy may be a significant risk factor for the fetus.[17] Fetal exposure to very high levels of cholesterol and its oxidative products has been found to cause programming of arterial cells and a predisposition to atherosclerosis in later life.[1] On the other hand, studies have also demonstrated adverse fetal effects because of decreased exogenous cholesterol. In Caucasian women, a greater incidence of prematurity has been observed with lower cholesterol levels.[18] Altered expression of LDL-R and SCARB1 in placenta is detected in intrauterine growth retardation (IUGR).[19]

Maternal Considerations

Although lipid and lipoprotein levels have been studied during pregnancy in women with uncomplicated and complicated pregnancies, but currently no reference standards for lipids during pregnancy exist. Pregnancy represents an insulin resistance state reflected by the maternal lipid profiles. Lipid levels fall slightly by 6 weeks of gestation, followed by a rise during each trimester of pregnancy. Generally, cholesterol and TG levels do not exceed 250 mg/dL.[20] Estrogens increase TG levels by stimulating very-low-density lipoprotein (VLDL) production in liver and inhibiting hepatic and adipose lipoprotein lipase. However, this physiologic increase in lipids is a mechanism to handle increasing fetal demands for normal growth and development.[21]

Abnormally high TG levels during the first trimester have been found to be strongly associated with gestational hypertension, pre-eclampsia, preterm birth, and large for gestational age fetuses.[22,23] Pre-eclampsia is characterized by endothelial dysfunction provoked by raised TG and FFA levels. TG levels together with ApoB and small LDL particles are elevated in pre-eclampsia. However, it is currently uncertain whether ApoB or small LDL particles cause this endothelial dysfunction.[24] Conditions that result in abnormal lipid levels should be investigated and, if found, treated appropriately. One of the more common causes for high TG levels during pregnancy is medication use. Ideally, offending drugs should be identified and discontinued before conception. Elevated VLDL and chylomicron levels are thought to be secondary to a genetic predisposition. TG levels are typically very high, increasing the risk of acute pancreatitis. In severe hypertriglyceridemia, a reduction in fat calories to 15–20% daily is usually required. Even in the absence of overt diabetes, insulin therapy may be used in this context. Insulin mediated increase in lipoprotein lipase activity accelerates chylomicron degradation.

All lipid-lowering medications, with the exception of bile sequestrant and omega-3-fatty acids, should be discontinued before conception or promptly when pregnancy happens unexpectedly. Lifestyle modifications and glycemic control should be started where required. In pregnancy, elevated cholesterol levels can be safely treated with a bile acid sequestrant. The actual risk of congenital anomalies attributable to statin use is not well substantiated in human pregnancies. Nevertheless, until more information is available statin use in pregnancy should be conducted only in a research setting.[25] Severe hypertriglyceridemia induced pancreatitis can be treated with omega-3 fatty acids, parenteral nutrition, and other lipid-lowering agents, particularly gemfibrozil during the last trimester of pregnancy. Monitoring is recommended within 6 weeks of initiating treatment. Therapeutic plasma exchange (TPE) may be utilized for an urgent need of a rapid and effective lowering of TG levels to prevent a severe pancreatitis episode.[26] In pregnant women with FH, a lipid profile should be obtained before conception and each trimester. FH can be treated with bile acid sequestrants, preferably colesevelam. Ultimately, mipomersen and LDL apheresis may be required in pregnancy. Management should be done in a specialized center where facilities are available. It is recommended that the mother with FH or with dysmetabolic problems should be followed-up closely.

A detailed understanding of medication safety in pregnancy and lactation is necessary to ensure maternal and fetal safety. At present, statins are classified as category X, whereas ezetimibe, fibrates, niacin, cholestyramine, and omega-3 are category C. Colesevelam and mipomersen are category B.

Fetal Considerations

Since gestational dyslipidemia has traditionally been considered physiologic, with little clinical relevance, lipid and lipoproteins are not measured usually during pregnancy. However, the detection of fatty streaks in the aortas of 6-month-old fetuses of hypercholesterolemic mothers, and the finding of aortic atherosclerosis at autopsy of deceased children with normal cholesterol levels born to mothers with hypercholesterolemia, underscore the importance of correcting maternal dyslipidemia.[1,27]

Atherosclerosis is the foremost of several conditions for which a role of developmental programming was described. A number of factors may contribute to the developmental programming of the fetus.[28] In pregnancy, maternal nutrition, maternal stress, preeclampsia, hypertension, gestational diabetes, maternal smoking, obesity, dyslipidemia, metabolic syndrome and placental function, may be important influences. Although uncertain, growing evidence has shown that epigenetic programming of metabolism during embryonic or fetal development might be implicated.[29] For example, in ApoE-deficient mice maternal hypercholesterolemia causes the activation of genes involved in the process of cholesterol synthesis and LDL-R activity in adult offspring.[30] Quite a few animal studies have found that the genes involved in immune pathways and FA metabolism are unregulated in the offspring of hypercholesterolemic dams.[31] These findings indicate that an unfavorable maternal environment may alter the basic cellular programming of the fetus. Further work is needed to decipher the precise mechanisms through which maternal hypercholesterolemia changes this process.

In utero, the fetal handling of lipid metabolism occurs in a dynamic fashion. It is well-known that lipids are actively transported to the fetus and this transport appears to be variable depending on the different stages of pregnancy. In early gestation, the fetus seems to preferentially use lipids for the proper membrane formation and probably for protection. Subsequently, the excess fat may be deposited in the liver, depending on the gestational age and the liver maturity. Besides, fetal epicardial fat can be recognized early in gestation. Apparently, these mechanisms take place in an attempt to protect the fetal brain. The offspring of obese mothers carry an elevated risk of mortality in later life. Long-term studies have demonstrated that offspring of mothers with a greater BMI and waist circumference have higher TG levels and blood pressure and increased insulin resistance.[32]

STATINS AND PREGNANCY

3-hydroxy-3-methyl-glutaryl-coenzyme A reductase (HMG-CoA reductase) activity is essential for the normal placental development in mammals. Inhibition of HMG-CoA reductase by statin may disrupt membrane development, cellular proliferation and growth, metabolism, and protein glycosylation, which are crucial for normal development of placenta.[33] Kenis et al.[34] concluded that simvastatin induced dysfunction of placenta in the first trimester could explain the rise in miscarriages found in animal models. Forbes et al.[35] support these observations by demonstrating that pravastatin and cerivastatin inhibit the proliferative response of the human cytotrophoblast to IGF-I and -II with other in vitro findings. Some statin effects are mediated by the activation of peroxisome proliferator-activated receptors (PPARs). PPARγ is bound and activated by many natural ligands including polyunsaturated FAs, eicosanoids, and oxidized LDL.[36] The heterodimer PPARγ/retinoid X receptor alpha (RXRα) is predominantly expressed in human trophoblasts and plays a key role in the differentiation of the human trophoblast toward the villous and the extravillous differentiation pathways. Its activation in the villous trophoblast stimulates the expression and secretion of the hormones required in pregnancy and for development of fetus.[37] Statins increase PPARγ expression in human macrophages and in murine endothelium. No animal study on the effects of statin on placental PPAR receptors has yet been published.

In rats, lovastatin administration at toxic doses for the dam (800 mg/kg/day) results in gastroschisis in the fetus, along with bone malformations, mainly in the vertebrae, ribs, and sternum.[38] An Food and Drug Administration (FDA) report supported these results affirming that fluvastatin in the pregnant rat causes fetal and bone malformations. In contrast, Tanase and Hirose,[39] reported that pravastatin 500 and 1,000 mg/kg administered to pregnant rats during organogenesis had no teratogenic effect. Dostal et al.[40] investigated the effects of incremental doses of atorvastatin (from 25 to 200 mg/kg) on the pregnant rat and fetus between day 6 and day 15 of gestation. The authors concluded that atorvastatin has no teratogenic effect in gestating rats when given at doses potentially toxic for the dam.

In 2004, Edison and Muenke[41] selected 52 of 178 cases of statin exposure in the first trimester of pregnancy after reviewing FDA data from 1987 to 2001. Fetal malformations were noted in 20 of these 52 cases, including 5 severe central nervous system malformations, one of which was holoprosencephaly. One case of VACTERL syndrome was described after lovastatin exposure between 6th and 11th week of gestation. In these 20 cases, all prescribed statins were lipophilic. No malformation was detected in the 14 pregnancies exposed to pravastatin (hydrophilic). Since the cases of holoprosencephaly and VACTERL syndrome were possibly linked to a defective cholesterol-dependent Sonic Hedgehog pathway, a pathophysiological mechanism is hypothesized by the authors, in which statins disrupt this signaling pathway through an effect on the fetal cholesterol metabolism.[42] This interpretation has been questioned and some authors suggest there is no established causal link between antenatal statin exposure and perturbation of the Sonic Hedgehog pathway.[43] No cases of Smith–Lemli–Opitz syndrome, caused by defective cholesterol biosynthesis in the fetus, were reported by Edison and Muenke. With three databases containing records from 1997 to 2003, Ofori et al.[44] investigated 153 cases of statin exposure during pregnancy. Statins administered in the first trimester of the three pregnancies complicated by malformations were lipophilic (lovastatin, simvastatin, and atorvastatin). No malformation was noted among the 11 pregnancies with exposure to pravastatin. No cases of holoprosencephaly or VACTERL syndrome were reported. Postmarketing surveillance of lovastatin and simvastatin showed no connection between early pregnancy exposure to statin and adverse pregnancy outcomes.[45] The Merck pharmacological vigilance database follow-up reports found that the rate of congenital anomalies in the exposed group (3.8%) was similar to the background population rate of 3%, and no specific patterns of anomalies were detected.[46] Nevertheless, lovastatin maintained X categorization, mainly because of a lack of indication to use it in pregnancy. The same designation has been given to subsequent statins, regardless of considerable differences in biochemical properties and lack of properly controlled studies to assess human teratogenicity. Recent epidemiologic data as well do not support teratogenicity concerns of statins in general and for pravastatin in particular.[47,48] Transplacental transfer of pravastatin is negligible.

Pre-eclampsia is known to be a multisystem disorder and theoretically, the pleiotropic effects of statins may be helpful in its treatment. Growing experimental evidence is there supporting the thought that statins could be of use in management of pre-eclampsia. Given the potential of statin therapy for women with high-risk for pre-eclampsia, as well as the reassuring safety profile, the Obstetric–Fetal Pharmacology Research Unit Network has initiated a pilot, prospective, randomized trial of pravastatin for prevention of pre-eclampsia (Clinicaltrials.gov Identifier NCT01717586).

POSTPARTUM CONSIDERATIONS

During the postpartum period, follow-up of women with dyslipidemia includes close observation, mostly for those who experienced pre-eclampsia and/or diabetes. Moreover, a history of pre-eclampsia doubles the long-term CVD risk in the mother.[49] A history of gestational diabetes increases the 10-year risk for developing overt type 2 diabetes mellitus to nearly 40%. The prevalence of significant and treatable dyslipidemia is approximately one-third in these populations.[50] The understanding of maternal dyslipidemia and its effect on the future health and well-being of both the mother and her offspring is incomplete. However, mounting evidence suggests that clinicians should be more cautious in assessment and treatment of CVD risk factors during pregnancy.[51]

CONCLUSION

Considering the increasing prevalence of risk conditions for the development of lipid abnormalities in reproductive age women, routine detection plus appropriate and secure management of dyslipidemia before, during, and after pregnancy may provide a good opportunity to reduce the rising atherosclerotic burden of the growing population. Indications for statins are increasing and their use in women of childbearing age is also becoming more frequent. These women could benefit from ongoing statins, but due to the concerns about statin safety during pregnancy, the recommendations are to discontinue the treatment. However, several diseases like FH are coupled with a high-risk of death as a result of CVD. Therefore, large studies are needed to determine the effects of hypercholesterolemia during pregnancy on CVD risk and the benefit of continuing statins during pregnancy in such high-risk patients. Albeit their widespread use in adults at risk for CVD, and even though statins have a known pharmacokinetic properties and favorable safety profile the data on the maternal and fetal safety after exposure during pregnancy are inadequate. The teratogenic risk related to the use of statin is imprecise because the existing data are conflicting but, statins remain contraindicated in pregnancy. Studies assessing the transplacental passage, benefits and safety of statins during pregnancy are urgently needed. Nevertheless, altered maternal-fetal pharmacokinetics and pharmacodynamics, issues related to fetal risk, along with regulatory challenges are there and have to be diligently considered in the progress of research in this field.

> **KEY POINTS**
>
> - Lipid disorders commonly associated with pregnancy are two major clinical disorders: Severe hypertriglyceridemia is a potent risk factor for acute pancreatitis and elevated cholesterol levels promote atherosclerosis.
> - Dyslipidemia during pregnancy is found to be associated with pre-eclampsia, preterm birth, and gestational diabetes, and the offspring of these mothers show an increased risk of atherosclerosis progression.
> - Routine assessment of supraphysiological changes in lipid levels will identify high-risk pregnancies for maternal and fetal after-effects which require more rigorous follow-up and care.
> - Drug-based lipid management strategies will only be applicable if maternal safety during pregnancy and fetal safety; both short and long-term—are first well understood.
> - Dyslipidemia in pregnancy has been undervalued and underevaluated in basic and clinical research. Nevertheless, this situation is changing and the field is expanding.

REFERENCES

1. Napoli C, D'Armiento FP, Mancini FP, Mancini FP, Balestrieri ML, Della Ragione F, et al. Fatty streak formation occurs in human fetal aortas and is greatly enhanced by maternal hypercholesterolemia. Intimal accumulation of low-density lipoprotein and its oxidation precede monocyte recruitment into early atherosclerotic lesions. J Clin Invest. 1997;100:2680-90.
2. Ford ES, Li C, Pearson WS, Zhao G, Mokdad AH. Trends in hypercholesterolemia, treatment and control among United States adults. Int J Cardiol. 2008;140:226-35.
3. Henshaw SK. Unintended pregnancy in the United States. Fam Plann Perspect. 1998;30:24-9.
4. Salganicoff A, Ranji U, Beamesderfer A, Kurani N. Women and Health Care in the Early Years of the ACA: Key Findings from the 2013 Kaiser Women's Health Survey – Preventive Services – 8590. California: KFF; 2014.
5. Alvarez JJ, Montelongo A, Iglesias A, Lasunción MA, Herrera E. Longitudinal study on lipoprotein profile, high-density lipoprotein subclass, and post heparin lipases during gestation in women. J Lipid Res. 1996;37:299-308.
6. Piechota W, Staszewski A. Reference ranges of lipids and apolipoproteins in pregnancy. Eur J Obstet Gynecol Reprod Biol. 1992;45:27-35.
7. Winkler K, Wetzka B, Hoffmann MM, Friedrich I, Kinner M, Baumstark MW, et al. Low-density lipoprotein (LDL) subfractions during pregnancy: accumulation of buoyant LDL with advancing gestation. J Clin Endocrinol Metab. 2000;85:4543-50.
8. Hornstra G, Al MD, van Houwelingen AC, Drongelen MM. Essential fatty acids in pregnancy and early human development. Eur J Obstet Gynecol Reprod Biol. 1995;61:57-62.
9. Shekhawat P, Bennett MJ, Sadovsky Y, Nelson DM, Rakheja D, Strauss AW. Human placenta metabolizes fatty acids: implications for fetal fatty acid oxidation disorders and maternal liver diseases. Am J Physiol Endocrinol Metab. 2003;284:E1098-105.
10. Neuringer M, Connor WE. n-3 fatty acids in the brain and retina: evidence for their essentiality. Nutr Rev. 1986;44:285-94.
11. Bourre JM, Durand G, Pascal G, Youyou A. Brain cell and tissue recovery in rats made deficient in n-3 fatty acids by alteration of dietary fat. J Nutr. 1989;119:15-22.
12. Crawford MA, Doyle W, Drury P, Lennon A, Costeloe K, Leighfield M. n-6 and n-3 fatty acids during early human development. J Intern Med Suppl. 1989;731:159-69.
13. Allen KG, Harris MA. The role of n-3 fatty acids in gestation and parturition. Exp Biol Med (Maywood). 2001;226:498-506.
14. Woollett LA. Where does fetal and embryonic cholesterol originate and what does it do? Annu Rev Nutr. 2008;28:97-114.
15. Woollett LA. Review: transport of maternal cholesterol to the fetal circulation. Placenta. 2011;32(Suppl 2):218-21.

16. Yoshida S, Wada Y. Transfer of maternal cholesterol to embryo and fetus in pregnant mice. J Lipid Res. 2005;46:2168-74.
17. Narverud I, Iversen PO, Aukrust P, Halvorsen B, Ueland T, Johansen SG, et al. Maternal familial hypercholesterolaemia (FH) confers altered haemostatic profile in offspring with and without FH. Thromb Res. 2013;131:178-82.
18. Edison RJ, Muenke M. Gestational exposure to lovastatin followed by cardiac malformation misclassified as holoprosencephaly [erratum]. N Engl J Med. 2005;352:2759.
19. Wadsack C, Tabano S, Maier A, Hiden U, Alvino G, Cozzi V, et al. Intrauterine growth restriction is associated with alterations in placental lipoprotein receptors and maternal lipoprotein composition. Am J Physiol Endocrinol Metab. 2007;292:E476-84.
20. Potter JM, Nestel PJ. The hyperlipidemia of pregnancy in normal and complicated pregnancies. Am J Obstet Gynecol. 1979;133:165-70.
21. Herrera E. Metabolic adaptations in pregnancy and their implications for the availability of substrates to the fetus. Eur J Clin Nutr. 2000;54(Suppl 1):S47-51.
22. Wiznitzer A, Mayer A, Novack V, Sheiner E, Gilutz H, Malhotra A, et al. Association of lipid levels during gestation with preeclampsia and gestational diabetes mellitus: a population-based study. Am J Obstet Gynecol. 2009;20:482-8.
23. Barbour LA, Hernandez TL. Maternal Non-glycemic Contributors to Fetal Growth in Obesity and Gestational Diabetes: spotlight on Lipids. Curr Diab Rep. 2018;18(6):37.
24. Hubel CA, Roberts JM, Taylor RN, Musci TJ, Rogers GM, McLaughlin MK. Lipid peroxidation in pregnancy: new perspectives on preeclampsia. Am J Obstet Gynecol. 1989;161:1025-34.
25. Cleary KL, Roney K, Costantine M. Challenges of studying drugs in pregnancy for off-label indications: pravastatin for preeclampsia prevention. Semin Perinatol. 2014;38:523-7.
26. Russi G. Severe dyslipidemia in pregnancy: the role of therapeutic apheresis. Transfus Apher Sci. 2015;53:283-7.
27. Napoli C, Glass CK, Witztum JL, Deutsch R, D'Armiento FP, Palinski W. Influence of maternal hypercholesterolaemia during pregnancy on progression of early atherosclerotic lesions in childhood: fate of Early Lesions in Children (FELIC) study. Lancet. 1999;354:1234-41.
28. Hanson M, Godfrey KM, Lillycrop KA, Burdge GC, Gluckman PD. Developmental plasticity and developmental origins of non-communicable disease: theoretical considerations and epigenetic mechanisms. Prog Biophys Mol Biol. 2011;106:272-80.
29. DeRuiter MC, Alkemade FE, Gittenberger-de Groot AC, Poelmann RE, Havekes LM, van Dijk KW. Maternal transmission of risk for atherosclerosis. Curr Opin Lipidol. 2008;19:333-7.
30. Reymer PW, Groenemeyer BE, van de Burg R, Kastelein JJ. Apolipoprotein E genotyping on agarose gels. Clin Chem. 1995;41:1046-7.
31. Palinski W. Maternal-fetal cholesterol transport in the placenta: good, bad, and target for modulation. Circ Res. 2009;104:569-71.
32. Hochner H, Friedlander Y, Calderon-Margalit R, Meiner V, Sagy Y, Avgil-Tsadok M, et al. Associations of maternal prepregnancy body mass index and gestational weight gain with adult offspring cardiometabolic risk factors: the Jerusalem Perinatal Family Follow-up Study. Circulation. 2012;125:1381-9.
33. Kazmin A, Garcia-Bournissen F, Koren G. Risks of statin use during pregnancy: a systematic review. J Obstet Gynaecol Can. 2007;29:906-8.
34. Kenis I, Tartakover-Matalong S, Cherepnin N, Drucker L, Fishman A, Pomeranz M, et al. Simvastatin has deleterious effects on human first trimester placental explants. Hum Reprod. 2005;20:2866-72.
35. Forbes K, Hurst LM, Aplin JD, Aplin JA, Westwooda M. Statins are detrimental to human placental development and function; use of statins during early pregnancy is inadvisable. J Cell Mol Med. 2008;12:2295-6.
36. Fournier T, Guibourdenche J, Handschuh K, Tsatsaris V, Rauwel B, Davrinche C, et al. PPARgamma and human trophoblast differentiation. J Reprod Immunol. 2011;90:41-9.
37. Tarrade A, Schoonjans K, Guibourdenche J, Bidart JM, Vidaud M, Auwerx J, et al. PPAR gamma/RXR alpha heterodimers are involved in human CG beta synthesis and human trophoblast differentiation. Endocrinology. 2001;142:4504-14.
38. Minsker DH, MacDonald JS, Robertson RT, Bokelman DL. Mevalonate supplementation in pregnant rats suppresses the teratogenicity of mevinolinic acid, an inhibitor of 3-hydroxy-3-methylglutaryl-coenzyme a reductase. Teratology. 1983;28:449-56.

39. Tanase TM, Hirose K. Reproduction study of pravastatin sodium administered during the period of fetal organogenesis in rats. Jpn Pharmacol Ther. 1987;15:4983-94.
40. Dostal LA, Schardein JL, Anderson JA. Developmental toxicity of the HMG-CoA reductase inhibitor, atorvastatin, in rats and rabbits. Teratology. 1994;50:387-94.
41. Edison RJ, Muenke M. Central nervous system and limb anomalies in case reports of first-trimester statin exposure. N Engl J Med. 2004;350:1579-82.
42. Edison RJ, Muenke M. Mechanistic and epidemiologic considerations in the evaluation of adverse birth outcomes following gestational exposure to statins. Am J Med Genet A. 2004;131:287-98.
43. Gibb H, Scialli AR. Statin drugs and congenital anomalies. Am J Med Genet A. 2005;135:230-1; author reply 232-4.
44. Ofori B, Rey E, Berard A. Risk of congenital anomalies in pregnant users of statin drugs. Br J Clin Pharmacol. 2007;64:496-509.
45. Manson JM, Freyssinges C, Ducrocq MB, Stephenson WP. Postmarketing surveillance of lovastatin and simvastatin exposure during pregnancy. Reprod Toxicol. 1996;10:439-46.
46. Pollack PS, Shields KE, Burnett DM, Osborne MJ, Cunningham ML, Stepanavage ME. Pregnancy outcomes after maternal exposure to simvastatin and lovastatin. Birth Defects Res A Clin Mol Teratol. 2005;73:888-96.
47. Petersen EE, Mitchell AA, Carey JC, Werler MM, LouikC, Rasmussen SA. Maternal exposure to statins and risk for birth defects: a case-series approach. Am J Med Genet A. 2008;146A:2701-5.
48. Taguchi N, Rubin ET, Hosokawa A, Choi J, Ying AY, Moretti ME, et al. Prenatal exposure to HMG-CoA reductase inhibitors: effects on fetal and neonatal outcomes. Reprod Toxicol. 2008;26:175-7.
49. Brown MC, Best KE, Pearce MS, Waugh J, Robson SC, Bell R. Cardiovascular disease risk in women with pre-eclampsia: systematic review and meta-analysis. Eur J Epidemiol. 2013;28:1-19.
50. Nerenberg K, Daskalopoulou SS, Dasgupta K. Gestational diabetes and hypertensive disorders of pregnancy as vascular risk signals: an overview and grading of the evidence. Can J Cardiol. 2014;30:765-73.
51. Sattar N, Greer IA. Pregnancy complications and maternal cardiovascular risk: opportunities for intervention and screening? BMJ. 2002;325:157-60.

CHAPTER

29

Drug Treatment of Dyslipidemia in Older Adults

Ruby Raphael, Beatrice Anne

ABSTRACT

Statin therapy in elderly and older adults is still a clinical dilemma. As the age advances atherosclerotic changes in the vessel wall occurs. It often leads to increased risk of cardio vascular disease (CVD) and other conditions like dementia and frailty. Few epidemiological studies have shown that low cholesterol in old age is associated with worse prognosis which further results in concerns in the management. Due to lack of the evidence regarding the effect of statins in older adults for primary prevention, this often results in low adherence and discontinuation. This is a brief review on the present evidence and recommendations in management of dyslipidemia in the elderly.

INTRODUCTION

As the mean life expectancy is increasing, there is a parallel increase in percentage of elderly population in the society.[1] Studies have shown that elderly population are at a higher risk of cardiovascular events. The newer cardiovascular disease (CVD) events and most of the coronary deaths occur in elderly people (≥65 years). An elevated low-density lipoprotein cholesterol (LDL-C) and low high-density lipoprotein cholesterol (HDL-C) can be predictive as well as major risk factor for the development of coronary artery disease (CAD) especially in elderly population.[2] In a cardiovascular health study, among 5,201 adults aged >65 years or older, 3,584 individuals without clinical CAD were found to have subclinical CVD.[3,4] Several randomized controlled trials (RCTs) have proved that reduction in serum cholesterol can reduce cardiovascular morbidity and mortality.[3,5] Only few subjects aged >75 years were included in the RCTs reviewed and the data regarding use of statins in older adults for primary prevention in atherosclerotic cardiovascular disease (ASCVD) are sparse.[5,6] Studies on high intensity statin therapy in individuals >75 years is still lacking. Hence, the use of statin for primary prevention in older adults is still a dilemma due to clinical heterogeneity and lack of clear guidelines.

Manual of Lipidology

DEFINITION

According to WHO, older age group can be classified in to elderly and oldest-old; the term "elderly" is generally referred to persons aged 65 years and above and oldest-old are aged 80 years and older.[7] By 2050 it is expected that the worldwide proportion of elderly will increase by 16.4% and the older adult population is expected to triple between 2015 and 2050 from 126.5 to 446.6 million. There could be 16% increase in centenarians by 2050.[7,8]

ROLE OF AGING AND ATHEROSCLEROSIS (FIGS. 1 AND 2)

The process of aging is influenced by various environmental factors, genetic, lifestyle and disease conditions. It can be defined as progressive impairment in function, leading to failure in adaptive response to stress and increasing risk of age-related diseases.[7] Atherosclerosis is linked with premature aging and a reduced cell proliferation, apoptosis, increased DNA damage, telomere dysfunction, and epigenetic modifications.[9]

LIPOPROTEINS AND AGING

The triglyceride (TG) levels increase with age. The peak TG is seen men of age 50–59 and in women of age 60–69 years. Apolipoprotein B (ApoB) also increases with age and it is

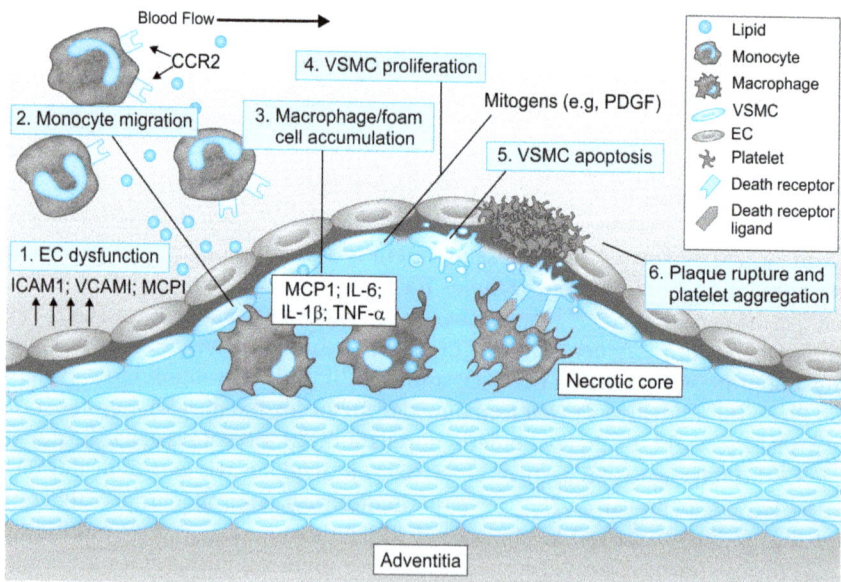

(EC: endothelial cell; ICAM-1: intracellular adhesion molecule; IL-1β: interleukin; MCP-1: monocyte chemotactic protein; TNF-α: tumor necrosis factor; VCAM-1 Vascular Cell Adhesion Molecule-1; VSMC: vascular smooth muscle cell-1)

FIG. 1: Atherogenesis and unstable plaque.
Source: Adapted from Wang JC, Martin B. Aging and Atherosclerosis. Circ Res. 2012;111(2):245-59.

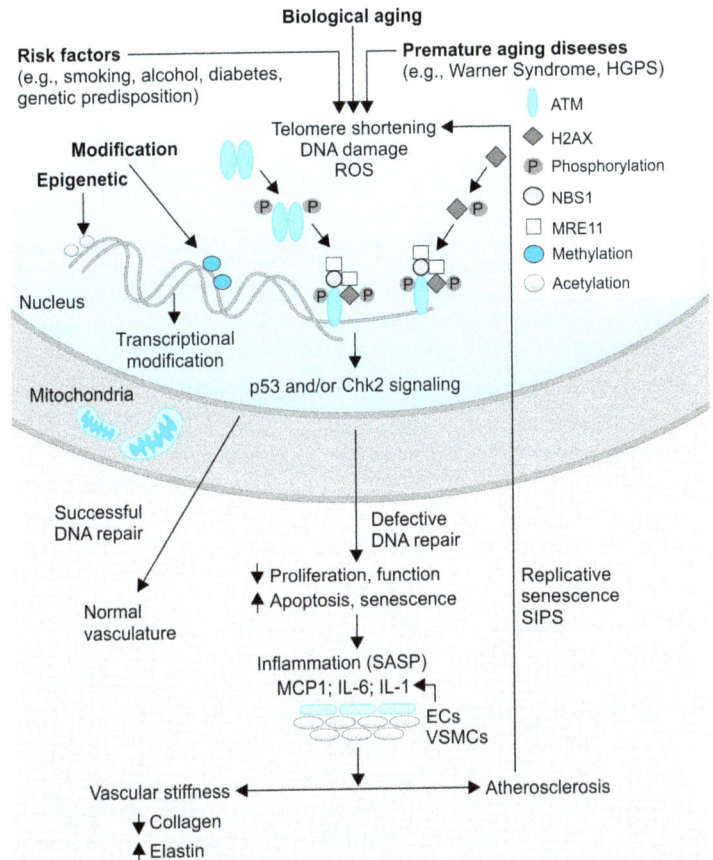

FIG. 2: Model showing cellular senescence in atherosclerosis.
Source: Adapted from Wang JC, Martin B. Aging and Atherosclerosis. Circ Res. 2012;111(2):245-59.

associated with an increased prevalence of small dense LDL-C. However, HDL-C does not seem to increase with age. The decrease in concentration of atherogenic particles with age could be due to poor socioeconomic status, comorbidity, attrition, and frailty.[4] There is an apparent paradox of low cholesterol in old age being associated with increased mortality. This can be explained by changes in the cholesterol metabolism[10], or as a marker of terminal decline[11] or subclinical disease such as cancer [12].

The increased interleukin levels[13] in these conditions can lower the serum LDL-C level by an increase in the LDL receptors. The distinction between therapeutic cholesterol lowering and these endogenous mechanisms of cholesterol lowering in the background of the inflammatory state, needs to be clearly defined in order to understand the true change in mortality risk due to any interventions.

IS THERE A NEED TO INITIATE LIPID-LOWERING AGENTS IN OLDER ADULTS FOR PRIMARY PREVENTION (TABLE 1)?

In a meta-analysis of statin trials, it was found that 1 mmol/L reduction in LDL-C can result in one-fifth reduction of major ASCVD events in a year and decline in total mortality by 10%. The absolute risk of ASCVD is higher in older adults and initiating statin therapy could be beneficial.[1,6]

TABLE 1: An arbitrary categorization of the elderly age group based on current available evidence for lipid lowering therapy.[4]

Age	Evidence for benefit of cholesterol lowering	Decision making process
65–74 years	Strong evidence	After CVD risk evaluation in patients without known CVD
75 years or older	No direct evidence or lack of evidence	After evaluation of ASCVD, or non-ASCVD, comorbidities, lifestyle, socioeconomic status, overall condition of the patient

(ASCVD: atherosclerotic cardiovascular disease; CVD: cardiovascular disease)

STUDIES SHOWING BENEFIT OF LIPID LOWERING AGENTS IN OLDER ADULTS

The PROSPER (Prospective Study of Pravastatin in the Elderly at Risk) Trial:[14] A Randomized, double blind placebo-controlled trial evaluated effect of statins in 5,804 elderly patients with mean age of 75 years and with or at high risk of developing CAD. Subjects were randomized to pravastatin, 40 mg, or placebo. After a mean follow-up of 3.2 years, it was found that pravastatin resulted in 15% relative reduction in risk of the primary endpoint (coronary death, nonfatal myocardial infarction (MI), or fatal and nonfatal).

The JUPITER (Justification for the Use of Statins in Primary Prevention: An Intervention Trial Evaluating Rosuvastatin Trial):[15] A randomized double-blind placebo-controlled study to investigate use of rosuvastatin in apparently healthy persons aged 50 years without hyperlipidemia but with increased levels of high-sensitivity C-reactive protein, rosuvastatin significantly reduced the incidence of major cardiovascular events in both older (>65 years) and younger (50–65 years) individuals. The statin therapy was shown to be protective in the older age group [hazard ratio 0.61, (95% CI 0.46–0.82)], but rate of all-cause death did not differ significantly between both the groups.

Patient and Provider Assessment of Lipid Management (PALM) Registry:[16] Adults aged >75 years and ≤75 years who were eligible for primary or secondary prevention with statin were assessed for tolerance of therapy and patient reported symptoms were surveyed in these age groups. A total 6,717 people were enrolled, out of which 1,704 subjects were >75 years old. For primary prevention, use of any statin or high dose statin did not vary between the groups whereas for secondary prevention, older patients were less likely to receive high-intensity statin. It was found that statins were similarly tolerated in both the groups.

Drug Treatment of Dyslipidemia in Older Adults

The HOPE-3 (Heart Outcomes Prevention Evaluation) Trial:[17] In this trial, the benefit of statins was studied in an intermediate risk ethnically diverse population without cardiovascular disease. Subjects aged 55 years or older with one or two risk factors with intermediate ASCVD risk on rosuvastatin were compared with placebo. In half of the study subjects 12,704 (65 years or older), composite outcome (death due to cardiovascular causes and nonfatal MI or stroke) with statin therapy was significantly lower compared to placebo [hazard ratio 0.75 (95% CI 0.61–0.93)].

The STAREE (Statin Therapy for Reducing Events in the Elderly) Trial-(NCT020991223):[18] A randomized placebo-controlled study to assess the effect of statin (atorvastatin 40 mg) versus placebo on overall survival or disability free survival in healthy elderly people (≥70 years). This is an ongoing trial which could give more insight on primary prevention in older adults and is likely to be completed by December, 2023.

WHAT DO THE GUIDELINES RECOMMEND?

The ACC/AHA 2018 Guideline on Management of Blood Cholesterol (Tables 2 to 4)[19]

TABLE 2: The ACC/AHA task force clinical practice guidelines for primary prevention.[19]

Primary prevention		
	Recommendations	LOE$
Adults with diabetes		
Age 65–75 years	Regardless of estimated 10 years ASCVD risk**, a moderate intensity statin therapy is indicated	A
	With an LDL–C level between 70 and 189 mg/dL, it is reasonable to assess the 10-year risk of ASCVD	B-NR
	In the presence of multiple ASCVD risk factors, it is reasonable to prescribe high intensity statin therapy with the aim to reduce LDL-C levels by 50% or more	B-R
	If the 10-year ASCVD risk is 20% or more, it may be reasonable to add ezetimibe to maximally tolerated statin therapy to reduce LDL-C levels by 50% or more	C-LD
Age >75 years	It is reasonable to continue statin therapy if the patient is already on statins	B-NR
	In those not on statin therapy, it may be reasonable to initiate statin therapy after a clinician patient discussion of potential benefits and risks	C-LD

Continued

Continued

TABLE 2: The ACC/AHA task force clinical practice guidelines for primary prevention.[19]

Primary prevention

	Recommendations	LOE$
Adults without diabetes		
Age 65–75 years	LDL-C 70–189 mg/dL + intermediate risk category, moderate intensity statin recommended LDL–C should be reduced by 30% or more and in high-risk patients, it should be reduced by 50% or more	A
	In intermediate risk adults, risk enhancing factors* favor initiation or intensification of statin therapy	B-R
	In intermediate risk adults or selected borderline risk adults, if the decision about statin therapy remains uncertain, it is reasonable to use CAC score	B-NR
	In those not tolerating high intensity statins, ezetimibe or bile acid sequestrants can be added to moderate intensity statin therapy	B-R
	In those at borderline risk with risk enhancing factors*, initiate moderate intensity statin therapy	B-R
Age >75 years	LDL-C level of 70–189 mg/dL (1.7–4.8 mmol/L), initiate a moderate-intensity statin	B-R
	It may be reasonable to stop statin therapy when functional decline (physical or cognitive), multimorbidity, frailty, or reduced life expectancy limits the potential benefits of statin therapy	B-R
	In adults 76–80 years of age with an LDL-C level of 70–189 mg/dL (1.7–4.8 mmol/L), it may be reasonable to measure CAC to reclassify those with a CAC score of zero to avoid statin therapy	B-R

(ASCVD: atherosclerotic cardiovascular disease; CAC: coronary artery calcium; LDL-C: low-density lipoprotein cholesterol)

*Risk enhancing factors[19]
- Family history of premature ASCVD (males age <55 years; females age <65 years).
- Primary hypercholesterolemia [LDL-C, 160–189 mg/dL (4.1–4.8 mmol/L); non–HDL-C 190–219 mg/dL (4.9–5.6 mmol/L)].
- Metabolic syndrome.
- Chronic kidney disease.
- Chronic inflammatory conditions such as psoriasis, rheumatoid arthritis (RA), or HIV/AIDS.
- History of premature menopause (before age 40 years) and history of pregnancy-associated conditions that increase later ASCVD risk such as preeclampsia.
- High-risk race/ethnicities (e.g., South Asian ancestry).
- Lipid/biomarkers: Associated with increased ASCVD risk- *Persistently (optimally, 3 determinations) elevated, primary hypertriglyceridemia (≥175 mg/dL); Elevated high-sensitivity C-reactive protein (≥2.0 mg/L); elevated Lp(a); (≥50 mg/dL or ≥125 nmol/L); elevated ApoB ≥130 mg/dL; ABI <0.9.

$LOE: level of evidence: Level A: high quality evidence from more than 1 RCT, meta-analyses of high-quality RCTs, one or more RCTs Corroborated by high-quality registry studies. Level B-R: Moderate quality. Evidence from 1 or more RCTs, Meta-analysis of moderate quality RCTs, Level C-LD: Randomized or nonrandomized observational or registry studies with limitations of design or execution, Meta-analysis of such studies, physiological or mechanistic studies in human subjects.[19]

TABLE 3: Treatment decisions based on CAC in intermediate risk patients.[19]

CAC score	Recommendation
Zero	Withhold statin therapy and reassess in 5–10 years, as long as higher risk conditions are absent (diabetes mellitus, family history of premature CHD, cigarette smoking)
1–99	Initiate statin therapy for patients ≥55 years of age
100 or higher or in the 75th percentile or higher	Initiate statin therapy

(CHD: coronary heart disease; CAC: coronary artery calcium)

TABLE 4: The ACC/AHA task force clinical practice guidelines for secondary prevention.[19]

Secondary prevention		
Age	Recommendation	LOE[$]
65–75 years		
Patients with clinical ASCVD	High intensity statin therapy initiated or continued for 50% or greater reduction in LDL-C levels	A
Patients with clinical ASCVD in whom high-intensity statin therapy is contraindicated or who experience statin-associated side effects	Moderate intensity statin therapy initiated or continued for 30–49% reduction in LDL-C levels	A
>75 years		
Patients with clinical ASCVD	Initiate moderate- or high-intensity statin therapy after evaluation of ASCVD risk reduction, adverse effects, and drug–drug interactions, patient frailty and patient preferences	B-R
Patients tolerating high-intensity statin therapy	Reasonable to continue high-intensity statin therapy after evaluation of ASCVD risk reduction, adverse effects, and drug-drug interactions, patient frailty and patient preferences	C-LD
Patients with clinical ASCVD on maximally tolerated statin therapy and LDL-C level 70 mg/dL or higher (≥1.8 mmol/L)	Ezetimibe may be added to the maximally tolerated statin therapy	B-R
Patients with heart failure with reduced ejection fraction attributable to ischemic heart disease who have a reasonable life expectancy (3–5 years) and statin because of ASCVD	Consider initiation of moderate-intensity statin therapy to reduce the occurrence of ASCVD events	B-R

(ASCVD: atherosclerotic cardiovascular disease; LDL-C: low-density lipoprotein cholesterol)

**10 year ASCVD can be assessed and classed as low risk (<5%), borderline risk (5% to <7.5%) intermediate risk (≥7.5% to <20%), high risk ≥20%, Very high risk – history of multiple major ASCD events or 1 major ASCVD event and multiple risk factors.[19]

[$]LOE: level of evidence: Level A: high quality evidence from more than 1 RCT, meta-analyses of high-quality RCTs, one or more RCTs Corroborated by high-quality registry studies. Level B-R: Moderate quality. Evidence from 1 or more RCTs, Meta-analysis of moderate quality RCTs, Level C-LD: Randomized or nonrandomized observational or registry studies with limitations of design or execution, Meta-analysis of such studies, physiological or mechanistic studies in human subjects.[19]

Euroean Society of Cardiology 2019 Recommendations for Older People Aged More than 65 Years.[20]
The Task Force for the management of dyslipidemias of the ESC (European Society of Cardiology) and EAS (European Atherosclerosis Society) recommendations:
- For older people with ASCVD—statins recommended.
- For primary prevention (according to the level of risk)—Statins recommended.
- Statin should be started at low dose if there is significant renal impairment or drug interactions and then tritiated.

PHARMACOLOGICAL MANAGEMENT

Statin Therapy

It can be divided into three categories (**Table 5**).[19]

Clinical implications: The major concern in older patients is statin-associated adverse effects. Due to comorbidities, frailty, poly pharmacy, and drug-drug interactions, statins should be administered with caution. One of the main reasons for discontinuation of statins are myalgias.[18] The statin associated muscle symptoms such as myalgias (CK normal), myositis/myopathy (CK > ULN), rhabdomyolysis (CK > 10 × ULN + renal injury), statin associated autoimmune myopathy (SAAM) are infrequent or rare (<1%). Myopathy and rhabdomyolysis results in one death per million prescriptions and the risk can increase with the dose.[20,21] Hepatotoxicity (transaminase elevation 3x ULN) is seen in one case per million person-years of use or infrequent. LFT should be assessed at the base-line and thereafter when clinically indicated. Careful monitoring required when it is administered with other concomitant drugs that interferes with hepatic uptake or statin metabolism. Another concern is its effect on cognition; randomized trials are still not conclusive.[18,22] In a meta-analysis of 23 RCTs, 29,012 participants of all age group, adverse cognitive outcomes were rarely reported. Further analysis of the cognitive data in 14 studies with 27,643 participants did not show any significant adverse effects on cognitive in statin users.[23,24] A meta-analysis of 30 observational studies which included 9,162,509 participants showed that patients with statin had lower all caused dementia risk than those without statin therapy with risk ratio 0.83, 95% CI 0.79–0.87. Hence, this study suggested that use of statin is associated with decreased risk of dementia.[25] In a cohort study by Lee S-E et al. followed 17,798 statin nonusers and 22,366 statin users for a duration of 7.66 ± 3.21 years found that 7.64% of statin users developed diabetes mellitus (DM) compared to 5.68% statin nonusers.[26]

TABLE 5: Categorization of statin therapy.			
Statins	**High intensity**	**Moderate intensity**	**Low intensity**
	• Atorvastatin 40–80 mg • Rosuvastatin 20–40 mg	• Atorvastatin 10–20 mg • Rosuvastatin 5–10 mg • Simvastatin 20–40 mg • Pravastatin 40–80 mg • Lovastatin 40–80 mg • Fluvastatin XL 80 mg • Fluvastatin 40 mg BID • Pitavastatin 1–4 mg	• Simvastatin 10 mg • Pravastatin 10–20 mg • Lovastatin 20 mg • Fluvastatin 20–40 mg
LDL-C lowering	≥50%	30–49%	<30%

(LDL-C: low-density lipoprotein cholesterol)

Nonstatin Therapy

The most commonly recommended non-statin agent is ezetimibe. It is used as monotherapy in patients with increased LDL-C who are intolerant to statins. If combined with statins, it can further reduce LDL-C. The concurrent use of ezetimibe and bile-acid sequestrants should be avoided because bile-acid sequestrants can inhibit the absorption of ezetimibe. It can lower LDL-C levels by 13–20% and low incidence of side effects. High intensity statins plus ezetimibe can reduce LDL-C up to 65%.[20] Bile acid sequestrants such as cholestyramine, colestipol, and colesevelam reduce LDL-C levels by 15–30%. These are associated with GI complaints (bloating, dyspepsia, constipation) and can cause severe hypertriglyceridemia. Cholestyramine and colestipol can interfere with the absorption of many drugs. It should be administered 1 hour before or 3-4 hours after. The newer class of lipid lowering agents, proprotein-convertase-subtilisin/kexin type 9 (PCSK9) inhibitors such as alirocumab, evolocumab when used as an adjunct to diet and maximally tolerated statin regimen, has been shown to reduce LDL-C levels by 43–75%.[6,7] PCSK9 inhibitor plus high intensity statin plus ezetimibe can result in 85% average LDL-C reduction.[20]

CONCLUSION

Due to the lack of evidence and clear-cut guidelines on the management of dyslipidemia in elderly, the risk versus benefit should be assessed and a decision on use of lipid lowering agent should be taken on a case- by- case basis. Such interventions could further decrease the pill burden and avoid polypharmacy in the elderly.

KEY POINTS

- Advocate heart-healthy lifestyle.
- Lipid profile should be assessed 4–12 weeks after initiation or dose adjustment and should be repeated every 3–12 months as required.
- Statin therapy is generally safe and well tolerated in older adults, and ongoing treatment should be continued after 75 years, however, individual assessment and shared decision should be made to improve adherence.
- More research is warranted in elderly population to mitigate this knowledge gap.

REFERENCES

1. Strandberg TE. Role of statin therapy in primary prevention of cardiovascular disease in elderly patients. Curr Atheroscler Rep. 2019;21(8).
2. Grundy Scott M. Primary prevention of coronary heart disease. Circulation. 1999;100(9):988-98.
3. Karalis DG. Intensive lowering of low-density lipoprotein cholesterol levels for primary prevention of coronary artery disease. Mayo Clinic Proceedings. 2009;84(4):345-52.
4. Kuller LH, Shemanski L, Psaty BM, Borhani NO, Gardin J, Haan MN, et al. Subclinical disease as an independent risk factor for cardiovascular disease. Circulation. 1995;92(4):720-6.
5. Streja D, Streja E. Management of dyslipidemia in the elderly. In: Feingold KR, Anawalt B, Boyce A, Chrousos G, Dungan K, Grossman A, et al. (Eds). South Dartmouth (MA): MDText.com Inc.; 2000.
6. Cholesterol Treatment Trialists' Collaboration. Efficacy and safety of statin therapy in older people: a meta-analysis of individual participant data from 28 randomised controlled trials. Lancet. 2019;393(10170):407-15.
7. World Health Organization. Men, ageing and health: Achieving health across the life span. [online] Available from https://apps.who.int/iris/bitstream/handle/10665/66941/WHO_NMH_NPH_01.2.pdf;jsessionid=3256BC67B13F9C754281CD04F7C7272D?sequence=1. [Last accessed July, 2020].
8. National Institutes of Health (NIH). (2016) World's older population grows dramatically. [online] Available from https://www.nih.gov/news-events/news-releases/worlds-older-population-grows-dramatically. [Last accessed July, 2020]
9. Wang JC, Martin B. Aging and Atherosclerosis. Circ Res. 2012;111(2):245-59.
10. Tilvis RS, Valvanne JN, Strandberg TE, Miettinen TA. Prognostic significance of serum cholesterol, lathosterol, and sitosterol in old age; a 17-year population study. Ann Med. 2011;43:292-301.

11. Charlton J, Ravindrarajah R, Hamada S, Jackson SH, Gulliford MC. Trajectory of total cholesterol in the last years of life over age 80 years: cohort study of 99,758 participants. J Gerontol A Biol Sci Med Sci. 2018;73(8):1083-9.
12. Ahn J, Lim U, Weinstein SJ, Schatzkin A, Hayes RB, Virtamo J, et al. Prediagnostic total and high-density lipoprotein cholesterol and risk of cancer. Cancer Epidemiol Biomark Prev. 2009;18:2814-21.
13. DiNicolantonio JJ, McCarty MF. Is interleukin-6 the link between low LDL cholesterol and increased non-cardiovascular mortality in the elderly? Open Heart. 2018;5:000789.
14. Ford I, Blauw GJ, Murphy MB, Shepherd J, Cobbe SM, Bollen EL, et al. A Prospective Study of Pravastatin in the Elderly at Risk (PROSPER): Screening Experience and Baseline Characteristics. Curr Control Trials Cardiovasc Med. 2002;3(1):8.
15. Ridker PM, Danielson E, Fonseca FA, Genest J, Gotto AM Jr, Kastelein JJP, et al. Rosuvastatin to prevent vascular events in men and women with elevated C-reactive protein. N Engl J Med. 2008;359(21):2195-207.
16. Nanna MG, Navar AM, Wang TY, Mi X, Virani SS, Louie MJ, et al. Statin use and adverse effects among adults >75 years of age: Insights From the Patient and Provider Assessment of Lipid Management (PALM) Registry. J Am Heart Assoc. 2018;7(10):e008546.
17. Yusuf S, Bosch J, Dagenais G, Zhu J, Xavier D, Liu L, et al. HOPE-3 Investigators. Cholesterol lowering in intermediaterisk persons without cardiovascular disease. N Engl J Med. 2016;374(21):2021-31.
18. Gurwitz JH, Go AS, Fortmann SP. Statins for Primary Prevention in Older Adults. JAMA. 2016;316(19):1971-2.
19. Grundy SM, Stone NJ, Bailey AL, Beam C, Birtcher KK, Blumenthal RS, et al. 2018 AHA/ACC/AACVPR/AAPA/ABC/ACPM/ADA/AGS/APhA/ASPC/NLA/PCNA Guideline on the Management of Blood Cholesterol. J Am Coll Cardiol. 2019;73(24):e285-350.
20. Mach F, Baigent C, Catapano AL, Koskinas KC, Casula M, Badimon L, et al. 2019 ESC/EAS Guidelines for the management of dyslipidaemias: lipid modification to reduce cardiovascular risk. Eur Heart J. 2020;41(1):111-88.
21. Ito MK, Maki KC, Brinton EA, Cohen JD, Jacobson TA. Muscle symptoms in statin users, associations with cytochrome P450, and membrane transporter inhibitor use: a subanalysis of the USAGE study. J Clin Lipidol. 2014;8(1):69-76.
22. Gurgle HE Donald K. Blumenthal drug therapy for dyslipidemias. In: Brunton LL, Hilal-Dandan R, Knollman BC. Goodman & Gilman's the Pharmacological Basis of Therapeutics, 13th edition. New York: McGraw-Hill; 2018. pp. 606-18.
23. Schultz BG, Patten DK, Berlau DJ. The role of statins in both cognitive impairment and protection against dementia: a tale of two mechanisms. Transl Neurodegener. 2018;7.
24. Ott BR, Daiello LA, Dahabreh IJ, Springate BA, Bixby K, Murali M, et al. Do statins impair cognition? a systematic review and meta-analysis of randomized controlled trials. J Gen Intern Med. 2015;30(3):348-58.
25. Poly TN, Islam MM, Walther BA, Yang H-C, Wu C-C, Lin M-C, et al. Association between Use of Statin and Risk of Dementia: A Meta-Analysis of Observational Studies. Neuroepidemiology. 2020;54(3):214-26.
26. Lee S-E, Sung JM, Cho I-J, Kim HC, Chang H-J. Risk of new-onset diabetes among patients treated with statins according to hypertension and gender: Results from a nationwide health-screening cohort. PLoS ONE. 2018;13(4):e0195459.

CHAPTER
30

Dyslipidemia in Human Immunodeficiency Virus

Ranajit Bari

ABSTRACT

Human immunodeficiency virus (HIV) infected patients are at risk for cardiovascular disease due to various risk factors. Lipid disorders in these patients can be due to disease process or drugs. Drug naïve patients with HIV/acquired immunodeficiency syndrome (AIDS) usually present with increase of triglyceride and decrease of total cholesterol, low density lipoprotein-cholesterol (LDL-C), and high density lipoprotein-cholesterol (HDL-C) levels. Alteration of lipid metabolism varies according to different antiretroviral therapy (ART) regimens. Protease inhibitors (PIs) such as indinavir, lopinavir are usually associated with high levels of cholesterol, triglyceride, LDL-C and low level HDL-C. Apparently atazanavir is associated with more favorable lipid profile. Nucleoside reverse transcriptase inhibitors (NRTIs) such as zidovudine, didanosine, stavudine cause hypertriglyceridemia and lipoatrophy, whereas tenofovir results minimal alteration of cholesterol and triglyceride. Although non-nucleoside reverse transcriptase inhibitors (NNRTIs) are associated with increased cholesterol and triglyceride, nevirapine favorably cause high HDL-C. Therefore, not only infection itself but distinct ART regimens also responsible for the alteration of lipid parameters. Management of dyslipidemia includes clinical and biochemical evaluation for risk stratification, lifestyle modification and lipid lowering drugs with proper consideration of drug interactions.

INTRODUCTION

Human immunodeficiency virus (HIV) is a chronic infective disease with increase morbidity and mortality due to different HIV related conditions.[1] Combined anti-retroviral therapy (c-ART) has significantly improved the morbidity and mortality of HIV affected individuals.[2] As a result of longer survival and ART, various clinical aspects of these group of patients have developed, e.g., insulin resistance, altered glucose homeostasis, lipid disorders and fat redistribution.[3] Dyslipidemia, a common metabolic complication of HIV/acquired immunodeficiency syndrome (AIDS) can be associated with disease itself or antiretroviral therapy (ART). Mounting evidence indicates that lipid abnormalities are associated with increased cardiovascular morbidity and mortality. These epidemiologic findings have prompted increased need for assessment and modification of cardiovascular risk as an integral part of routine HIV care.

LIPID CHANGES AND LIPODYSTROPHY SYNDROME IN HIV

In a study by Grunfeld et al.[4] Patients with HIV infection without AIDS, had decreased total cholesterol and HDL-C and those with AIDS had elevated triglyceride and free fatty acid levels. Total cholesterol, HDL-C, LDL-C declined following HIV seroconversion, as shown in MACS (multicenter AIIDS cohort study).[5] After initiation of highly active ART, the total cholesterol and LDL-C increased but HDL-C level remained low. Combined ART associated dyslipidemia usually associated with elevated total cholesterol, triglyceride, LDL-C, very low density lipoprotein (VLDL), apolipoprotein B (ApoB) and low level of HDL-C.[6] Though it was the first lipid disorders described in association with protease inhibitors (PIs), later it was also observed in patients receiving nucleoside reverse transcriptase inhibitors (NRTIs) and non-nucleoside reverse transcriptase inhibitors (NNRTIs).[7,8]

The terminology "lipodystrophy syndrome associated with HIV" was first described in 1998 and is characterized by dyslipidemia, centripetal adiposity (lipohypertrophy), lipoatrophy and metabolic syndrome.[3] Lipohypertrophy includes fat deposition in abdomen (omental, retroperitoneal mesenteric), dorsocervical area along with increased breast size.[9] On the other hand lipoatrophy is characterized by loss of subcutaneous fat in gluteal and facial region, upper and lower limbs with vascular prominence. There is also decreased periorbital and temporal fat double line nasolabial fold.[10] Lipoatrophy is commonly associated with use of NRTIs (Zidovudine, Stavudine).

Antiretroviral therapy induced dyslipidemia most commonly seen with PIs. Almost 70–80% patients receiving PIs had some kind of lipid abnormality. PI induced dyslipidemia is characterized by hypertriglyceridemia, elevated LDL-C, low HDL-C with accumulation of apolipoprotein E (ApoE) and apolipoprotein C-III (Apo C-III).[11,12] Lipid disorders are associated with almost all PIs (saquinavir, indinavir, nelfinavir, and ritonavir) except atazanavir containing regimen, which showed improvement of lipid parameters. DAD (Data collection on adverse events of anti-HIV drugs) study group, in their prospective, observational study on 23,437 patients, showed that use of PIs was associated with increased risk of myocardial infarction. Although the relative risk was slightly lower after adjustment of established cardiovascular risk factors, other drug classes and lipid levels.[13]

Body fat alterations as well as hypertriglyceridemia also have been shown to be associated with use of NRTIs.[14] Lipid alterations are less evident with tenofovir containing NRTI combination than others. When zidovudine + lamivudine or didanosine + lamivudine were compared with tenofovir + lamivudine, the later therapy showed lower total cholesterol, LDL-C and triglyceride than former.[15] on the other hand, use of NRTIs such as abacavir, didanosine were shown to be an independent risk factor for myocardial infarction as per DAD study.[13]

Among NNRTIs, use of nevirapine showed better lipid profile mainly because of elevated HDL-C, whereas elevated total cholesterol and triglyceride were observed with efavirenz containing regimen.[16]

EPIDEMIOLOGY OF DYSLIPIDEMIA IN HIV

Prevalence of dyslipidemia varies from 30 to 80% among HIV infected individuals on c-ART depending on the diagnostic criteria and drug combination.[17] Most common form of ART induced dyslipidemia was hypertriglyceridemia (40–80%) followed by hypercholesterolemia (10–50%).[18] This disorder has been reported in all age group though children are more vulnerable than young adults because of status of their growing organs and longer exposure to ART.[19]

PATHOPHYSIOLOGY OF DYSLIPIDEMIA IN HIV

In patients with HIV/AIDS before receiving ART, dyslipidemia is typically characterized by high triglyceride, low HDL-C which mostly due to chronic inflammatory state and insulin resistance. Other factors such as delayed lipid clearance and increased hepatic synthesis of VLDL are also playing significant role.[9] Proinflammatory cytokines such as tumor necrosis factor-alpha (TNF-α), and interleukin-6 (IL-6) appear to promote lipid peroxidation, endothelial dysfunction, platelet aggregation and activation of reactive oxygen species.[20] Low CD4 T lymphocyte count is considered as major atherosclerotic risk factor in HIV infected patients. The lower the CD4 T-cell count, higher the level of triglyceride, the lower the level of LDL-C and total cholesterol. HDL-C concentration remains to be low irrespective of CD4 T lymphocyte status.[4,21]

Among the ARTs, PIs are generally associated with dyslipidemia, which is highest with ritonavir and other boosted PIs. There are several theories to explain the mechanism of PIs induced dyslipidemia, although none of them is able to cover all the aspects. Carr et al. proposed a theory of molecular structural homology of HIV protease with two human proteins, cellular retinoic acid binding protein-1 (CRABP-1) and protein related to LDL receptor (LDL-R). Therefore multiple steps of lipid metabolism are being inhibited by PIs. Activation of 9 cis retinoic acid would be impaired with binding of PI with CRABP-1. In the same way, inhibition of receptor for Peroxisome Proliferator activated receptor–gamma (PPARγ) would lead to alterations of adipocyte differentiation and insulin sensitivity. Another theory proposed that PIs inhibit the activity of lipoprotein lipase by binding to LDL-receptor.

This may lead to impaired capturing of chylomicrons and decreased hepatic clearance of triglyceride.[22] Similarly, lipogenesis including increased production of VLDL-C could be related to increased activity of sterol regulatory element binding protein-1 (SRBP-1).[23]

Almost all PI (except Atazanavir) appear to inhibit glucose transporter 4 (GLUT4), thereby reduction of insulin mediated glucose uptake and dyslipidemia.[24,25] NRTIs reduce synthesis of mitochondrial DNA, which leads to enzyme deficiency for oxidative phosphorylation and apoptosis of subcutaneous adipocytes.[26]

MANAGEMENT OF DYSLIPIDEMIA IN HIV

The guidelines by ACTG (AIDS Clinical Trial Group) and IDSA-HIVMA (Infectious Disease Society of America-HIV Medicines Associations) suggest management of dyslipidemia in HIV infected patients in a way similar to NCEP (National Cholesterol Education Program) guideline.[27] They recommended, lipid should be measured at baseline, 3 and 6 months after starting ART. Framingham Risk score for 10-year coronary heart disease (CHD) risk should be calculated for HIV infected patients having two or more traditional risk factor.

Consideration for lipid lowering drugs should be made if target lipid levels are achieved after lifestyle modification. ART should be modified wherever possible. Statins as the first-line therapy should be started for elevated non-HDL-C with TG level between 200 and 500 mg/dL as recommended in IDSA guideline. Fibrates should be considered when TG is >500 mg/dL which is consistent with guidelines by AHA (American Heart Association) and NCEP for general population.[28,29] Assessment and management of cardiovascular risk should be implemented for all patients with HIV infection according to WHO antiretroviral guideline 2016.

So in summary, guideline for management of dyslipidemia in HIV infected person is same as recommended in general population.

According to NCEP ATP III and IDSA/ACTG guidelines, lifestyle modification including diet and exercise should be implemented to all patients. Other modifiable risk factors such as hypertension, diabetes mellitus, obesity, smoking, sedentary lifestyle need to be addressed.

Stepwise Approach for Management of Dyslipidemia in HIV
Step 1: Assess Cardiovascular Risk

- *History and physical examination:* A detail history (**Box 1**) and physical examination (**Box 2**) should be taken in each patient of HIV to assess the cardiovascular risk. History must include details of ART particularly PIs and NNRTI.

BOX 1 | History.

- Age >45 years in men or >55 years in female
- History of coronary artery disease, peripheral vascular disease, aortic aneurysm
- History of dyslipidemia prior to diagnosis and treatment of HIV
- Family history of premature coronary artery disease in first degree relative (in male <55 years and in female <65 years of age)
- Family history of diabetes or lipid disorders
- History of occupation, food habits, exercise and lifestyle
- Drug history including HRT, antipsychotic, antihypertensive medications
- Current medications for HIV
- History of cigarette smoking, alcohol abuse, or illicit drug use
- History of comorbidities such as hypothyroidism, kidney disease, obstructive sleep apnea syndrome, polycystic ovarian syndrome

(HRT: hormone replacement therapy)

BOX 2 | Physical examinations.

- BMI, waist circumference, blood pressure
- Peripheral pulses and arterial bruit
- Stigmata of dyslipidemia: Tendon xanthoma, xanthelasma (due to elevated LDL-C), eruptive xanthoma (elevated TG)
- Presence of lipoatrophy or lipohypertrophy
- Clinical evidence of hypothyroidism

(BMI: body mass index; LDL-C: low-density lipoprotein-cholesterol; TG: triglycerides)

> **BOX 3** | **Laboratory workup.**
> - Fasting lipid profile
> - Calculate non-HDL-C (ApoB and LDL-P not routinely recommended)
> - Lipoprotein(a) (in case of patients with premature CVD, family history of premature CVD, or familial hypercholesterolemia)
> - hs-CRP
> - TSH
> - AST and ALT—baseline testing before starting lipid medication and evaluation for obstructive liver disease
> - Serum creatinine kinase
> - Serum creatinine and calculation of GFR
> - Fasting glucose and/or HbA1c
>
> (AST: aspartate transaminase; ALT: alanine transaminase; CVD: cardiovascular disease; GFR: glomerular filtration rate; HbA1c: glycosylated hemoglobin; hs-CRP: high-sensitivity C-reactive protein; HDL: high-density lipoprotein; LDL: low-density lipoprotein; TSH: thyroid stimulating hormone)

- *Laboratory evaluation (Box 3)*: Lipid examination should include fasting total cholesterol, HDL-C, triglyceride, calculated LDL-C. Direct LDL-C measurement may be considered only when calculated LDL-C is inaccurate, e.g., triglyceride >400 mg/dL. However, laboratories equipped for direct LDL-C measurement are limited. Hence routine recommendation does not include direct LDL-C measurement.[30]

Step 2: Therapeutic Intervention

- *Lifestyle modification:* Lifestyle modification should be advised and implemented to all patients with HIV dyslipidemia. Therapeutic lifestyle management should include dietary modification (cholesterol intake <200 mg/day, saturated fat intake <7% of total calorie), increased physical activity, weight management and smoking cessation. Higher frequency of cigarette smoking, alcohol abuse, illicit drug use and unhealthy food habits impose these individuals at risk for metabolic complications.
- *Medications:*
  - Hydroxymethylglutaryl Co-A reductase inhibitors: Statins are widely studied and most commonly used drugs for management of dyslipidemia. Apart from primary LDL-C lowering effect, they have modest effect on elevating HDL-C and lowering triglycerides. However selection of statin should be taken into account considering efficacy, safety and potential drug-drug interaction. Pitavastatin is least likely to interact with ART. Atorvastatin or rosuvastatin are also reasonable choice, but interaction with ART and side effects need to be considered. Simvastatin and lovastatin should be avoided. Though pravastatin has least drug interaction, but is less potent than others.
  - Fibrates: Fenofibrate is effective in reduction of triglyceride. It also has minimal interaction with ART.
  - Bile acid sequestrants: They include colestipol, cholestyramine, and colesevelam. These drugs are usually avoided as they cause decreased ART absorption.
  - Ezetimibe: It inhibits cholesterol absorption from intestinal brush border. It neither has interaction with CYP3A, nor any drug interaction with ART.

There are some important strategies should be considered for HIV infected patients on ART while choosing them for lipid lowering therapy:
- For patients on PIs, rosuvastatin is better than atorvastatin as lipid lowering drug.
- Fenofibrate are effective in reducing triglycerides and it does not have significant interaction with ARTs.
- Fish oil can be safely used as an alternative to fenofibrate for triglyceride reduction.
- Niacin also reduces triglycerides, but may cause impaired glucose tolerance.
- Bile acid sequestrants may impair ART absorption, so should be avoided.
- Use of ezetimibe along with pravastatin helps further reduction of total cholesterol and LDL-C.

Drug Interaction with Statins and Antiretroviral Therapy (Box 4)

Some important drug interactions should be remembered while selecting statins to those patients on ART.

Physicians and healthcare workers involved in treatment of HIV infected patients are primarily focused on management of infection, reduction of viral load, improvement of CD4 count. Prevention of cardiovascular disease is now becoming an increasing priority in these individuals with advancement of ART and life expectancy.

BOX 4 | Drug interaction with statins and antiretroviral therapy (ART).

- PIs, NNRTIs and some statins are metabolized by CYP3A4
- Atorvastatin can be used with PIs but in low doses
- Simvastatin and lovastatin are contraindicated with PIs due to risk of serious rhabdomyolysis
- Pravastatin can be given with PIs but may be less effective due to induction of enzymes that metabolize it. However its levels increase five-fold with darunavir
- Pitavastatin can be given with ritonavir-boosted atazanavir (no data with other PIs)
- Rosuvastatin concentration increases with lopinavir/ritonavir and fosamprenavir/ritonavir but not with atazanavir/ritonavir
- Efavirenz decreases concentrations of simvastatin, atorvastatin and pravastatin
- Raltegravir does not interact with statins
- Maraviroc is a CYP3A inducer and may interact with statins

(PIs: protease inhibitors; NNRTIs: non-nucleoside reverse transcriptase inhibitors)

CONCLUSION

HIV, being primarily an infective disease, can be managed by different ART regimens. But metabolic consequences of chronic HIV infection are growing concern now a days, of which dyslipidemia is the leading one. Apart from routine HIV care, these individuals should also be screened for cardiovascular risk factors periodically. Careful selection of ART regimen, proper implementation of lifestyle measures and judicious initiation of lipid lowering drugs may help to improve dyslipidemia and cardiovascular sequelae in these patients.

KEY POINTS

- HIV infected patients are at high cardiovascular risk due to multiple factors.
- Dyslipidemia in HIV infected patients can be due to disease and ART.
- Particular lipid disorder and body fat alteration depend on distinct ART regimen.
- There is increased need for thorough cardiovascular assessment and modification of risk factors for each HIV infected patient as a part of routine HIV care.

REFERENCES

1. Husain NO, Ahmed MH. Managing dyslipidemia in HIV/AIDS patients: challenges and solutions. HIV AIDS (Auckl).. 2015;7:1-10.
2. Palella FJ Jr, Delaney KM, Moorman AC, Loveless MO, Fuhrer J, Satten GA, et al. Declining morbidity and mortality among patients with advanced human immunodeficiency virus infection. HIV Outpatient Study Investigators. N Engl J Med. 1998;338:853-60.
3. Carr A, Sâmaras K, Burton S, Law M, Freund J, Chisholm DJ, et al. A syndrome of peripheral lipodystrophy, hyperlipidemia and insulin resistance in patients receiving HIV protease inhibitors. AIDS. 1998;12:51-8.
4. Grunfeld C, Pang M, Doerrler W, Shigenaga JK, Jensen P, Feingold KR. Lipids, lipoproteins, triglyceride clearance, and cytokines in human immunodeficiency virus infection and the acquired immunodeficiency syndrome. J Clin Endocrinol Metab. 1992;74:1045-52.
5. Riddler SA, Smit E, Cole SR, Li R, Chmiel JS, Dobs A, et al. Impact of HIV infection and HAART on serum lipids in men. JAMA. 2003;289:2978-82.
6. Dronda F. Cardiovascular risk in patients with chronic HIV-1 infection: a controversy with therapeutic, clinical and prognostic implications. Enferm Infecc Microbiol Clin. 2004;22:40-5.
7. Fantoni M, Autore C, Del Borgo C. Drugs and cardiotoxicity in HIV and AIDS. Ann N Y Acad Sci. 2001;946:179-99.
8. Nguemaïm NF, Mbuagbaw J, Nkoa T, Alemnji G, Této G, Fanhi TC, et al. Serum lipid profile in highly active antiretroviral therapy-naïve HIV-infected patients in Cameroon: a case-control study. HIV Med. 2010;11:353-9.
9. Grinspoon S, Carr A. Cardiovascular risk and body fat abnormalities in HIV-infected adults. N Engl J Med. 2005;352:48-62.
10. Safrin S, Grunfeld C. Fat Distribution and metabolic changes in patients with HIV infection. AIDS. 1999;13:2493-505.
11. Fauvel J, Bonnet E, Ruidavets JB, Ferrières J, Toffoletti A, Massip P, et al. An interaction between apo C-III variants and protease inhibitors contributes to high triglyceride/low HDL levels in treated HIV patients. AIDS. 2001;15:2397-406.
12. Calza L, Manfredi R, Colangeli V, Pocaterra D, Rosseti N, Pavoni M, et al. Efficacy and safety of atazanavir-ritonavir plus abacavir-lamivudine or tenofovir-emtricitabine in patients with hyperlipidaemia switched from a stable protease inhibitor-based regimen including one thymidine analogue. AIDS Patient Care STDS. 2009;23:691-7.
13. Friis-Møller N, Reiss P, Sabin CA, Weber R, Monforte A, El-Sadr W, et al. Class of antiretroviral drugs and the risk of myocardial infarction. N Engl J Med. 2007;356:1723-35.
14. Galli M, Ridolfo AL, Adorni F, Gervasoni C, Ravasio L, Corsico L, et al. Body habitus changes and metabolic alterations in protease inhibitor-naïve HIV-1-infected patients treated with two nucleoside reverse transcriptase inhibitors. J Acquir Immune Defic Syndr. 2002;29:21-31.
15. Crane HM, Grunfeld C, Willig JH, Mugavero MJ, Van Rompaey S, Moore R, et al. Impact of NRTIs on lipid levels among a large HIV-infected cohort initiating antiretroviral therapy in clinical care. AIDS. 2011;25:185-95.
16. Williams P, Wu J, Cohn S, Koletar S, McCutchan J, Murphy R, et al. Improvement in lipid profiles over 6 years of follow-up in adults with AIDS and immune reconstitution. HIV Med. 2009;10:290-301.
17. Troll JG. Approach to dyslipidemia, lipodystrophy, and cardiovascular risk in patients with HIV infection. Curr Atheroscler Rep. 2011;13:51-6.
18. Carr A, Samaras K, Thorisdottir A, Kaufmann GR, Chisholm DJ, Cooper DA. Diagnosis, prediction, and natural course of HIV-1 protease-inhibitor-associated lipodystrophy, hyperlipidaemia, and diabetes mellitus: a cohort study. Lancet. 1999;353:2093-9.

19. Leonard EG, McComsey GA. Metabolic complications of antiretroviral therapy in children. Pediatr Infect Dis J. 2003;22:77-84.
20. Ducobu J, Payen MC. Lipids and AIDS. Rev Med Brux. 2000;21(1):11-7.
21. Zangerle R, Sarcletti M, Gallati H, Reibnegger G, Wachter H, Fuchs D. Decreased plasma concentrations of HDL cholesterol in HIV-infected individuals are associated with immune activation. J Acquir Immune Defic Syndr. 1994;7:1149-56.
22. Mooser V, Carr A. Antiretroviral therapy-associated hyperlipidaemia in HIV disease. Curr Opin Lipidol. 2001;12:313-9.
23. Sprinz E, Lazzaretti RK, Kuhmmer R, Ribeiro JP. Dyslipidemia in HIV-infected individuals. Braz J Infect Dis. 2010;14(6):575-88.
24. Noor MA. The role of protease inhibitors in the pathogenesis of HIV-associated insulin resistance: cellular mechanisms and clinical implications. Curr HIV/AIDS Rep. 2007;4:126-34.
25. Hertel J, Struthers H, Horj CB, Hruz PW. A structural basis for the acute effects of HIV protease inhibitors on GLUT4 intrinsic activity. J Biol Chem. 2004;279:55147-52.
26. Nolan D, Hammond E, Martin A, Taylor L, Herrmann S, McKinnon E, et al. Mitochondrial DNA depletion and morphologic changes in adipocytes associated with nucleoside reverse transcriptase inhibitor therapy. AIDS. 2003;17:1329-38.
27. Aberg JA, Gallant JE, Ghanem KG, Emmanuel P, Zingman BS, Horberg MA. Primary care guidelines for the management of persons infected with HIV: 2013 update by the HIV medicine association of the Infectious Diseases Society of America. Clin Infect Dis. 2014;58(1):e1-e34.
28. Expert Panel on Detection, Evaluation, and Treatment of High Blood Cholesterol in Adults. Executive summary of the third report of the National Cholesterol Education Program (NCEP) Expert Panel on Detection, Evaluation, and Treatment of High Blood Cholesterol in Adults (Adult Treatment Panel III). JAMA. 2001;285:2486-97.
29. Stone NJ, Robinson J, Lichtenstein AH, Merz CN, Blum CB, Eckel RH, et al. 2013 ACC/AHA guideline on the treatment of blood cholesterol to reduce atherosclerotic cardiovascular risk in adults: a report of the American College of Cardiology/American Heart Association Task Force on Practice Guidelines. J Am Coll Cardiol. 2013;129:46-8.
30. Baruch L, Agarwal S, Gupta B, Haynos A, Johnson S, Kelly-Johnson K, et al. Is directly measured lowdensity lipoprotein clinically equivalent to calculated low-density lipoprotein. J Clin Lipidol. 2010;4(4):259-64.

Index

Page numbers followed by *f* refer to figure, *fc* refer to flowchart, and *t* refer to table.

A

Abetalipoproteinemia 39, 43, 123, 124*t*
Acquired immunodeficiency
 syndrome 214, 297
Acute coronary syndrome 84, 90, 163, 164, 182, 222
Acyl-CoA cholesterol acyl transferase 7
Adeno-associated virus 8 250
Adenosine triphosphate 7, 125, 164, 191
 binding cassette
  subfamily G member 85, 88
  transporter 14, 82, 85, 88, 90, 125
Adipocyte triglyceride lipase 96, 199
Adiponectin 137
Advanced glycosylation end products 12
Advanced lipoprotein tests 123
Alanine aminotransferase 68, 156, 223, 232, 273
Alanine transaminase 301
Alirocumab 164
Alpha-linolenic acid 55
Alpha-tocopherol, beta-carotene cancer
 prevention study 57*f*–59*f*
American Academy of Pediatrics 267
American Association of Clinical
 Endocrinologists 101, 102, 151, 162, 206, 213, 222
American College of Cardiology 117, 204, 212, 214, 219, 223, 225, 259, 263*t*
American College of Clinical Endocrinology
 Recommendations 163*t*
American College of Endocrinology 206
 Guidelines for Management of Dyslipidemia
  and Prevention of Atherosclerosis 102
American Diabetes Association 205
 Standards of Medical Care in Diabetes 102
American Heart Association 53, 54, 59, 117, 204, 212, 214, 219, 223, 225, 229, 259, 263*t*, 300
Amiodarone 192
Amlodipine 192
Anacetrapib 182
Angiopoietin-like protein 3 250
 inhibitors 101, 164, 204

Anglo-Scandinavian Cardiac Outcomes Trial 217*t*
Ankle-brachial index 214
Antiretroviral protease inhibitors 118
Antiretroviral therapy 297, 298, 302
Antisense oligonucleotides 166, 185
 mechanism of action of 34*f*
Antisense therapy 34
Apheresis 237
Apolipoproteins 4, 5, 132, 213
 A1 77, 82, 90, 182, 184
  transcriptional upregulators 183
 B 11, 23, 163, 214, 222
  particles 25*f*
  synthesis inhibitor 164
 C2 82
 C3 82
  inhibitors 101
  functions of 5*t*
  measurement of 132
  types of 5*t*, 24
Arterial blood pressure lowering drugs
 study 157
Arterial disease, peripheral 161
Aspartate
 aminotransferase 273
 transaminase 301
Atazanavir 299
Atherogenesis 288*f*
Atherogenic diabetic dyslipidemia 198, 200, 207
Atheroma
 complications 15
 evolution 15
Atherosclerosis 11, 16, 18*f*, 20*f*, 21, 23, 24, 68, 69, 74–76, 109, 288, 289*f*
 coronary 176
 mechanism of 11, 31*f*
 multiethnic study of 215, 222
 oxidation theory of 17
 postprandial hypothesis of 18
 response-to-retention hypothesis of 16*f*, 17
 role in 24

initiation of 12
progression of 14
subclinical 64, 69
Atherosclerotic cardiovascular disease 27, 29, 73, 74, 78t, 87, 94, 100, 108, 136, 145, 161, 170, 173, 189, 200, 211, 214, 216, 219-221, 222t, 225, 246, 250, 263, 266, 287, 290, 292
 development of 180
 prevention of 108
 risk 30, 32, 94, 231t
  factors of 73
Atherosclerotic lesion formation 12fc
Atherosclerotic vascular disease 81, 91
Atherothrombosis, role in 15
Atorvastatin 137, 157, 192, 206, 218, 224, 232, 272
Autoimmune disorders 119
Autosomal recessive mutations 94

B

Basic lipidology 1
Bempedoic acid 164, 202
Beta-quantification 130
Bezafibrate Infarction Prevention study 172t, 203
Bile acid 202
 sequestrants 164, 165, 195, 202
Blood
 cholesterol, management of 291
 lipids, abnormal 161
 pressure 65, 273
  systolic 216
Body mass index 68, 155, 229, 270, 273t, 300
Bromodomain 185

C

Canadian Cardiovascular Society 213, 214, 219, 225
Cancer 53, 223
Carbohydrates 54, 56, 58f, 148
 types of 60f
Cardiovascular death 201, 228
Cardiovascular disease 8, 18, 23, 25f, 32, 53, 56, 59, 65, 81, 82, 89, 94, 103, 161, 170, 172, 182, 218t, 228, 259-261, 266, 278, 287, 290, 301
 risk factors, prevalence of 279
 therapies 162
Cardiovascular events 162, 170
Cardiovascular Health Integrated Lifestyle Diet-2 271f
Cardiovascular outcome trials 253
Cardiovascular risk 81
 factors 211
 reduction 162

Care lipid analysis, point of 132
Carotid intima medial thickness 68
Cerebrotendinous xanthomatosis 123
Cerebrovascular accident 216
Cerebrovascular disease 182
Cholestasis 222
 severe 119
Cholesterol 38, 42, 86, 127, 148, 261, 279
 absorption inhibitor 164
 efflux capacity 69, 85f, 86
 elevated 38
 metabolism 7
 remnant 94
 rich lipoproteins 237
 treatment trials 228
Cholesteryl esters 3, 4, 8, 13, 25, 66, 85
 storage disease 39, 41, 122
 transfer 181
  protein 5, 19, 47, 66, 82, 84, 85, 90, 96, 125, 141, 167, 180, 181, 184, 185, 199, 250
Cholestyramine 164, 165
Chondroitin sulfate 25
Chronic inflammatory
 disease 69
 disorders 89
Chronic kidney disease 45, 73, 88, 118, 163, 194, 220, 222, 224, 224t, 230, 231, 259-263
 grades of 260t
Chylomicronemia 123
Chylomicrons 3, 4, 17, 24, 146, 261
Citrate-dextrose 243
Clarithromycin 192
Clinical lipidology 51
Collaborative Atorvastatin Diabetes Study 201
Coronary artery calcium 292
 score 221
Coronary artery disease 43, 66, 87, 111, 146, 156, 176, 183, 224, 287
Coronary revascularization 201
C-reactive protein 65
Creatine kinase 190, 273
Cushing's syndrome 118, 119
Cyclosporine 192
Cytochrome p450 16
Cytosolic lipid droplets 86

D

Dalcetrapib 182
Danazol 192
De novo synthesis 7
Dermatan sulfate 25
Desmosomal protein desmocollin-1 186
Dextran sulfate 239
 liposorber 239

Diabetes mellitus 44, 45, 53, 73, 136, 163, 198, 219, 222, 225, 261
　risk of development of 136
　type 1 90, 205
　type 2 8, 21, 82, 89, 94, 121f, 154, 166, 182, 198, 212, 216, 267
Diabetic dyslipidemia 198
　management of 200, 204
Diacylglycerol acyltransferase 250
Dietary cholesterol 148, 271
Dietary fats 53, 55, 56, 59
　guidance, evolution of 54
Dietary fibers 149
Diltiazem 192
Dimethylarginine, symmetric 88
Docosahexaenoic acid 99, 171, 176t, 203
Double filtration plasmapheresis 239
Drug
　interaction 192, 302
　therapy 200, 270, 272
Dysbetalipoproteinemia 94
　familial 38, 42, 45, 119, 124
Dyslipidemias 37, 64, 74, 77, 78t, 115, 116, 161, 198, 211, 212, 226, 247t, 266, 270t, 271, 278, 297
　atherogenic 64, 102fc, 212
　drug treatment of 287
　epidemiology of 299
　management of 259, 299, 300
　mixed 272
　pathophysiology of 299
　patterns of 70
　severe 246
Dyslipoproteinemias, monogenic 47

E

Eicosapentaenoic acid 55, 99, 167, 171, 174, 176, 203
Electrophoresis 128
Endogenous lipid transport pathway 6
Endothelial cell 288, 289
Endothelial derived hyperpolarization factors 16
Endothelial lipase 82
Endothelial nitric oxide synthase 16, 88, 90
End-stage renal disease 224, 225
Enzyme 85
Epidermal growth factor-A 252
Erythromycin 192
Escherichia coli 252
Estimated glomerular filtration rate 172, 214, 224, 230

European Atherosclerosis Society 190, 213, 214, 219, 223, 225, 259
European Society of Cardiology 213, 214, 219, 223, 225, 259
Evacetrapib 182
Evolocumab 164
Exercise 150, 193
　regimens 151
Exogenous lipid transport pathway 5
Extracellular matrix 12
Ezetimibe 158, 164, 195, 201

F

Familial chylomicronemia
　score 42fc
　syndrome 38, 41, 42
Familial hepatic lipase deficiency 38
Familial hypercholesterolemia, Simon Broome diagnostic criteria for 41t
Farsenoid X receptor 6
Fasting lipid-profile 271
Fasting plasma glucose 223
Fats 54
　types of 60f
Fatty acids 5, 13, 54, 56, 56t, 96, 198, 199
　monounsaturated 55, 57, 58, 59f, 148
　nonesterified 95, 96, 198, 199
　omega-3 polyunsaturated 158
　polyunsaturated 53-56, 58, 60, 148
　saturated 53, 56, 58, 58f, 60
　short-chain 55
　triglycerides consist of 95
Fatty liver 64, 68
　index 69
Fatty streak 12
Fenofibrate Intervention and Event Lowering in Diabetes
　study 166, 172
　trial 171, 203
Fibrates 82, 98, 166, 171, 203
　randomized control trials of 172t
　role of 171
Fibrinogen 77
Fluvastatin 206, 224, 272
Foam cells, formation of 13, 14f
Food and Drug Administration 57, 171, 250, 282
　approved statins 272t
Framingham Heart Study 81, 228
Fredrickson and Levy classification 38t
Free cholesterol 4
Free fatty acid 5, 19, 65, 66, 155, 199
Friedewald equation 130

G

Gemcabene 164
Gemfibrozil 192
Gene-silencing techniques 251
Genetic 30
    lipoprotein disorders 45*fc*
    polymorphisms 192
Glitazars 99, 203
Global burden of metabolic risk factors study 212
Glomerular filtration rate 260, 301
    basis of 260*t*
Glomerulosclerosis, focal segmental 241
Glucagon like peptide-1 97
Glucose
    lowering medications 97
    transporter 4 137, 151
Glycosaminoglycan 16
Glycosylated hemoglobin 216, 223, 301
Glycosylphosphatidylinositol anchored high-density lipoprotein binding protein 1 125*t*
Greek Atorvastatin and Coronary Heart Disease Evaluation Study 157

H

Hashimoto's thyroiditis, chronic 119*f*
Heart
    disease
        chronic 31
        congenital 195
        coronary 26, 39, 42, 56, 57*f*-59*f*, 67, 161, 170, 172, 217, 299
        ischemic 259
    Outcomes Prevention Evaluation Trial 291
    Protection Study 217
Helsinki Heart Study 172*t*
Hemodialysis 263
Heparin 239
    sulfate proteoglycans 14
Hepatic effects 222
Hepatic lipase 43, 82, 96, 151, 199
    deficiency 43, 124
Hepatitis 119, 222
Hepatocellular carcinoma 154
High-density lipoproteins 3-5, 8, 11, 14, 20, 58, 65, 66, 73, 74, 81, 85, 87, 88*f*, 90, 91, 96, 109, 127, 129, 145, 146, 161, 176, 180, 182, 182*t*, 185, 185*t*, 199, 239, 245, 259, 261, 263, 279, 301
    cholesterol 23, 37, 39, 45, 55, 67, 72, 75, 77, 78, 81, 82, 84, 88, 90, 108, 110, 116, 122*f*, 125, 128, 155, 162, 163, 166, 172, 176, 180, 212-214, 216, 218-220, 222, 225, 234, 267, 269, 271
    high levels of 47
    low levels of 46
    role of 109
    elimination of 85*f*
    function 87
    infusions 182
    metabolism, genetic disorders of 46
    raising therapies, mechanism of action of 184*f*
    role of 19
    structural diversity 87
    targeted therapeutics 181
    trials of 185*t*
High-intensity statin 194, 232
    therapy 206, 206*t*
High-low-density lipoproteins cholesterol 212
High-sensitivity C-reactive protein 74, 216, 231, 301
High-triglycerides 198
Hirsutism 119
Homocysteine 77
Homogeneous assays 129, 131
Hormone sensitive lipase 95, 96, 198, 199
Human immunodeficiency virus 214, 231, 297
    infection 118
    protease inhibitors 192
Human leukocyte antigen 192
Human lipoprotein metabolism 3
Human plasma lipoproteins 73, 73*t*, 74*t*
Hyperalphalipoproteinemia, familial 47
Hypercholesterolemia 38, 45*fc*, 48, 117, 241, 280
    autosomal dominant 38, 40, 124, 246
    autosomal recessive 38, 40, 124
    familial 38-40, 121*f*, 122*f*, 124, 162, 202, 231, 237, 240, 266, 272, 280
    heterozygous familial 40, 45, 163, 222
    homozygous familial 40, 45, 250
    prevalence of 212*t*
Hyperchylomicronemia 48
Hyperglycemia 76, 136, 199
Hyperinsulinemia 76
Hyperlipidemia 38, 49, 145
    familial combined 39, 43, 45, 46, 94, 122*f*, 123, 124, 270
    postprandial 17
Hypertension 73, 260
Hypertriglyceridemia 38, 42, 44, 45, 45*fc*, 76, 299
    familial 43, 94, 125
    isolated 44*fc*
    moderate 101
    polygenic causes of 43
    severe 117, 118, 120, 171

Index

Hypoalphalipoproteinemia
  genetic forms of 46
  primary 39, 46
Hypobetalipoproteinemia, familial 39, 44, 124
Hypothyroidism 118, 119, 194
  primary 119f

I

Immunization 252
Immunoglobulin G 239
Indian Council of Medical Research-India Diabetes Study 212
Inflammatory diseases 231
Insulin
  receptor substrate-1 137
  resistance 64, 66
Intercellular adhesion molecule 13, 20
Intermediate-density lipoproteins 3, 4, 11, 25, 38, 66, 72-74, 95, 109, 129, 132, 146, 184, 199, 261
  cholesterol 132
International Lipid Expert Panel 190
International Lipid Guidelines 228
Intima-media thickness 220
Intracellular adhesion molecule 288, 289
Intrauterine malnutrition 260t
Itraconazole 192

J

Japan EPA Lipid Intervention Study 173

K

Ketoconazole 192
Kidney disease 261, 262
  chronic 45, 73, 88, 118, 163, 194, 220, 222, 224, 224t, 230, 231, 259-263
  outcomes quality initiative 261

L

Lecithin-cholesterol acyltransferase 5, 8, 82, 85, 90, 125, 151, 184
  deficiency 46, 120
Linoleic acid 56
Lipase maturation factor 1 125
Lipid 115, 145, 150, 279
  direct adsorption of 239
  disorders 39, 41, 42, 115
  levels 267t
  lowering
    arm 217
    drugs, effect of 154
    therapy 245

management 234fc, 290
measurements and mathematical equations 109
oxidation product 17
  role of 17
parameter 78
profile 115
  abnormal 124t
research clinics-coronary primary prevention trial 165
Lipodystrophy 119
  partial 120f
  syndrome 298
Lipo-oxygenase controls collagen 17
Lipoproteins 11, 17, 21, 24, 45, 82, 145, 146, 214, 239, 288
  A 4, 14, 20, 25, 29, 30f-32f, 35, 73-75, 109, 131
    function 29
    structure 29
    role of 20
  apheresis 237, 238, 241fc
  associated phospholipase A2 77
  atherogenic 237
  classification of 129f, 261t
  disorders
    classification of 37
    Fredrickson and Levy classification of 38t
  hemoperfusion 239
  isolated elevation of 241
  laboratory assessment of 127
  lipase 4, 12, 18, 44, 82, 92, 96, 118, 125, 198, 199, 250, 272
    deficiency 38, 123
    hydrolyzes 95
    role of 19
  metabolism 4, 6f, 9
    disorders of 37
  modification of 13
  receptor related protein 5
  residual 15
  types of 24, 127
Liposorber system 238f, 239t
Liver
  disease, chronic 119, 193
  lipid disease of 154
  X receptor 85, 86, 185
Lomitapide 164, 202
Lovastatin 192, 206, 218, 232, 272
Low-carbohydrate diet 150
Low-density lipoproteins 3-5, 11, 14-17, 25, 29, 39, 48, 65, 66, 73, 74, 78, 82, 85, 88, 94, 96, 102, 103, 109, 127, 129, 145, 146, 182, 184, 199, 204, 237-239, 245, 250, 261, 263, 301

apheresis 237
cholesterol 12, 23, 26, 39, 55, 72, 74, 77, 82, 84, 108, 110, 110f, 119f, 121f, 122f, 125, 130, 161-164, 194, 214, 217-219, 221, 222, 225, 231, 232, 234, 241, 245, 252, 267, 269, 271, 273, 287, 292, 294, 297, 300
   raised 270
   role of 16
receptor 14, 41, 85, 251
Low-dose statin 194
Low-high-density lipoproteins 95, 103
   cholesterol 120f, 198
Low-intensity statin 232
Lymphocytes, T-regulatory 14
Lysosomal-acid lipase deficiency 41, 48, 122

M

Magnetic resonance spectroscopy 68
Major adverse cardiovascular events 95, 200, 228
Major cholesterol guidelines, comparative analysis of 211
Matrix metalloproteinases 15
   regulation of 17
Medical Research Council 60
Mediterranean diet 149
Messenger ribonucleic acid 166
Metabolic adaptation 278
Metabolic syndrome 64, 65f, 65t, 216, 231, 298
   atherothrombosis intervention in 100, 165, 173
Metalloproteinase-1, tissue inhibitor of 17
Microsomal transfer protein inhibitor 164
Microsomal triacylglycerol transfer protein 6
Microsomal triglyceride transfer protein 43, 202, 250
Mipomersen 26, 164, 166, 202
Moderate-intensity statin 232
   therapy 206, 206t
Monoclonal antibodies 251, 252
Monocyte
   chemoattractant protein-1 13, 20
   chemotactic protein 288, 289
Monogenic lipid disorders 39
   treatment of 48
Monogenic lipoprotein disorders 37, 44fc, 45fc
   classification of 38t
Mononeuritis multiplex 123
Monounsaturated fat 271
Muscle
   inflammation 190
   weakness 190
Myalgia 190

Myocardial infarction 31, 73, 161, 171, 176, 191, 201, 211, 213t, 216, 217, 229, 260, 262
Myonecrosis 190
Myopathy 190
Myositis 190

N

National Cholesterol Education Program 54, 65, 132, 259
   Adult Treatment Panel 204, 262
   Recommendations 132
National Health and Nutrition Examination Surveys 127, 267
National Institute for Health and Clinical Excellence 259, 263
National Lipid Association 190, 263t
National Nutrition Monitoring Bureau 54, 56
National Obesity and Metabolic Syndrome Summit 229
Nefazodone 192
Nephelometry 31
Nephrotic syndrome 119
Neuromuscular disorders 193
Next-generation peroxisome proliferator-activated receptor agonists 167
Niacin 82, 100, 164, 165, 173, 203
Nicotinamide adenine dinucleotide phosphate oxidase and xanthine oxidase 16
Nicotinic acid 100, 164
Nitric oxide levels 16
Nonalcoholic fatty liver disease 9, 64, 68, 154, 223, 250
Noncommunicable diseases 53
Non-high-density lipoproteins 72
   cholesterol 76, 108, 110f
Nonstatin therapy 294
Nuclear magnetic resonance 45, 87, 131
Nucleoside reverse transcriptase inhibitors 297

O

Obesity 21, 198
Omega-3 carboxylic acids 171
Omega-3 fatty acids 99, 149, 167, 170, 176, 176t, 203
Oral estrogens 200
Oxidized low-density lipoproteins 14, 15, 131

P

Pancreatitis, acute 117
Paraoxonase-1 88
Pediatric dyslipidemia 267
Percutaneous coronary intervention 230

Peripheral tissues 198
Peripheral vascular disease 182, 242
Peritoneal dialysis 263
Peroxisome proliferator-activated receptors 157, 166, 203, 282
Pharmacokinetics 253
Phospholipase domain containing 3 protein 68
Phospholipids 4, 5, 13
  transfer protein 6, 82, 85
Phytoestrogens 150
Pitavastatin 206, 218, 272
Plant stanols 149
Plant sterols 148
Plasma
  high-density lipoproteins cholesterol 81
  lipoproteins 3, 72
    properties of 4t
  remnant lipoproteins 11, 17
Plasmat-futura 239
Plasmat-secura 239
Plasminogen
  activator inhibitor 1 15, 65
  structure 30f
Polyacrylate-coated polyacrylamide beads 239
Polyunsaturated fat 60f
Portfolio diet 150
Posaconazole 192
Pravastatin 206, 218, 224, 232, 272
Progressive inflammatory disease 69
Proprotein convertase subtilisin/kexin type 9 40, 164, 165, 201, 245
  deficiency 44
  inhibitors 195
  targeted therapy 165
Protein 4, 148
  kinase A 139
Proteoglycans 25
  sub-endothelial 12
Psoriasis 231

R

Radial immunodiffusion 132
Radioimmunoassay 132
Randomized controlled trials 58f, 162, 191, 245, 268
Ranolazine 192
Refractory angina 242
Refsum's disease 242
Respiratory diseases, chronic 53
Retinoids 200
Reverse cholesterol transport 4, 8f, 85f, 86, 88, 180
  pathway 7, 81
  quantification of 86

Rhabdomyolysis 190
Rheumatoid arthritis 214, 231
Ribonucleic acid 87, 164
  small-interfering 166, 251
Rosuvastatin 206, 218, 224, 232, 272

S

SAMS Clinical Index 190
Saturated fat 148
  placement of 60f
Scavenger receptor
  A1 14
  B1 14, 82
Senile sensory-neural hearing loss 242
Serum amyloid A 88, 89
Simvastatin 192, 206, 218, 224, 232, 272
Single-nucleotide polymorphism 192
Single-spin density gradient ultracentrifugation 128
Sitosterolemia 38, 41, 124
  diagnosis of 124
Skeletal muscle function 191
Small-dense low-density lipoproteins 131
  cholesterol 76, 155
Small-molecule inhibitors 252
Smooth muscle cells 15
Sodium 148
  glucose cotransporter-2 97
Sphingosine-1-phosphate receptors 88
Stanols 148
Statin 98, 156, 190, 200, 281
  characteristics 192
  diabetogenicity 136
  induced diabetes, mechanism of 139fc
  intensity of 232t
  intolerance 189, 190
    epidemiology of 191
    mechanism of 191
    pathogenesis of 191
  rechallenge 194
  safety 274
  therapy 191, 217, 294
    categorization of 294t
  withdrawal 194
Stearoyl-CoA desaturase-1 inhibitors 159
Steatohepatitis, nonalcoholic 223
Sterols 149
  regulatory-element-binding protein 7
Stroke 161, 201
Sulfonylureas 97

T

Tangier disease 39t, 46, 121
Telithromycin 192

Tendon xanthomas 122f
Therapeutic lifestyle change diet 147
  dietary recommendations of 148t
Therapeutic lipidology 116, 143
Therapeutic plasma exchange 280
Thiazide diuretics 200
Thyroid
  hormone receptor-beta 164t
  stimulating hormone 123, 301
Torcetrapib 182
Total cardiovascular disease 213
Total cholesterol 23, 45, 58, 76-78, 127
Trans fatty acids 53, 55, 57, 58f, 147
Transaminitis 190
  statin-associated 190
Triglycerides 3, 23, 25, 38, 39, 41, 42, 44, 45,
  65-67, 82, 94, 96, 97, 102, 103, 109, 110, 110f,
  122f, 127, 128, 145, 155, 162, 164, 166, 170,
  172, 176, 199, 234, 239, 261, 267, 271, 272,
  288, 300
  elevated 94
  lowering therapies 101, 204
  management of 102fc
  metabolism 95
  rich lipoproteins 11, 17, 18, 66, 75, 96, 163,
    199, 199f
    metabolism 96f
    remnants 18
  role of 17
  transfer of 181
Tuberous xanthomas 121f
Tumor necrosis factor 288f, 289f, 299
  alpha 19, 65, 87

U

United Kingdom Prospective Diabetes Study
  216
United States Food and Drug Administration
  270
United States Preventive Services Task Force
  212, 214f, 219t, 225t, 268
Unstable angina 73, 201
Uric acid 77

V

Vascular cell adhesion molecule 13, 20, 88, 288,
  289
Vascular smooth muscle cells 14, 17, 86, 288,
  289
Verapamil 192t
Very-low-density lipoproteins 3, 5, 11, 25, 38,
  39, 41, 45, 65, 66, 72-74, 85, 96, 109, 127,
  129, 146, 199, 155, 184, 198, 229, 261
  cholesterol 82, 116, 139
  remnants 66
Veterans Affairs High-Density Lipoprotein
  Cholesterol Intervention Trial 98, 171, 172,
  203
Vitamin E 123

W

Wolman's disease 122

X

Xanthelasma palpebrarum 122f

EU GSPR Authorised Reprsentative
Logos Europe, 9 rue Nicolas Poussin
1700, La Rochelle, France
Phone: +33 (0) 6 67 93 73 78
E-mail: contact@logoseurope.eu

www.ingramcontent.com/pod-product-compliance
Ingram Content Group UK Ltd.
Pitfield, Milton Keynes, MK11 3LW, UK
UKHW050428150426
5217IPUK00019B/1285